Rivers Beyond Borders

This book explores the intricate challenges and potential solutions in managing transboundary rivers in South Asia, a densely populated and ecologically diverse region. It examines the significant impacts of the growing climate crisis, disaster risks, population growth, and socio-political conflicts on transboundary river basins, which pose substantial risks to large areas of the Global South. The book highlights the critical need for multi-sectoral, multi-disciplinary, and cross-country perspectives to formulate collaborative policies that support sustainable development goals. Through specific case studies, the book addresses hydro-politics and flood hazards in countries like India, China, Pakistan, Nepal, Bhutan, and Bangladesh.

Features

- Provides a holistic explanation of transboundary water governance and integrates perspectives from environmental science, geography, hydrology, political science, international relations studies, law, water governance and sociology.
- Includes detailed case studies and practical examples from different countries to help understand theoretical concepts through real-world scenarios.
- Utilizes advanced technologies such as machine learning, AI, and deep learning for data analysis and predictive modelling and provides cutting-edge tools for predicting water-related challenges and improving management strategies.
- Incorporates spatial maps and geostatistical diagrams to visually enhance comprehension of complex data analysis and spatial relationships.
- Features contributions by renowned scholars, policymakers, and practitioners in the field of water governance, providing cutting-edge research, expert analyses, and authoritative perspectives.

Rivers Beyond Borders is an excellent reference for students, researchers, and academics in environmental science, geography, hydrology, political science, law, sociology, and South Asian studies; policymakers, government officials, and professionals in water resource management and international relations; and practitioners in non-governmental organizations (NGOs), international development agencies, and environmental advocacy groups.

Rivers Beyond Borders

Environmental Sustainability in South Asia's Transboundary Waters

Edited by
Mehebub Sahana

CRC Press
Taylor & Francis Group
Boca Raton London New York

CRC Press is an imprint of the
Taylor & Francis Group, an **informa** business

Designed cover image: © Getty Images

First edition published 2026
by CRC Press
2385 NW Executive Center Drive, Suite 320, Boca Raton FL 33431

and by CRC Press
4 Park Square, Milton Park, Abingdon, Oxon, OX14 4RN

CRC Press is an imprint of Taylor & Francis Group, LLC

ISBN: 978-1-032-98629-6 (hbk)
ISBN: 978-1-032-98630-2 (pbk)
ISBN: 978-1-003-59969-2 (ebk)

DOI: 10.1201/9781003599692

Typeset in Times
by KnowledgeWorks Global Ltd.

Contents

Foreword

It gives me immense pleasure to write the foreword to the book 'Rivers Beyond Borders," edited by Dr. Mehebub Sahana of Manchester University, UK. It is a compilation of 14 papers contributed by experts in Geography, Environmental Science, Juridical Science, International Studies, and Political Science. The hydro-politics in South Asia have become more complicated after the terrorists killing 26 innocent tourists at Pahalgam and India started 'Operation Sindur' in retaliation against Pakistan on 7 May 2025. Further Government of India kept in abeyance the Indus Water Treaty. While the Indo-Bangladesh Ganges water treaty(1996) will come to an end in 2026 and the issue of sharing the Teesta water remains unresolved, China has started a mighty hydropower project on the Brahmaputra (Yarlung Zanbo) with five cascading dams, possibly the largest in the world, the hydro-politics in South Asia is now an important and debated topic. It is widely admitted that the countries need to withdraw water from rivers but it is important to understand the rational meeting point which defines how much water may be exploited by the upper riparian States, keeping in mind the demand of downstream countries and the ecosystem services.

The political boundaries of South Asia, delineating China, India, Pakistan, Nepal, Bhutan, and Bangladesh, were drawn across river basins with a total lack of understanding of how holistic ecohydrology functions within the Indus, Ganga, and Brahmaputra basins. It was quite impossible to avoid the bifurcation of the rivers while the Radcliffe Commission drew the border between India and Pakistan in 1947 based on the numerical majority of Muslim and Non-Muslim communities then living in the contiguous areas of the subcontinent. More than 4,000 kilometres long border between India and China was a notional demarcation, evolving along the ceasefire line established after the Indo-China war in 1962. The endeavour to delineate the India-Nepal border was initiated in 1816 and continued till 1941, when the boundary pillars were finally put in place. The India-Bhutan border was first established in 1865 and reaffirmed through subsequent treaties, including that in 1949, which was again negotiated and finally accepted in 2007.

Three mighty rivers, the Indus, Ganga, and Brahmaputra, have their sources in the Himalaya, which is truly described as the 'water tower' of South Asia. South Asia experiences temporally skewed and spatially uneven rainfall, contributing to the flow of approximately 1433 BCM water from the Himalayan slopes through the Indus-Ganga-Brahmaputra system. This water, rich in sediment, historically sustained the agrarian economies of the vast alluvial plains extending from Pakistan through India to Bangladesh. In the backdrop of skewed annual rainfall, mostly concentrated during June to September, the challenge of water management has been to store the monsoon flow in dams/ reservoirs and transfer the same to the non-monsoon season. The rivers are tamed in this subcontinent either for irrigation or to exploit energy. The structures built for the above-noted purpose led to longitudinal disconnection of the rivers, downstream desiccation, and interrupted their ability to transfer sediment load downstream. Further, caused aquatic habitat fragmentation and loss of biodiversity. The flow was altered without considering the crucial ecological flow necessary for the ecosystem services and the well-being of millions of people living in the mighty Indus-Ganga-Brahmaputra plain. The situation is further complicated by both the Himalaya and the downstream IGB plain being immensely impacted by climate change.

No State has ownership over a transboundary river, but has the right to use water rationally and equitably. Unfortunately, the prevalent approach of upper riparian States often goes beyond 'right to use' and treats the flowing water as stock, as if having exclusive ownership. Any unilateral venture or declaration of stopping the river at the border and diverting water to any other direction clearly violates principles of 'equitable and reasonable utilisation' and protection of downstream ecology, proclaimed in the UN Convention on the Law of the Non-Navigational Uses of International Watercourses (1997). The flowing water should not be used to consolidate political power and dominate over the lower riparian States.

In this backdrop, the 'Rivers Beyond Border' will definitely be considered as an outstanding dossier, which will help to develop hydro-wisdom in South Asia. The knowledge about the delicate hydrology of the Indus-Ganga-Brahmaputra system has to be made available in the public domain for the greater common good. I congratulate Dr. Sahana and all the authors for their research and contribution towards ensuring peace in this subcontinent.

Dr. Kalyan Rudra
Chairman, West Bengal Pollution Control Board, India
Adjunct Professor, Vidyasagar University,
West Medinipur, West Bengal

1.

Preface

River basins and water resources are central to achieving sustainable development, positioning transboundary cooperation between upper and lower riparian regions as a cornerstone of Agenda 2030 and the Sustainable Development Goals (SDGs). In South Asia—home to around a quarter of the world's population—shared rivers are not merely ecological lifelines but are also deeply intertwined with political, economic, and cultural realities. The complex politics surrounding these shared waters emphasise the urgent need for collaborative governance, informed diplomacy, and evidence-based policy interventions. Against this backdrop, *Rivers Beyond Borders: Environmental Sustainability in South Asia's Transboundary Waters* explores the pressing challenges and emerging opportunities in the governance of shared rivers within one of the world's most densely populated and ecologically diverse regions.

This edited volume examines how rapid population growth, intensifying climate change, and recurring geopolitical tensions shape the management of transboundary river basins in South Asia. These pressures pose substantial risks across the region, particularly in climate-sensitive areas such as the Himalayan watershed, the Ganga-Brahmaputra-Meghna (GBM) basin, and the Indus river basin. Large-scale river diversions, hydropower projects, and unregulated infrastructure development continue to influence downstream water availability, alter ecological balances, and heighten disputes over water sharing. Using grounded case studies and real-world examples from India, Nepal, Bhutan, Bangladesh, and Pakistan, this volume highlights both the challenges and opportunities for cooperative and sustainable water governance.

The book draws on insights from three major international events: the *International Conference on Transboundary Rivers of South Asia* (7–8 May 2024, North South University, Dhaka), the *Interdisciplinary Workshop on Urban Flooding and Transboundary River Basins in South Asia* (7–8 May 2024, Indian Institute of Science, Bengaluru), and the *International Symposium on Himalayan River Systems* (16–17 April 2025, Kathmandu University, Nepal). Together, these events involved over 200 participants and featured more than 100 presentations across diverse disciplines. Fourteen high-quality chapters were selected from this body of work, rigorously reviewed, and refined to form the edited volume. A limitation is the absence of contributions from Pakistani authors; despite repeated invitations, geopolitical and travel constraints prevented their participation, underscoring the book's central theme: the urgent need for enhanced regional cooperation across politically divided landscapes.

A central theme of the book is the importance of multi-sectoral, multidisciplinary, and cross-border approaches to managing shared river basins. Sustainable development in South Asia depends on addressing the ecological, socio-economic, and political dimensions of river governance. The chapters highlight the integration of hydrology, environmental science, geography, political science, law, and international relations to understand the drivers, risks, and impacts of transboundary water challenges. Readers will encounter detailed analyses of hydro-politics, flood and drought management, and institutional frameworks governing the GBM and Indus river systems. Drawing on real-world examples—such as Himalayan dam construction, sedimentation, riverbank erosion, and groundwater–surface water interactions—the book offers evidence-based insights for policymakers and practitioners.

The management of transboundary rivers is far more than an academic concern; it directly affects the daily lives of millions across South Asia. For instance, farmers in India, Bangladesh, and Nepal rely heavily on shared rivers such as the GBM for irrigation. Effective transboundary cooperation ensures reliable water availability, supporting food security and rural livelihoods, whereas mismanagement can lead to reduced flows, water stress, and declining crop yields. Flood management provides another illustrative example: from Khyber Pakhtunkhwa in the eastern Himalayas to highland regions in Nepal and the Bengal delta, severe monsoon flooding occurs annually. Coordinated measures—such as joint early warning systems, synchronised dam releases, and cross-border disaster planning—significantly reduce losses, protecting both human life and property. These real-world examples highlight why robust transboundary water governance is essential for regional resilience and human security.

The urgency of addressing these challenges is heightened by climate change, which accelerates glacial melt, alters rainfall patterns, and increases the frequency and severity of extreme weather events. Existing water-sharing frameworks, often decades old, are ill-equipped to address contemporary hydrological realities or the socio-political complexities of shared river management. Moreover, integrated, region-specific research bridging ecological science, infrastructure development, and geopolitics remains limited.

This book addresses these gaps, offering holistic, interdisciplinary, and actionable insights into the governance of South Asia's transboundary rivers. Each chapter presents rigorous analyses of hydro-politics, climate hazards, legal frameworks, governance structures, river health, and geopolitical tensions, while also proposing practical solutions. Policy relevance is a key strength, providing guidance on institutional strengthening, basin-wide cooperation, and climate adaptation strategies. Forward-looking discussions explore how South Asian nations can collectively advance sustainable development and water security. With its accessible style and strong empirical foundation, the volume is invaluable to scholars, students, policymakers, practitioners, environmental activists, and concerned citizens. Whether new to the field or deeply engaged in water governance, readers gain the conceptual tools and real-world insights necessary to navigate the complexities of shared rivers.

By combining scientific analysis, legal insights, political perspectives, and community-level observations, *Rivers Beyond Borders* equips readers with the knowledge required to engage meaningfully with the future of South Asia's rivers. Through its detailed case studies and policy recommendations, the book aims to inspire collaborative, sustainable, and equitable water governance across one of the world's most critical and vulnerable river systems.

Mehebub Sahana
Manchester, UK
20th October 2025

About the Editor

Mehebub Sahana is a Leverhulme Fellow and Lecturer in GIS in the Department of Geography at The University of Manchester, UK. He is an environmental geographer with an interest in analysing land-use changes, and his work focuses on landscape alterations in the post-colonial era and their consequences for sustainability in developing countries. His current research interests include the transboundary river basin management, landscape ecology, socio-ecological resilience, hydro-climatic extremes, and the socio-political implications of land-use dynamics in the Global South.

Previously, he was a Lecturer at the Indira Gandhi Conservation Monitoring Centre (IGCMC), WWF-India, New Delhi. He is a highly cited, globally recognised researcher, and his work has been supported by projects funded by the Leverhulme Trust (UK, 2022–25), the International Science Partnership Fund (2024–25), the British Council (2025–26), and ICSSR, India. He is a Fellow of the Higher Education Academy (UK) and a member of the British Academy Early Career Researcher Network. He actively engages with public and policy audiences through opinion pieces, editorials, and commentary in various media outlets.

Contributors

Sk Ajim Ali
The Global Development Institute
School of Environment, Education and Development
University of Manchester
Manchester, United Kingdom

Ammu S Anil
International Studies
MMAJ Academy of International Studies
Jamia Millia Islamia
New Delhi, India

Md. Tawhidul Aziz
National River Conservation Commission
Bangladesh

Japungsar Basumatary
Department of International Relations
Sikkim University
Gangtok, India

Debojyoti Das
Sustainability (Education), Geography, University of Birmingham, UK.

Udyan Devkota
Independent Researcher
Kathmandu, Nepal

Md Kutubuddin Dhali
Jadavpur University
Jadavpur, Kolkata, India

Genevieve Donnellon-May
Oxford Global Society
Oxford, United Kingdom

Chhimi Dorji
Consultant Water Resources
ChhimiD Consulting
Thimphu, Bhutan

Sadia Afrose Esha
Department of Soil, Water, and Environment
University of Dhaka
Dhaka, Bangladesh

Bishwajit Ghosh
Department of Law
Green University of Bangladesh
Kanchon, Bangladesh

Kausik Ghosh
Department of Geography
Faculty of Science
Vidyasagar University
Paschim Medinipur, West Bengal, India

Md. Ashraful Haque
National River Conservation Commission
Dhaka, Bangladesh

Sufia Khanom
Bangladesh Institute of International and Strategic Studies (BIISS)
Dhaka, Bangladesh

Tek Raj Koirala
South Asian University
Maidan Garhi, New Delhi, India

Sakib Mahmud
National River Conservation Commission
Dhaka, Bangladesh

Neeraj Singh Manhas
Department of Political Science
The Maharaja Sayajirao University of Baroda, Pratapgunj, Vadodara, Gujarat
India

Radhika Mulay
Centre for Sustainability, Environment, and Climate Change
FLAME University
Pune, Maharashtra, India

Vijay Paranjpye
Centre for Sustainability, Environment, and Climate Change
FLAME University
Pune, Maharashtra, India

Siddharth Prabhakar
Delhi High Court
Delhi, India

Naresh Prajapati
School of Law, Bennett University,
Greater Noida, Uttar Pradesh, India

Ahmad Shams Qadir
National River Conservation Commission
Bangladesh

Dipmala Roka
Department of International Relations
Sikkim University
Gangtok, India

Mehebub Sahana
Department of Geography
School of Environment, Education and Development
University of Manchester
Manchester, United Kingdom

Md. Abdus Samad
Department of History, Jagannath University
Dhaka, Bangladesh

Nyamat Sekhon
Rajiv Gandhi National University of Law
Patiala, Punjab, India
and
National Law University
Delhi, India

Kumkum Sharma
Delhi High Court
Delhi, India

Md Abu Bakar Siddique
Department of Law
Bangladesh University of Professionals (BUP)
Dhaka, Bangladesh

Taranjot Singh
Department of Laws
Panjab University
Chandigarh, India

Deki Pem Tamang
ChhimiD Consulting
Bhutan

Bibek Dhoj Thapa
NIICE
Kathmandu, Nepal
and
The Institute of South Asian Studies
Sichuan University
Chengdu, China

1 Transboundary River Conflicts, Cooperation, and Environmental Sustainability in South Asia

Mehebub Sahana, Sk Ajim Ali,
Md Kutubuddin Dhali, and Md. Abdus Samad

INTRODUCTION

Water is a fundamental resource for human survival and socio-economic development. In South Asia, however, water resources have emerged not only as a foundation for growth but also as a source of both conflict and cooperation. The region is home to some of the world's largest and most critical transboundary river systems (Figure 1.1). Rivers such as the Ganges–Brahmaputra–Meghna (GBM) and the Indus shape the lives, cultures, and religions of millions of people, with the GBM and Indus basins together supporting nearly one-tenth (1,000 million) of the global population (Shamsudduha, 2021). Following the post-colonial redrawing of national borders, these rivers came to traverse multiple countries, linking riparian states including China, India, Pakistan, Bangladesh, Nepal, and Bhutan, which today face both shared challenges and opportunities (Huda, 2017). Within this context, transboundary river governance is inextricably tied to geopolitical dynamics, environmental sustainability, and socio-economic development (Williams, 2020). Cooperation between upstream and downstream riparian states is central to Agenda 2030 and the Sustainable Development Goals (SDGs), as effective transboundary water governance is crucial for sustainable development, reducing regional disparities, and ensuring equitable access to resources.

Globally, there are 286 international transboundary river basins (TRBs), collectively covering 47.1% of the Earth's land surface, with nearly half of the world's population residing within a TRB, a transboundary lake basin, or above a shared aquifer. South Asian TRBs are vital to agrarian economies, industrial growth, urban development, and regional sustainability (Sahana et al 2024). Yet increasing population, rapid urbanisation, agricultural intensification, climate change, and rising water demand have intensified pressures on these systems. While they offer immense potential for cooperation, they are also sites of recurring disputes (Karthe et al., 2015). Historical tensions—such as those over the Indus between India and Pakistan,

DOI: 10.1201/9781003599692-1

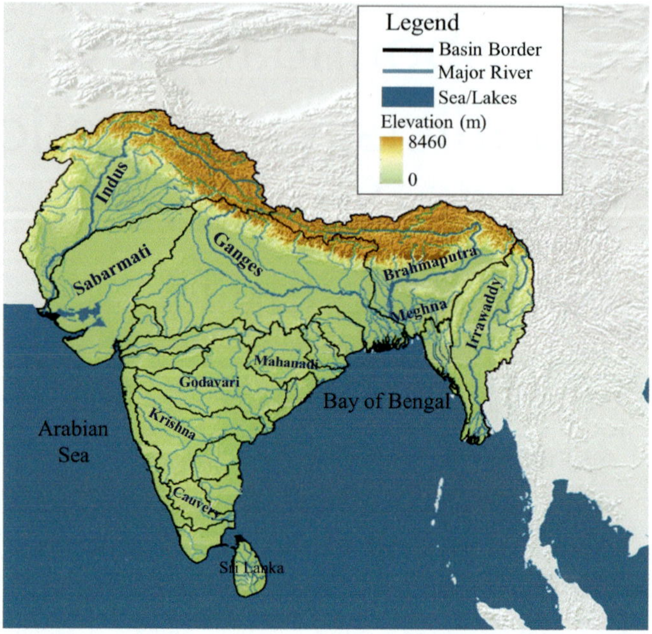

FIGURE 1.1 Transboundary river basins of South Asia, highlighting their linkages with major hydro-climatic zones and associated challenges. (*Source:* Prepared by authors.)

the Yarlung Tsangpo/Brahmaputra between India and China, the Ganges and Teesta between India and Bangladesh, and the Mahakali and Kosi between Nepal and India—highlight the complexity of governing shared waters within politically sensitive contexts.

This study examines the complex dynamics of conflict, cooperation, and environmental sustainability in relation to TRBs in South Asia, with the overarching aim of identifying pathways towards equitable and sustainable water governance. The fair and sustainable management of these shared resources is crucial not only for national and regional stability but also for securing global water security and advancing environmental protection. Given that the world relies heavily on TRBs, their governance carries major implications for achieving the SDGs (Albrecht and Gerlak, 2022). The inherently political nature of shared waters underscores the urgent need for transboundary cooperation and diplomacy, which remain complex yet vital aspects of the international development agenda. To address these challenges, interdisciplinary research is needed to better understand both the underlying causes of conflict and the consequences of governance choices.

South Asian TRBs, such as the Indus, Ganges and the Brahmaputra, face acute pressures linked to large-scale water diversions, food and energy security concerns, and contested international cooperation (Baten and Titumir, 2016; Katyaini et al., 2020). Several scholars highlight the importance of effective governance in these contexts. Uprety and Salman (2011) emphasise the role of international water law in addressing disputes, while Swain (2016) stresses the political economy underpinning water-related conflicts. Rahaman (2009) discusses the competing priorities between hydropower development and water sharing, illustrating the difficulties riparian states face when balancing national interests. Likewise, Hanasz (2017) introduces the concept of hydro-diplomacy as a potentially transformative mechanism for reducing tensions. Despite these valuable contributions, gaps remain in terms of understanding the socio-environmental implications of transboundary disputes and designing cooperative frameworks that prioritise ecological sustainability.

Existing scholarship has largely centred on bilateral agreements and conflict resolution mechanisms. The Indus Waters Treaty (IWT) (1960) and the Ganges Water Sharing Treaty (1996) are often celebrated as landmarks of water diplomacy. However, their limitations are becoming increasingly visible. These treaties seldom incorporate ecological dimensions, seasonal variability, or the realities of climate change, leaving downstream populations vulnerable to both water-scarcity, drought and floods. Moreover, asymmetries of power complicate the cooperative landscape. China and India's dominant role in most basins frequently results in unilateral decisions, fuelling resentment among smaller states such as Bangladesh, Bhutan and Nepal. This power imbalance often marginalises the perspectives of vulnerable river-dependent communities, whose voices are rarely incorporated into negotiations.

Another key gap concerns the insufficient prioritisation of environmental sustainability. To often, political and economic imperatives overshadow ecological needs (Albrecht and Gerlak, 2022). Large-scale dams and barrages have altered river flows, damaged ecosystems, and accelerated biodiversity decline, while climate change is introducing new uncertainties, including Himalayan glacier retreat and changing monsoon patterns. These challenges are poorly reflected in existing policy frameworks. Ongoing dam-building projects in upstream areas, combined with water diversions and barrage operations in the downstream plains, have already triggered several disputes. The Koshi River conflict between India and Nepal and India's disagreements with Bangladesh over the Teesta and Farakka barrages and abeyance of indus water treaty exemplify failures of current water governance structures.

The ecological consequences are stark. Many rivers across the region are beginning to dry seasonally, with recurring drought conditions during the summer months (Crausbay et al., 2020; Ojima, 2021). Furthermore, many small tributaries and distributaries within the GBM and Indus river systems are disappearing, disrupting surface water hydrology and groundwater recharge. These changes have serious consequences, intensifying monsoon flooding and increasing the frequency and severity of summer droughts across South Asia. Paradoxically, these periods of low flow are often followed by devastating monsoon floods, highlighting the fragile and unpredictable hydrological balance (Moser, 2010). Demographic pressures exacerbate the situation. South Asia's population, recorded at 1.68 billion in 2010, is projected to rise by 32% to 2.22 billion by 2040 (Chaudhary et al., 2024). This growth will drastically increase demand for water in agriculture, food production, hydropower generation, and domestic use. Expanding industries, rapid urbanisation, and changing consumption patterns further intensify demand. This widening gap between supply and demand has already created recurrent water crises, aggravating social and political tensions.

Given these dynamics, the Himalayan region and the adjoining Ganges–Brahmaputra delta represent one of the world's most significant areas for future research and policy innovation (Paszkowski et al., 2021; Rudra, 2018a). The environmental and social stakes are extraordinarily high, but so too are the opportunities for cooperation. While the challenges are daunting, local communities often demonstrate resilience through mutual aid, social cohesion, and creative adaptation. Harnessing such community-level innovation is essential for building resilience and designing governance frameworks that integrate ecological, social, and political priorities. Ultimately, achieving the SDGs in South Asia will be difficult without addressing the entrenched problems of transboundary water governance. Equitable, ecologically grounded cooperation across scales—local, national, and regional—is urgently required. Only through such efforts can South Asia manage its shared waters in ways that reduce conflict, safeguard ecosystems, and promote sustainable development for generations to come.

CLIMATE CHANGE AND THE NEED FOR TRANSBOUNDARY RIVER COOPERATION IN SOUTH ASIA

The GBM and Indus River systems are among the most significant transboundary rivers in the world. Despite their critical importance, there is no regional, multilateral transboundary governance framework, and existing bilateral treaties often contain ambiguous language, rendering them largely ineffective in resolving water-related conflicts. Agreements enacted by South Asian governments to address water disputes are frequently flawed, failing to address the fundamental challenges inherent in managing shared water resources. For instance, India and Pakistan have long-standing disputes over the Indus River and its tributaries. Pakistan has accused India of storing and diverting waters from the western rivers, which are allocated under the IWT, with the intention of establishing hydro-hegemony over downstream Pakistan. According to Khan et al. (2022), the IWT has historically represented a geographic water divide, providing little incentive for cooperative water-sharing. Cross-border terrorism has further strained the treaty, leaving the effectiveness of the IWT in abeyance as of April 2025.

Similarly, the Ganges Water Treaty, signed in 1996 between Bangladesh and India to resolve the Farakka Barrage dispute, is due to expire in 2026 and faces increasing political and environmental pressures. Management of the Teesta River, which is shared between the two countries, remains a source of tension, compounded by the absence of clear mechanisms for resolving disputes if the parties cannot agree. In Nepal, agreements such as the Sarada Agreement (1920), Kosi Agreement (1954), Gandak Agreement (1959), and Mahakali Treaty (1996) have historically favoured India, leaving many Nepalese stakeholders dissatisfied with their perceived share of water resources. In Afghanistan, the construction of 12 dams along the Kabul River, with Indian assistance, has raised concerns in Pakistan regarding downstream water diversion. Additionally, China's plans to construct world's large-scale dams on the Brahmaputra headwaters in Tibet have generated significant anxiety over potential ecological, social, and hydrological impacts downstream in India and Bangladesh.

Climate change has further amplified the urgency of transboundary water governance in South Asia. In recent decades, the Himalayan region has experienced increasingly devastating floods, while major rivers such as the Ganges are simultaneously undergoing accelerated drying. The disappearance of smaller tributaries and distributaries signals a looming large-scale hydrological crisis if the Ganges itself continues on this declining trajectory. At the same time, the degradation of these smaller river systems (tributaries and distributaries) is intensifying challenges related to transboundary water sharing and governance across national borders. Effective cooperation and management of these transboundary rivers are therefore essential. A collaborative framework enables real-time data sharing, joint early warning systems, coordinated disaster management, and sustainable development strategies for shared water resources. Strengthening regional institutions, promoting scientific collaboration, and fostering political trust are critical to achieving efficient river management. A well-integrated approach to transboundary water governance can transform these shared rivers from potential sources of conflict into instruments of regional peace and prosperity. This study aims to address existing research gaps by providing a comprehensive analysis of conflicts and cooperation surrounding transboundary rivers in South Asia through the lens of environmental sustainability. Its objectives are threefold: to identify major conflicts among riparian countries, including India, Bangladesh, Nepal, Pakistan, and China; to assess the impact of these conflicts on environmental sustainability; and to demonstrate how enhanced cooperation over transboundary rivers can contribute to sustainable development and the achievement of the SDGs in South Asian countries.

MATERIALS AND METHOD

This study employs a mixed-methods approach, combining qualitative desk-based research on agreements, treaties, and grey literature with quantitative methodologies, including existing environmental data on TRBs and geospatial data visualisation. The aim is to analyse transboundary river conflicts, cooperation frameworks, and environmental sustainability in South Asia. The research focuses on major TRBs, including the large basins such as the GBM basin and the Indus basin, as well as medium and small sized basins like the Teesta, Mahakali, Atrai and Koshi/Saptakoshi (within the GBM system) and the Ravi, Beas, and Sutlej (within the Indus system). These basins have been selected due to their geopolitical significance, histories of water-sharing disputes, and ecological importance. The study examines interactions between upstream and downstream riparian states, particularly India, Bangladesh, Nepal, Bhutan, and Pakistan. The study primarily relies on secondary data, both Statistical and geospatial and a literature-based analysis. Archival documents and historical records detailing transboundary water issues and conflicts were collected from government archives, including the International Centre for Integrated Mountain Development (ICIMOD) and the United Nations Environment Programme (UNEP), as well as diplomatic correspondence, government websites, government reports, daily news reports, and relevant historical publications. Policy documents and grey literature on water-sharing agreements, such as the IWT (1960) and the Ganges Water Sharing Treaty (1996), were reviewed to analyse institutional mechanisms for conflict resolution and cooperation. Comparative analyses of selected transboundary river disputes and cooperation frameworks were conducted to assess water availability and transboundary impacts and identify best practices and policy gaps. Data on climate change, disaster resilience, water pollution, and the occurrence of natural hazards and disasters were obtained from various government websites and environmental ministry portals across South Asian countries.

MAJOR TRANSBOUNDARY ISSUES AND CONFLICTS IN SOUTH ASIA

Himalayan glaciers, often referred to as the 'water towers of Asia', feed the Himalayan rivers, which serve as the principal source of water for South Asian countries and have historically sustained some of the world's most significant civilisations in this region. However, post-colonial South Asia faces profound transboundary water challenges arising from the complex river systems originating in the Himalayas, coupled with growing water demands, hydropower expansion, dam construction, shifting political boundaries, and escalating geopolitical tensions (Figure 1.2). For the past century, the geopolitics of South Asian countries have been shaped by critical concerns such as water sharing, hydropower projects, climate change, cross-border pollution, inadequate agreements and cooperation, sociopolitical conflicts, insufficient mechanisms for data sharing, unplanned development, and rapid population growth. These tensions have intensified in the 21st century, as water-related disasters have become more frequent and are closely linked to water infrastructure.

Following the end of colonial rule and the emergence of independent states, the history of transboundary water disputes in South Asia—including China, India, Bangladesh, Pakistan, Bhutan, and Nepal—began to unfold. Water sharing for agriculture, industrial processes, and urban expansion soon became one of the most pressing challenges for these new republics, each of which had previously been governed by external powers. Major disputes have since centred on the Indus and GBM river systems. Notable examples include India–Pakistan: the IWT; India–Bangladesh: the Ganges and Teesta Water Disputes and the Kushiyara Water Dispute; India–Nepal:

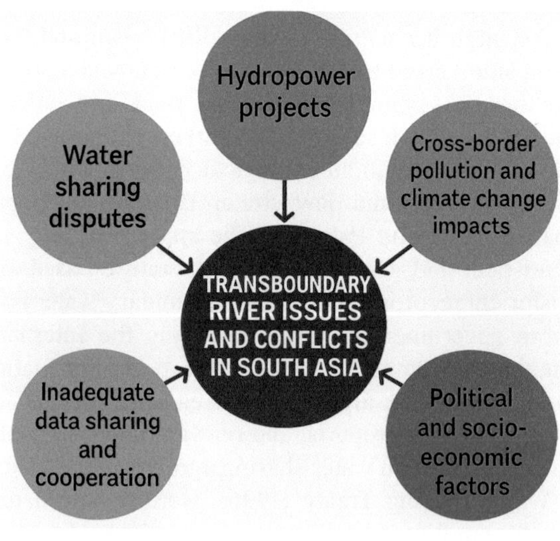

FIGURE 1.2 Major transboundary issues and conflicts in South Asia, highlighting water-sharing disputes, hydropower development challenges, cross-border pollution, climate change impacts, political and socio-economic tensions, and gaps in data sharing and cooperation, with examples from the Indus, Ganga, Brahmaputra, and other key river basins. (*Source:* Prepared by authors.)

Mahakali, Koshi, and Gandak Disputes; India–China: Brahmaputra mega-dam projects and water-sharing concerns; Bangladesh–Myanmar: Naf River and the Rohingya Crisis; and Bhutan–India: hydropower development and river management.

Water Sharing Disputes

India and Pakistan both gained independence from British colonial rule in 1947. Almost immediately, disputes arose between the two countries, one of the most significant being related to the sharing of the waters of the river Indus (Alam, 2002; Kalair et al., 2019; Mustafa, 2007). The first Indo-Pakistan war of 1947–1948 was intrinsically linked to disagreements over Indus water use and access (Munir et al., 2021). From independence until 1960, the two nations remained in near-constant conflict over water resources, ultimately leading to the signing of the IWT. The IWT, mediated by the World Bank, aimed to resolve disputes by partitioning rather than sharing river waters. It allocated the three eastern rivers (Ravi, Beas, and Sutlej) to India and the three western rivers (Indus, Chenab, and Jhelum) to Pakistan. Although considered a milestone in transboundary water diplomacy, the treaty did not eliminate disputes. Pakistan has repeatedly accused India of seeking hydrohegemony by storing or diverting waters from the western rivers through hydroelectric projects, even when technically permissible under the IWT framework. In response, the Permanent Indus Commission (PIC) was created, with two commissioners from each side tasked with resolving issues and meeting annually (Condon et al., 2014). The 118th meeting of the PIC took place on 30–31 May 2022. Yet tensions persist, and in 2025 India announced the suspension of the IWT, citing Pakistan's alleged support for cross-border terrorism, further destabilising already fragile water relations.

The Indo-Bangladesh water disputes similarly reflect a legacy of partition and post-colonial geopolitics. After Bangladesh's independence in 1971, the two neighbours sought to stabilise relations through the India-Bangladesh Treaty of Friendship, Cooperation and Peace (Haines, 2014). Yet the construction of the Farakka Barrage by India during 1962–1972 soon emerged as a major source of contention. Commissioned in 1975, the barrage was designed to divert water to maintain the navigability of Kolkata port and support the National Thermal Power Corporation at Farakka (Adel, 2013; Kawser and Samad, 2016). However, Bangladesh accused India of excessive withdrawal of dry season flows, resulting in declining water availability, environmental degradation in the Sundarbans, increasing salinity, and water contamination, with knock-on impacts on agriculture, fisheries, and public health (Huq and Easher, 2021). To address these disputes, the 30-year Ganges Water Sharing Treaty was signed in 1996 (Vinayaraj, 2009). While hailed as progress at the time, the agreement has not resolved tensions. As the treaty approaches expiry in 2026, political mistrust remains high. In addition to Farakka, unresolved disagreements over the Teesta River Agreement and the Kushiyara Water Sharing Agreement have exacerbated bilateral strains (Rahman and Rahaman, 2018; Rudra, 2018b). With 54 rivers flowing into Bangladesh from India, most of which have been dammed or diverted upstream, Dhaka continues to accuse New Delhi of inequitable water distribution.

Further east, the India–China water-sharing dispute over the Brahmaputra represents one of South Asia's most complex transboundary challenges. Originating as the

Yarlung Tsangpo in Tibet, the river flows through China into India's Arunachal Pradesh and Assam, before joining Bangladesh. While most of the basin area lies in China, it is largely arid, whereas the Indian segment receives heavy monsoon rainfall. Tensions have escalated due to China's hydropower development, including the Zangmu Dam and proposed mega-projects at the Great Bend in Tibet (Giordano and Wahal, 2024). Beijing insists these are run-of-river projects that do not significantly alter flows, yet India remains concerned about reduced downstream water availability, ecological disruptions, and long-term diversion under schemes such as the South-North Water Diversion Project (Bukhari, 2024). Another point of contention is China's limited sharing of hydrological data. Although a bilateral agreement ensures the provision of monsoon-season flow data, the lack of real-time information during critical events such as flash floods leaves India vulnerable, particularly in Assam and Arunachal Pradesh (Deka et al., 2023). The absence of a legally binding agreement on Brahmaputra waters, compounded by broader Sino-Indian border tensions, means that water-sharing discussions remain sporadic, fragmented, and overshadowed by strategic rivalry (Feng et al., 2019). India–China water relations have further deteriorated following China's decision to construct what is set to be the world's largest dam on the Great Bend of the Yarlung Tsangpo River, just upstream of its entry into India. These tensions are compounded by ongoing geopolitical disputes between the two countries, including recent confrontations over control of the Galwan River valley in the Ladakh region.

India's water disputes with Nepal are equally long-standing. The Mahakali, Koshi, and Gandak rivers are vital for irrigation, hydropower, and flood management across the two countries. Several treaties have been signed, including the Koshi Agreement (1954), the Gandak Agreement (1959), and the Mahakali Treaty (1996) (Bagale, 2020). Nevertheless, these arrangements have often fuelled rather than resolved discontent. For example, the Koshi Agreement authorised the Koshi Barrage for flood control and irrigation, but repeated embankment breaches, notably in 2008, 2017, and 2023, displaced thousands in Nepal and millions in India's Bihar state. Nepal has argued that India has failed to adequately maintain infrastructure, increasing flood vulnerability on both sides of the border. The Gandak Agreement also remains controversial, with Nepal contending that India has not fulfilled commitments to guarantee irrigation supplies, while mismanagement of canals has restricted Nepalese farmers' access (Gupta et al., 2021; Vasani, 2023). The Mahakali Treaty, intended to provide equal benefits through the Pancheshwar Multipurpose Project (PMP), has been plagued by delays. Nepal maintains that India's reluctance to implement the PMP reflects asymmetrical power dynamics, undermining its hydropower and development goals. Despite regular dialogues, progress remains slow, with bureaucratic and political obstacles preventing the creation of genuinely cooperative water-sharing mechanisms.

In addition to these major disputes, Afghanistan and Pakistan have their own tensions. With Indian assistance, Afghanistan has pursued dam construction on the Kabul River—up to 12 projects are planned—which flows into Pakistan. Islamabad fears these dams will reduce its downstream supply, further straining already scarce resources. Similarly, Bhutan and India, despite generally cordial relations, face challenges over hydropower cooperation, while Bangladesh and Myanmar have clashed over the Naf River, exacerbated by the Rohingya refugee crisis. Underlying all these disputes is the intensifying influence of climate change. South Asia, heavily reliant on Himalayan-fed rivers, faces unprecedented pressures due to glacial retreat, shifting monsoon patterns, and increasing extremes of flood and drought (Dar and Dar, 2022). The Himalayas are warming faster than the global average, threatening the long-term sustainability of river flows. Seasonal water shortages in the Ganges, coupled with catastrophic floods in the Brahmaputra and Koshi basins, underscore the urgent need for adaptive, cooperative frameworks. Without enhanced cooperation, unilateral dam building, unchecked withdrawals, and inadequate data sharing are likely to exacerbate conflict and undermine regional stability.

The persistence of disputes highlights several structural weaknesses (Figure 1.3). Firstly, many treaties, such as the IWT and Ganges Treaty, reflect mid-20th-century concerns but inadequately address 21st-century challenges like climate change, disaster resilience, changing monsoon patterns, ecological degradation, and environmental sustainability. Secondly, asymmetries of power—particularly India and China's dominant riparian position—have generated mistrust among smaller neighbours such as Bangladesh, Bhutan and Nepal. Other political conflicts have also affected water governance, particularly the wars and prolonged tensions between India and Pakistan, which have placed sustained pressure on existing water treaties. Thirdly, there is a notable absence of multilateral regional frameworks. Most agreements are bilateral and narrowly focused, leaving broader basin-wide cooperation underdeveloped. This fragmented approach hinders effective management of shared resources. Furthermore, most existing agreements and treaties are largely

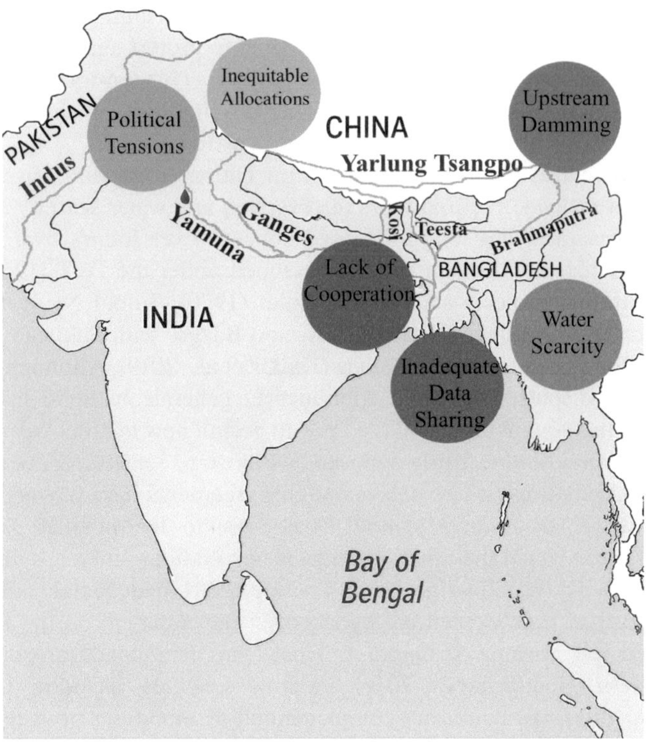

FIGURE 1.3 Graphical representation of transboundary water-sharing disputes in South Asia, focusing on the Indus, Ganges, and Brahmaputra basins. (*Source:* Prepared by authors.)

focused on major river basins, with little consideration given to smaller tributaries. The disappearance of these tributary systems, along with the lack of tributary-scale documentation and multi-level discussion, represents a critical gap in South Asian transboundary water governance and an urgent area for policy and research attention. To overcome these challenges, stronger regional institutions, scientific collaboration, and inclusive governance mechanisms are needed. Cooperative frameworks should integrate real-time hydrological data sharing, joint disaster risk management, and ecological considerations into treaty design. Regional bodies could play a pivotal role in mediating disputes and promoting sustainable development. A well-integrated approach to transboundary water governance has the potential to transform rivers from sources of conflict into instruments of peace, resilience, and shared prosperity.

Hydropower Projects and the Origin of Disputes

The development of hydropower projects on transboundary rivers in South Asia has been both a catalyst for cooperation and a source of conflict among riparian nations, largely due to unequal water allocation, environmental degradation, and underlying geopolitical rivalries. India's first Prime Minister, Jawaharlal Nehru, famously described dams as the 'temples of modern India', encapsulating the post-colonial South Asian enthusiasm for harnessing Himalayan rivers for hydropower and modernisation. Yet, from the outset, resistance to such projects emerged. A notable example is the anti-dam mobilisation led by Maulana Abdul Hamid Khan Bhasani, who organised the Farakka Long March in May 1976 to protest against India's Farakka Barrage and its downstream impacts on Bangladesh. This tension reflects a broader pattern across South Asia, where hydropower development on shared river basins often creates friction between upstream and downstream states.

One of the central challenges is upstream nations exercising disproportionate control over river flows, fuelling fears of diversion and water scarcity downstream. Pakistan, for instance, has long expressed concern over India's hydropower projects on the Chenab and Jhelum rivers, developed under the IWT (1960). Disputes have arisen around projects such as the Salal (1970), Tulbul Navigation Project, Baglihar Dam, Kishenganga, Ratle, and Nimoo Bazgo, with Pakistan arguing that India's schemes violate treaty provisions (Kalair et al., 2019). Although a PIC exists to resolve such disputes, the issue continues to generate bilateral mistrust. Most recently, India has sought to amend the IWT to permit new hydropower development on the Chenab, an initiative firmly opposed by Pakistan. Tensions are equally evident in India–Nepal relations, where delays and disagreements have plagued joint initiatives such as the PMP on the Mahakali River. Despite the Mahakali Treaty (1996), concerns persist in Nepal that India dominates negotiations and reaps disproportionate benefits, leaving Nepali irrigation and power needs inadequately addressed.

Further north, China's extensive hydropower development on the Brahmaputra River—called the Yarlung Tsangpo in Tibet—has heightened anxieties in India and Bangladesh (Bhattacharya, 2018). Existing schemes, including the Zangmu, Dagu, Jiacha, and Jeixu dams, are complemented by ambitious projects such as the Motuo hydropower plant at the Great Bend, a mere 3 km from the Indian border (Mohammed and Basu, 2022). China maintains that these are largely run-of-river schemes, but downstream states fear reductions in water flow, altered flood regimes, and ecological disruption. The India–Bangladesh context demonstrates similar contestations. The Farakka Barrage, commissioned in 1975 to secure water for Kolkata Port and power plants, remains highly contentious. Bangladesh accuses India of excessive water withdrawal, which has intensified salinity, environmental degradation, and public health challenges in downstream areas, including the Sundarbans (Sen et al., 2022). Conversely, India opposes Bangladesh's proposed dam at Pangsha, arguing it would heighten flood risks on its territory.

Small-scale dam disputes, such as rubber dams in Bangladesh, are rarely discussed at the national or central policy level. For example, the Mohanpur rubber dam in Bangladesh and small concrete dams in Balurghat on the Atrai River illustrate how localized infrastructure interventions can generate significant social and hydrological tensions, yet remain largely overlooked in broader transboundary water governance debates. The hydropower development in South Asia exemplifies the delicate balance between energy security, ecological sustainability, and regional geopolitics. While projects promise development and cooperation, their unilateral implementation often exacerbates mistrust, intensifying transboundary water disputes. Figure 1.4 highlights both existing and planned hydropower schemes in South Asia, illustrating how energy ambitions are deeply entangled with regional water politics.

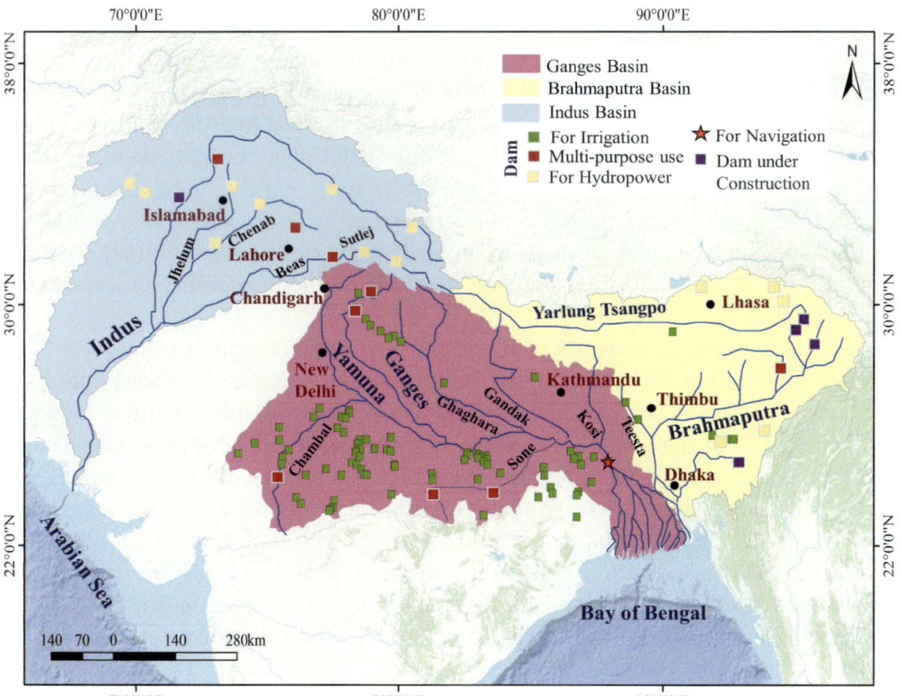

FIGURE 1.4 Transboundary hydropower projects in South Asia, illustrating upstream–downstream dynamics, contested dam construction, and associated disputes over water allocation, energy generation, and ecological impacts across the Indus, Ganges, and Brahmaputra. (*Source:* Prepared by authors.)

Cross-Border Water Pollution and Climate Change Impacts

Pollution of transboundary rivers in South Asia remains a significant concern for riparian countries (Biswas, 2011; Uprety and Salman, 2011). The region's large populations exert immense pressure on already scarce resources, with high water demand frequently resulting in contamination and scarcity (Sather et al., 2022). Rapid industrialisation and urban expansion have intensified water pollution, particularly from chemicals, untreated sewage, and industrial waste. Increasing urbanisation has also exacerbated river contamination, particularly through sewage disposal and the pervasive challenge of plastic waste. The river Ganga, for instance, flows through major urban centres and is heavily contaminated by sewage and waste inputs. Industrial waste from factories in the Hugli Industrial Belt and agricultural run-off from pesticides and insecticides further degrade water quality. Similar challenges are observed in the upstream Indus basin, which traverses extensive agricultural tracts, leading to chemical contamination of shared waters. Several recent incidents of cross-border water pollution have escalated into diplomatic concerns. The Bangladeshi government has raised alarms about the contamination of the cross-border Titas River. Likewise, microplastics in the Koshi River have emerged as a transboundary issue, with pollutants originating in China and travelling downstream through Nepal to India (Yang et al., 2021). Pakistan has also voiced concerns that untreated wastewater and industrial effluents discharged into the Hudiara drain in India contribute to the pollution of shared rivers, notably the Ravi.

In addition to anthropogenic pressures, climate change is compounding transboundary water challenges in South Asia. The region is experiencing accelerated glacier melt due to extreme warming, which is altering river hydrology. Rapid glacier retreat has increased river discharge, triggering GLOFs and flash floods in upstream regions. Downstream, these processes exacerbate recurrent riverine floods, salinisation of freshwater systems, and the submergence of deltas and coastal zones. Given South Asia's high population density and growing water demand, climate-induced changes further threaten water security by reducing the availability of uncontaminated freshwater.

Upstream countries such as China and India have the capacity to alter river flows through the construction of hydraulic infrastructure. Such interventions could significantly reduce, or even halt, downstream flows, creating vulnerabilities for riparian states. China's rapid construction of large-scale hydropower dams and barrages on rivers originating in Tibet remains a source of deep concern for its neighbours, particularly since Beijing has avoided entering into binding water treaties. India's own hydropower investments also generate tensions with Bangladesh over the Ganga and with Pakistan over the Indus. Transboundary water disputes are increasingly linked to climate stressors such as drought and summer desiccation. Both the Indian and Pakistani stretches of the Indus basin have reported recurrent water shortages during dry seasons. Bangladesh has also raised concerns about recurring summer droughts in its northern regions, attributing reduced flows in the Teesta and Ganga rivers to upstream withdrawals. These examples underscore the intersection of pollution, climate change, and geopolitical tensions in exacerbating cross-border water insecurity in South Asia.

Political and Socio-Economic Factors

The geopolitics of South Asia in relation to water carries deep historical and socio-economic legacies. The colonial era left behind extensive water engineering projects which reshaped the physical and political landscape of river systems (Williams, 2018). Partition and subsequent nation-building processes further influenced how water is perceived, managed, and contested in the region. Water diplomacy today continues to be shaped not only by geography and hydrology but also by the political histories and socio-economic realities of the countries involved (Imran et al., 2023). The region's demographic and economic pressures further intensify these challenges. South Asia is home to some of the most densely populated countries in the world, and millions of people depend directly on transboundary rivers for agriculture, fisheries, and their livelihoods. The vast floodplains of the Ganga-Brahmaputra and Indus basin, enriched by massive sediment flows, form the backbone of agro-economies, sustaining millions through fertile farmland (Bukhari, 2024). At the same time, rapid urbanisation and industrialisation across India, Pakistan, Bangladesh, and Nepal have driven increasing water demand, compounding pressures on already stressed systems. Political relations between riparian countries add another layer of complexity. India and Pakistan, for instance, have fought four wars (1947–1948, 1965, 1971, and 1999) and continue to be divided over Kashmir, border disputes, and cross-border terrorism. These conflicts inevitably spill over into water diplomacy, affecting the implementation and trustworthiness of agreements such as the IWT. India and China, too, have a fraught history, marked by the Sino-Indian War of 1962, the Nathu La and Cho La clashes of 1967, and recurring confrontations in Ladakh, Arunachal Pradesh, and Sikkim. More recently, political shifts in Bangladesh following the fall of the Hasina regime (2024) have strained its once-close relationship with India, with repercussions for bilateral water-sharing discussions. Socio-economic conditions across borders also heavily shape patterns of conflict and cooperation. Nationalist ideologies in particular have complicated diplomacy: in Pakistan, calls for a so-called 'water jihad' against India reflect the securitisation of water disputes, while in India, political voices demanding the cancellation of treaties with Pakistan demonstrate a rise in 'water nationalism'. Similar nationalist undertones affect India's relations with Bangladesh and Nepal. Such politicisation of water issues means that conflicts over rivers are often inseparable from broader disputes, making cooperative governance a persistent challenge.

Inadequate Data Sharing and Cooperation

The nations of South Asia are characterised by diverse political, social, and religious ideologies, which often create divisions and impede regional cooperation, particularly regarding the sharing of hydrological data (Hayden, 2002). Variations in economic development and military capacity further complicate collaboration, preventing the establishment of robust regional frameworks or organisations capable of facilitating meaningful data exchange. In some cases, more powerful states deliberately refrain from entering binding treaties with their neighbours, perceiving such commitments as potential constraints on national development strategies. Consequently, limited data sharing and weak communication channels between riparian countries exacerbate mistrust and hinder the effective management of transboundary river systems (Figure 1.5). These disputes are not solely determined by geography; rather, asymmetrical power relations sustain ongoing tensions. Given that the economies of many South Asian states remain heavily reliant on agriculture, disagreements over water allocation frequently dominate bilateral relations. Despite numerous treaties and agreements, many issues remain unresolved, and negotiations often stall. For instance, Bangladesh has repeatedly expressed concern that India does not provide real-time hydrological data on upstream diversions, barrage operations, or groundwater extraction, undermining effective flood and drought management. Similarly, during the 2017 Doklam stand-off, India reported that China suspended the sharing of Brahmaputra flow data, deepening mistrust. In the India–Nepal context, seasonal flow data for rivers such as the Gandak and Koshi, joint feasibility studies for hydropower projects, and water use agreements exist, yet critical operational and real-time data are often withheld, limiting the effectiveness of cooperative river basin management.

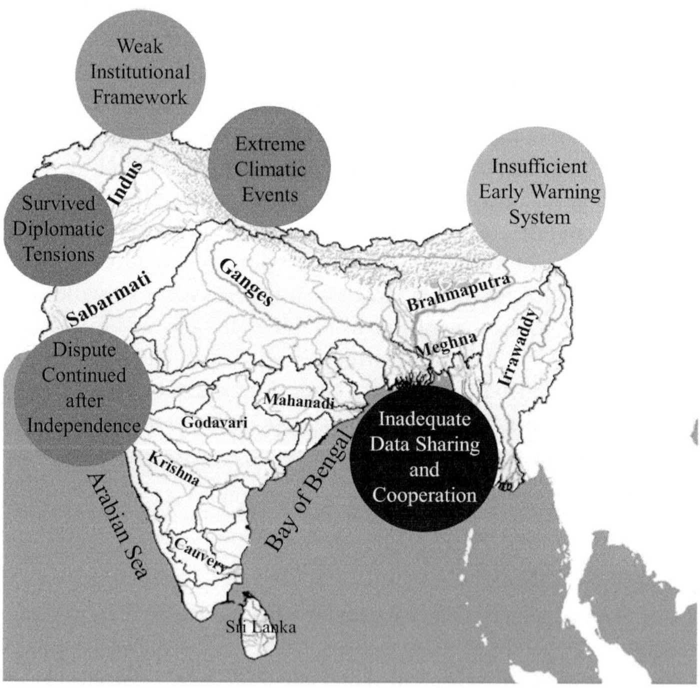

FIGURE 1.5 Inadequate data sharing and cooperation in South Asia, emphasising the limited exchange of hydrological information, lack of real-time forecasting and early warning, and the absence of legally binding agreements in the transboundary basins. (*Source:* Prepared by authors.)

TRANSBOUNDARY RIVER DISPUTES AND ENVIRONMENTAL CHALLENGES

The riverine ecosystems of border regions in South Asia have often been overlooked in policy and management decisions. However, geopolitical conflicts and unilateral developmental priorities have frequently resulted in substantial environmental degradation. Riparian nations in the region tend to prioritise economic growth and infrastructure development over sustainable approaches to river management, often ignoring long-term ecological consequences (Chen and Swain, 2014; Johnston, 2013). Rivers, by their very nature, are transboundary systems that flow across national borders, connecting upstream and downstream regions ecologically and hydrologically (Warner et al., 2008). Consequently, disruptions in one section of a river system can have cascading effects on downstream ecosystems, riparian livelihoods, and floodplain health. Flash floods in the hilly regions of South Asia provide stark examples of the environmental costs of mismanaged river systems. The 2021 flash flood in the Teesta River, Darjeeling Himalayan Foothill Region, West Bengal, was triggered by the excessive release of water from the upstream Teesta Barrage following intense rainfall. This event caused significant damage to the riparian ecosystem, altered river health, and impacted floodplain biodiversity. Similarly, uncoordinated dam releases, embankment failures, and alterations to river channels have repeatedly resulted in large-scale losses of life and property across the region. In parallel, the disappearance of small tributaries and distributaries has emerged as a critical and increasingly visible issue within South Asian transboundary river basin systems. This phenomenon demands urgent scholarly and policy attention, as it is closely linked to climate change impacts and poses serious risks to environmental sustainability and long-term water security across the region.

Rapid industrialisation, urbanisation, and population growth in South Asia have increased the demand for freshwater, while climate change and glacial retreat have reduced its availability. Upstream countries, leveraging their geographical advantage, frequently construct dams and hydraulic structures to divert river flow, exacerbating water scarcity downstream. For example, China's continuous construction of dams along the Yarlung Tsangpo River (Brahmaputra) poses significant concerns for India, particularly the North-Eastern states, as reduced downstream flows threaten agriculture and local water supply. Likewise, India's hydroelectric projects and diversion efforts on rivers such as the Teesta have created persistent water shortages in Bangladesh, affecting irrigation and the livelihoods of thousands of farmers, despite agreements such as the Teesta Water Treaty negotiations initiated in 1996.

Unilateral development and insufficient cooperation have broader ecological consequences. Continuous river diversions, dam construction, and excessive water extraction disrupt the natural flow regime, increase river pollution, and compromise sediment transport, leading to erosion, habitat loss, and ecosystem instability. For instance, India's Farakka Barrage on the Ganges has altered sediment deposition patterns, resulting in riverbank erosion and declining soil fertility in downstream Bangladesh. Similarly, floods along the Koshi River in 2008, caused by embankment breaches at the India-Nepal border, displaced millions and highlighted the consequences of uncoordinated cross-border river management (Gupta et al., 2021). Transboundary conflicts also accelerate biodiversity loss and wetland degradation.

Aquatic habitats are increasingly threatened, putting species such as the Ganges River Dolphin and migratory birds at risk. The Sundarbans mangrove forest, a UNESCO World Heritage Site shared by India and Bangladesh, experiences heightened salinity intrusion due to altered freshwater flows, largely resulting from upstream diversions like the Farakka Barrage (Sahana et al., 2022). Similarly, reduced flows in the Indus River have caused seawater intrusion in Pakistan's Indus Delta, leading to the loss of fertile agricultural land and fisheries (Janjua et al., 2021).

TRANSBOUNDARY RIVER COOPERATION FOR ENVIRONMENTAL SUSTAINABILITY

Transboundary river cooperation in South Asia is essential for achieving environmental sustainability, given that the region hosts several major rivers, including the Ganges, Brahmaputra, Indus, and Teesta, which traverse multiple national boundaries. Cooperation over these transboundary rivers is crucial for addressing pressing environmental challenges such as water scarcity, flood, drought, riverbank erosion, pollution, biodiversity loss, disappearance of small tributaries, climate change impacts, and ecosystem degradation. The hydrological and ecological sustainability of these basins depends on maintaining natural flow regimes, which are vital for preserving aquatic ecosystems, wetlands, and groundwater recharge. Scientific research highlights the importance of environmental flows (e-flows) in sustaining riverine biodiversity, fisheries, and sediment transport, all of which are critical for maintaining ecological integrity (Gebreegziabher et al., 2023). Collaborative frameworks among riparian nations facilitate the implementation of e-flow policies, ensuring downstream ecosystems receive adequate water supply, thereby preventing habitat degradation and species loss.

Water quality management represents another crucial dimension of transboundary river cooperation. Evidence indicates that pollution from agricultural run-off, industrial effluents, and untreated sewage significantly undermines river health in South Asia (Sarker et al., 2021). Coordinated approaches involving real-time water quality monitoring, data sharing, and stringent pollution control measures can enhance the condition of these rivers. The establishment of shared water quality standards and regional agreements can further enable joint action to address pollution hotspots, thus supporting environmental sustainability. Groundwater management is also a critical concern, as aquifers crossing national boundaries are often over-exploited. Scientific assessments of transboundary groundwater flow, recharge rates, and pollution levels provide a basis for developing joint aquifer management policies, preventing over-extraction and contamination. Collaborative groundwater recharge initiatives, including artificial recharge structures and afforestation programmes in catchment areas, can strengthen water security and reduce ecological stress.

Climate change has further exacerbated vulnerabilities in TRBs through alterations in precipitation patterns, glacial melt, and extreme weather events such as floods and droughts (Waseem and Chaudhry, 2024). Scientific hydrological modelling offers insights into future climate scenarios, informing adaptive water management strategies. Cooperation on climate resilience initiatives, including transboundary flood forecasting and early warning systems, reduces disaster risks and improves preparedness in downstream communities. The Integrated River Basin Management (IRBM), informed by scientific climate-risk assessments, ensures equitable water distribution while safeguarding ecological functions. Biodiversity conservation is closely intertwined with transboundary cooperation, as ecologically significant species, including the Ganges River dolphin and various migratory fish, depend on healthy, connected river systems. Habitat fragmentation caused by uncoordinated infrastructure development, such as dams and barrages, disrupts natural migration and breeding patterns. Modern scientific approaches, including environmental DNA (eDNA) monitoring, remote sensing, and habitat connectivity modelling, can guide conservation strategies to protect transboundary aquatic ecosystems.

Hydropower development provides both opportunities and challenges for environmental sustainability. While hydropower is a renewable energy source, uncoordinated dam construction has significant ecological consequences, including altered flow regimes, disrupted fish migration, and reduced sediment transport. Scientific impact assessments, including cumulative environmental impact studies and life-cycle analyses, inform sustainable hydropower planning that minimises ecological disruption while enhancing regional energy security. Sediment management is another vital aspect, as Himalayan rivers carry essential sediment loads that maintain deltaic ecosystems, prevent coastal erosion, and sustain agricultural fertility (Paszkowski et al., 2021). However, uncoordinated dam construction and excessive sand mining disrupt sediment flow, causing land subsidence and increasing flood risks. Coordinated sediment management among riparian nations is therefore critical to balance hydropower development with ecosystem preservation.

Globally, nations experience varying degrees of conflict over shared river resources, with disputes influenced by basin size, the number of riparian states, and geopolitical dynamics (Chen et al., 2016). In South Asia, research on TRBs has predominantly focused on geopolitical tensions and natural hazards, given their role as primary drivers of interstate conflict. In contrast, studies from Europe often emphasise policy frameworks and climate adaptation strategies (Kampragou et al., 2011). Despite widespread discussion of these issues, developing practical management strategies remains an urgent necessity. Effective cooperation mitigates over-exploitation, preserves natural flow and water quality, and sustains biodiversity. Resolving disputes fosters collaboration, reduces the likelihood of environmentally harmful conflicts, and promotes joint resource management, ensuring equitable access and economic stability. Collaborative management enables coordinated adaptation to climate change impacts, including water scarcity and extreme weather events. Such strategies require active engagement from organisations and associations at national, regional, and local levels. Additionally, integrating traditional knowledge and community-based water management practices has significant scientific value. Indigenous techniques, such as floodplain agriculture and wetland restoration, enhance ecosystem resilience (Gupta et al., 2024). Combining these traditional practices with modern scientific approaches through participatory governance frameworks can deliver effective, context-specific management solutions. Sustainable transboundary river management benefits communities reliant on these water resources. It ensures reliable water access, supports agriculture, and sustains socio-economic activities essential for livelihoods (Opiyo et al., 2024). Resolving conflicts builds trust among nations, facilitating cooperation that extends beyond water management to broader environmental policies and initiatives. Ultimately, collaborative approaches in South Asia's TRBs are indispensable for achieving ecological sustainability, socio-economic resilience, and long-term regional stability.

CONCLUSION

Transboundary river conflicts are widespread across South Asia, a region with the highest number of large rivers and numerous small tributaries that traverse multiple countries and share their water resources. Upstream nations often divert water through the construction of hydraulic structures, reducing the supply available to downstream regions. Such interventions significantly impact river health and alter natural flow regimes, with potentially severe consequences for the environment. The construction of dams and barrages has also contributed to flash floods during the monsoon season, further affecting riverine ecosystems. Countries such as Bangladesh and parts of India are experiencing water crises that threaten both the economy and local livelihoods.

Climate change is intensifying these challenges, with glacial melt, summer droughts, and prolonged dry periods increasing salinity in downstream river waters and creating significant water deficits. Conflicts between riparian nations, even after signing agreements and treaties, often fail to meet the demands of all parties, leading to tensions and, at times, conflict. Public mistrust, blame games, and rumours surrounding water-related disasters further exacerbate tensions in South Asia's TRBs. However, wars and conflicts are not a solution to the region's water challenges. Rapid population growth and developing economies are driving urbanisation and industrialisation, further straining the balance between water supply and demand, and causing substantial environmental degradation. There is therefore an urgent need for collaborative solutions. Governments, non-governmental organisations, and committees at national, regional, and local levels must take proactive steps and adopt strategies aligned with the SDGs. Integrated river basin management programmes, designed in consultation with environmental experts, can safeguard fragile ecosystems while supporting sustainable economic development, ensuring a balance between ecological preservation and growth.

REFERENCES

Adel, M. M. (2013). Upstream water piracy, the strongest weapon of cornering a downstream nation. Environment and Ecology Research, 1(3), 85–128.

Alam, U. Z. (2002). Questioning the water wars rationale: A case study of the Indus Waters Treaty. Geographical Journal, 168(4), 341–353.

Albrecht, T. R., & Gerlak, A. K. (2022). Beyond the basin: Water security in transboundary environments. Water Security, 17, 100–124.

Bagale, D. R. (2020). Nepal–India water cooperation: Consequences of mutuality or hegemony? Water Policy, 22(6), 1098–1108.

Baten, M. A., & Titumir, R. A. M. (2016). Environmental challenges of trans-boundary water resources management: The case of Bangladesh. Sustainable Water Resources Management, 2, 13–27.

Bhattacharya, S. (2018). China's hydro ambitions and the Brahmaputra. Institute for Defence Studies and Analyses. Available at: https://idsa.in/system/files/backgrounder/b-china-hydro-ambitions-and-the-brahmaputra.pdf (accessed 15 June 2020).

Biswas, A. K. (2011). Cooperation or conflict in transboundary water management: Case study of South Asia. Hydrological Sciences Journal, 56(4), 662–670.

Bukhari, S. R. H. (2024). Hydro-imperialism: China's water dominance and the transformation of global geopolitical power. Annals of Human and Social Sciences, 5(3), 411–425.

Chaudhary, M. A., Tanvir, M., & Choudhry, M. H. M. (2024). Navigating water security in South Asia: Towards regional cooperation and sustainable solutions. Annals of Human and Social Sciences, 5(2), 99–110.

Chen, H., & Swain, A. (2014). The Grand Ethiopian Renaissance Dam: Evaluating its sustainability standard and geopolitical significance. Energy Development Frontier, 3(1), 11–19.

Chen, J., Shi, H., Sivakumar, B., & Peart, M. R. (2016). Population, water, food, energy and dams. Renewable and Sustainable Energy Reviews, 56, 18–28.

Condon, M., Kriens, D., Lohani, A., & Sattar, E. (2014). Challenge and response in the Indus Basin. Water Policy, 16(S1), 58–86.

Crausbay, S. D., Betancourt, J., Bradford, J., Cartwright, J., Dennison, W. C., Dunham, J., … & Carter, S. (2020). Unfamiliar territory: Emerging themes for ecological drought research and management. One Earth, 3(3), 337–353.

Dar, J., & Dar, A. Q. (2022). Dominant patterns of seasonal precipitation variability in association with hydrological extremes over the North-west Himalayas. Environmental Science and Pollution Research, 29(47), 71593–71606.

Deka, A., Hazarika, N., Vij, S., Barua, A., & Fantini, E. (2023). Media reporting on conflicts and cooperation: What does it mean for the Brahmaputra basin? International Journal of Water Resources Development, 39(5), 819–845.

Feng, Y., Wang, W., & Liu, J. (2019). Dilemmas in and pathways to transboundary water cooperation between China and India on the Yaluzangbu-Brahmaputra River. Water, 11(10), 2096.

Gebreegziabher, G. A., Degefa, S., Furi, W., & Legesse, G. (2023). Evolution and concept of environmental flows (e-flows): Meta-analysis. Water Supply, 23(6), 2466–2490.

Giordano, M., & Wahal, A. (2024). Hydro-diplomacy on the Brahmaputra: Examining the interplay of water, territorial claims, and infrastructure. International Journal of Water Resources Development, 40(6), 1032–1052.

Gupta, N., Dahal, S., Kumar, A., Kumar, C., Kumar, M., Maharjan, A., … & Unni, A. (2021). Rich water, poor people: Potential for transboundary flood management between Nepal and India. Current Research in Environmental Sustainability, 3, 100031.

Gupta, R., Chembolu, V., Marjoribanks, T. I., & Dutta, S. (2024). Assessing the efficacy of hydro-ecological based wetland management approach for flood resilience of a large river catchment. Journal of Hydrology, 641, 131761.

Haines, D. (2014). Disputed rivers: Sovereignty, territory and state-making in South Asia, 1948–1951. Geopolitics, 19(3), 632–655.

Hanasz, P. (2017). Muddy waters: International actors and transboundary water cooperation in the Ganges-Brahmaputra Problemshed. Water Alternatives, 10(2), 459–474.

Hayden, R. (2002). Antagonistic tolerance: Competitive sharing of religious sites in South Asia and the Balkans. Current Anthropology, 43(2), 205–231.

Huda, M. S. (2017). Envisioning the future of cooperation on common rivers in South Asia: A cooperative security approach by Bangladesh and India to the Tipaimukh Dam. Water International, 42(1), 54–72.

Huq, H., & Easher, T. H. (2021). Coastal water: Wisdom, destruction, conflicts and contestation – A case of southwest coastal region of Bangladesh. Coastal Environments, 21, 1–14.

Imran, M., Mustafa, K., Mustafa, G., & Khan, M. U. (2023). Hydropolitics and conundrum of transboundary water issues: A case study of South Asia. Journal of Politics and International Studies, 7(1), 17–35.

Janjua, S., Hassan, I., Muhammad, S., Ahmed, S., & Ahmed, A. (2021). Water management in Pakistan's Indus Basin: Challenges and opportunities. Water Policy, 23(6), 1329–1343.

Johnston, B. R. (2013). Human needs and environmental rights to water: A biocultural systems approach to hydrodevelopment and management. Ecosphere, 4(3), 1–15.

Kalair, A. R., Abas, N., Hasan, Q. U., Kalair, E., Kalair, A., & Khan, N. (2019). Water, energy and food nexus of Indus Water Treaty: Water governance. Water-Energy Nexus, 2(1), 10–24.

Kampragou, E., Apostolaki, S., Manoli, E., Froebrich, J., & Assimacopoulos, D. (2011). Towards the harmonization of water-related policies for managing drought risks across the EU. Environmental Science & Policy, 14(7), 815–824.

Karthe, D., Chalov, S., & Borchardt, D. (2015). Water resources and their management in central Asia in the early twenty first century: Status, challenges and future prospects. Environmental Earth Sciences, 73, 487–499.

Katyaini, S., Barua, A., & Duarte, R. (2020). Science-policy interface on water scarcity in India: Giving 'visibility' to unsustainable virtual water flows (1996–2014). Journal of Cleaner Production, 275, 124059.

Kawser, M. A., & Samad, M. A. (2016). Political history of Farakka Barrage and its effects on environment in Bangladesh. Bandung, 3(1), 1–14.

Khan, R., Muzaffar, M., & Mustafa, G. (2022). Pakistan-India water conflict: A causal analysis. Annals of Social Sciences and Perspective, 3(1), 43–51.

Mohammed, A. N., & Basu, B. (2022). Geopolitics over Brahmaputra water: Dam constructions in Tibet and Arunachal Pradesh. Water and Energy International, 65(8), 28–34.

Moser, S. C. (2010). Communicating climate change: History, challenges, process and future directions. Wiley Interdisciplinary Reviews: Climate Change, 1(1), 31–53.

Munir, K., Khalid, I., & Shahrukh, W. (2021). Water conflict between Pakistan and India: Implications to regional peace and security. Journal of Humanities, Social and Management Sciences (JHSMS), 2(2), 255–270.

Mustafa, D. (2007). Social construction of hydropolitics: The geographical scales of water and security in the Indus Basin. Geographical Review, 97(4), 484–501.

Ojima, D. S. (2021). Climate resilient management in response to flash droughts in the US Northern great plains. Current Opinion in Environmental Sustainability, 48, 125–131.

Opiyo, S. B., Opinde, G., & Letema, S. (2024). A perspective of sustainable livelihood framework in analysis of sustainability of rural community livelihoods: Evidence from Migori River watershed in Kenya. International Journal of River Basin Management, 22(4), 627–643.

Paszkowski, A., Goodbred, S., Jr, Borgomeo, E., Khan, M. S. A., & Hall, J. W. (2021). Geomorphic change in the Ganges–Brahmaputra–Meghna delta. Nature Reviews Earth & Environment, 2(11), 763–780.

Rahaman, M. M. (2009). Principles of transboundary water resources management and Ganges treaties: An analysis. International Journal of Water Resources Development, 25(1), 159–173.

Rahman, M. M., & Rahaman, M. M. (2018). Impacts of Farakka barrage on hydrological flow of Ganges River and environment in Bangladesh. Sustainable Water Resources Management, 4(4), 767–780.

Rudra, K. (2018a). Rivers of the Ganga–Brahmaputra–Meghna Delta: An Overview. In *Rivers of the Ganga-Brahmaputra-Meghna Delta: A Fluvial Account of Bengal* (pp. 1–14). Springer International Publishing.

Rudra, K. (2018b). Evolution of the Bengal Basin. In *Rivers of the Ganga-Brahmaputra-Meghna Delta: A Fluvial Account of Bengal* (pp. 15–25). Springer International Publishing.

Sahana, M., Areendran, G., & Sajjad, H. (2022). Assessment of suitable habitat of mangrove species for prioritizing restoration in coastal ecosystem of Sundarban Biosphere Reserve, India. Scientific Reports, 12(1), 20997.

Sahana, M., Dhali, M. K., & Lindley, S. (2024). Global disparities in transboundary river research have implications for sustainable management. Communications Earth & Environment, 5(1), 786.

Sarker, B., Keya, K. N., Mahir, F. I., Nahiun, K. M., Shahida, S., & Khan, R. A. (2021). Surface and ground water pollution: Causes and effects of urbanization and industrialization in South asia. Scientific Review, 7(3), 32–41.

Sather, R., Antharam, S. M., & Catena, M. (2022). Water security in South Asian cities: A review of challenges and opportunities. CivilEng, 3(4), 873–894.

Sen, S. K., Al Nafi Khan, A. H., Dutta, S., Mortuza, A. A., & Sumaiya, U. (2022). Hydropower potentials in Bangladesh in context of current exploitation of energy sources: A comprehensive review. International Journal of Energy and Water Resources, 6(3), 413–435.

Shamsudduha, M. (2021). Groundwater storage dynamics in the Himalayan river basins and impacts of global change in the Anthropocene. In A. Mukherjee et al. (Eds.), *Global Groundwater* (pp. 47–63). Elsevier. https://doi.org/10.1016/B978-0-12-818172-0.00004-9.

Swain, A. (2016). Water and post-conflict peacebuilding. Hydrological Sciences Journal, 61(7), 1313–1322.

Uprety, K., & Salman, S. M. (2011). Legal aspects of sharing and management of transboundary waters in South Asia: Preventing conflicts and promoting cooperation. Hydrological Sciences Journal, 56(4), 641–661.

Vasani, H. (2023). International water law and hydropolitics: An enquiry into the water conflict between India and Nepal. Water International, 48(2), 259–281.

Vinayaraj, V. K. (2009). India as a threat: Bangladeshi Perceptions. South Asian Survey, 16(1), 101–118.

Warner, J., Wester, P., & Bolding, A. (2008). Going with the flow: River basins as the natural units for water management? Water Policy, 10(S2), 121–138.

Waseem, W., & Chaudhry, S. (2024). Assessing the geopolitical shifts in South Asia due to climate change: Analyzing the Brahmaputra River Basin (2010-2020). Journal of Development and Social Sciences, 5(2), 289–290.

Williams, J. M. (2018). Stagnant rivers: Transboundary water security in South and Southeast Asia. Water, 10(12), 1819.

Williams, J. M. (2020). Discourse inertia and the governance of transboundary rivers in Asia. Earth System Governance, 3, 100041.

Yang, L., Luo, W., Zhao, P., Zhang, Y., Kang, S., Giesy, J. P., & Zhang, F. (2021). Microplastics in the Koshi River, a remote alpine river crossing the Himalayas from China to Nepal. Environmental Pollution, 290, 118121.

2 Transboundary Water Dispute between Bangladesh and India and the Role of Bangladesh National River Conservation Commission

Prospects and Challenges

Md Abu Bakar Siddique and Bishwajit Ghosh

INTRODUCTION

Water sharing among upper and lower riparian states has been a common interstate crisis across the continents (Mianabadi et al., 2015). Water is an extremely politicized natural asset, yet we cannot survive without it for more than even two or three days (Buocz & Eisenberger, 2023). Furthermore, there currently doesn't exist any alternative to water, which keeps getting less accessible (Mianabadi et al., 2015). For nation states, including Bangladesh, securing freshwater supplies for the populace and economy has become a difficult and contentious problem (Begum, 2020). According to the National Water Resource Database, among 230 major rivers of Bangladesh, 54 of them (mainly the biggest ones) have been flowing in from India (Hanasz, 2014). Upstream regions frequently consume water to generate greater amounts of power, whereas nations that are downstream have to use diplomacy for acquiring water (Warner, 2004). Comparing socio-economic and political powers, Bangladesh has always been deprived of its fair share of river water as a downstream country (Naher et al., 2021).

In 1996, both India and Bangladesh accepted the Ganges Treaty, but water sharing is not being properly managed, and India is still dictating Ganga's water at its discretion (Rahman et al., 2019). Approximately 60% of the Ganges water has been withdrawn from Bangladesh's inland and aquatic ecosystems by India causing harm to the Ganges and its tributaries[1] (Adel, 2012). India has diverted, reserved, or

DOI: 10.1201/9781003599692-2

exploited water from transboundary rivers through dams, embankments, barrages, hydropower projects, and river linking projects[2] (Adel, 2001). Moreover, the abrupt discharge of surplus water through the Gajaldoba Barrage in the time of rainfall results in flooding and erosion of the bank inside Bangladesh (Afroz & Rahman, 2013; Higano & Islam, 2002). Furthermore, India's proposed Tipaimukh Dam poses threats to the Meghna River system (Baten & Titumir, 2016). Bangladesh's assessment suggests that the Tipaimukh Dam poses serious threats to the Meghna River system, notably the Surma and Kushiyara rivers in Sylhet. Bangladesh has urged India for details on the dam's layout but has not received timely responses (Bisht, 2012). As per the report of UN Bangladesh, more recently, Bangladesh asserted that a flood in 2024 devastated the lives and livelihoods of over 5 million people due to unregulated water circulation in the Dumboor Dam.[3]

However, to resolve such cross-border sharing of water disputes between Bangladesh and India in 1972, the Indo-Bangladesh Friendship, Cooperation and Peace Treaty was launched, and the Joint River Commission (JRC) was established. One of the mandates of this treaty was to ensure maximum benefit from the water resources by sharing water resources peacefully (Islam, 2005). Nevertheless, it expired in 1997 and was not renewed. Additionally, the Ganges water sharing treaty, signed in 1996, is arguably less effective due to missing provisions regarding data and information sharing, measures of consultation and dispute settlement (Arif, 2015). Besides, India and Bangladesh both voted in favour of the 1997 United Nations Convention on the Law of the Non-navigational Uses of International Watercourses (Watercourse Convention, 1997) during its adoption; however, neither country went on to sign the treaty (Islam, 2001).

On the other hand, new and emerging global environmental governance are being devised across the world to augment biodiversity, manage climate change, and reinterpret human ecosystem relationship (Pecharroman, 2018). Thus, concepts like 'Right of Nature' (RoN) have their roots in 1974, when University of Southern California law professor Christopher D. Stone first used the term 'legal personality' to describe an 'environmental entity' (Putzer et al., 2022). Nonetheless, granting legal citizenship to the environment would need a significant overhaul of the entire conversation surrounding environmental governance. Many countries have acknowledged the inherent value of rivers by classifying them as (1) living persons, (2) subjects of law, (3) legal persons, and (4) living entities (O'Donnell, 2020). In a similar vein, Bangladesh sought to protect its rivers, and the country's highest court, the High Court Division, recognized all the rivers of Bangladesh, including the Turag, as a distinct legal entity with inherent rights and capacities.[4] This decision was a first for Bangladesh, but follows a transnational line of judicial decisions granting legal personhood to rivers from across India and Colombia (Macpherson, 2019; O'Donnell, 2018).

After that, the court appointed the National River Conservation Commission (NRCC) to act as a guardian (in *loco parentis*) over all rivers, including the Turag, and ensure their protection and conservation. In these circumstances, NRCC is mandated to conserve and protect the rivers which are flowing through the territory of Bangladesh. However, if transboundary rivers have lost their natural flow and are in danger of going extinct because of political–human change of the river water before entering Bangladesh, the NRCC will be powerless to save those rivers within Bangladesh. Since the NRCC is tasked with the duty of *loco parentis* employing the public trust doctrine, it should be its responsibility to get involved in the TWD issue for this reason and for its own existence. The significance of this study revolves around the novelty of this research. Existing literature suggests that much legal research has been conducted before focusing on the national and international framework, but there has been less focus on the RoN movement, which has been elaborately discussed in this chapter. This movement advocates for the recognition of nature's intrinsic rights, challenging traditional legal paradigms and encouraging a shift towards more sustainable environmental practices. By examining this movement in depth, the study intends to address a major research gap in the corpus of existing scholarly work on law, ultimately contributing to a more comprehensive understanding of environmental jurisprudence. Hence, the chapter evaluates the prospects and challenges of the NRCC in resolving the dispute of joint river water sharing between Bangladesh and India, focusing on water sharing governance, the RoN movement, and potential challenges before posing concluding remarks.

METHODOLOGY

Research Approach

This study primarily employs a qualitative research methodology, integrating a doctrinal approach with discourse analysis. A theoretical framework has been developed to examine how contemporary TWDs can be addressed through emerging normative paradigms such as mutual cooperation and negotiated settlements. Traditionally, TWDs have emphasized state sovereignty, often at the expense of ecological integrity, treating rivers as mere resources rather than sentient, living systems. However, in the context of India and Bangladesh, both nations have begun to recognize rivers not only as shared resources but also as ecological entities with intrinsic value. Accordingly, this research investigates the legal and institutional barriers at both the domestic and global levels, adding to the evolving discourse on the RoN and its relevance to resolving TWDs in South Asia (Figure 2.1).

Sources And Interpretation Of Secondary Data

The chapter adopts a doctrinal lens, grounding its analysis in legal principles, statutes, judicial decisions, and scholarly literature. A comprehensive review of legal texts was undertaken to identify structural gaps in the water sharing frameworks of India and Bangladesh. Furthermore, the chapter investigates significant global and national frameworks including the UN Watercourses Convention (1997), the Stockholm Declaration (1972), the Rio Declaration on Environment and Development (1992), the Convention on Biological Diversity (1992), and the UN Framework Convention on Climate Change (1992) to explore the principle of non-harm and the broader goal of environmental sustainability in transboundary contexts. To deepen the analysis, the study engages with global discourses on the 'RoN', drawing on literature and case studies from Ecuador, Bolivia, Colombia, Australia, New Zealand, India, and Bangladesh. A comparative framework is used to assess the legal personhood of rivers in these jurisdictions, identifying typologies and implications for South Asia.

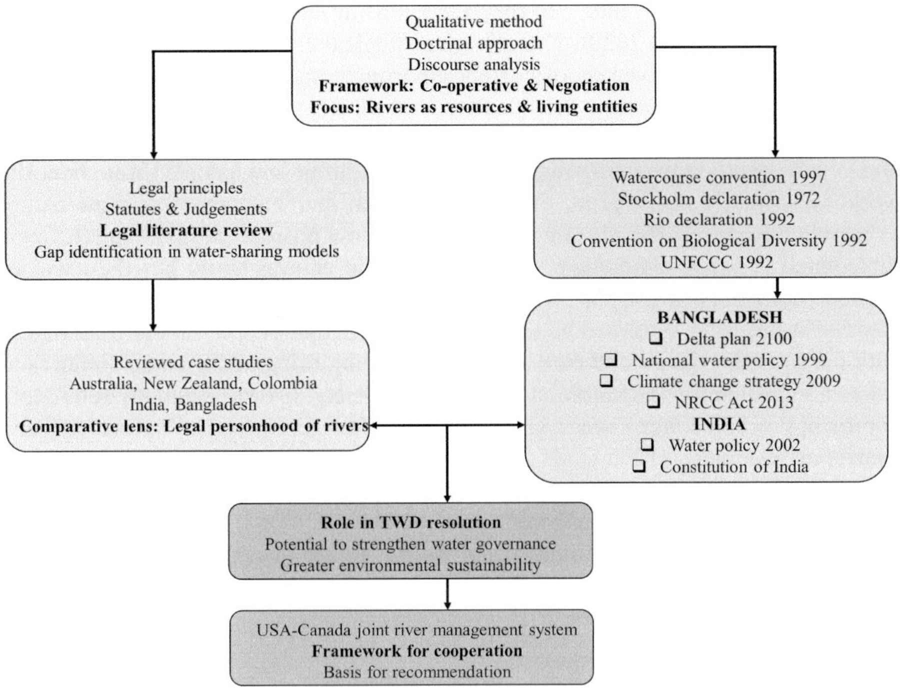

FIGURE 2.1 Data and methodological framework adopted in the present study. (*Source:* Prepared by authors.)

The research further explores institutional and legislative bottlenecks in both countries, focusing especially on the possible role of the Bangladesh NRCC in mediating TWDs. In Bangladesh, this includes a critical examination of the Bangladesh Delta Plan 2100, the Climate Change Strategy and Action Plan (2009), the National Water Policy (1999), the NRCC Act (2013), and Target 6.5 of Sustainable Development Goal 6. These instruments are analyzed to assess how NRCC can function within the existing governance architecture to promote environmental sustainability in the fragile transboundary river systems shared with India. Additionally, the Indian Water Policy (2002) and relevant provisions of the Constitution of India are briefly reviewed to identify structural constraints affecting bilateral water governance. Finally, the study offers recommendations for strengthening NRCC's role, drawing insights from the successful USA–Canada joint river management model, aligned with the theoretical framework guiding this research.

CONCEPTUAL FRAMEWORK

The RoN movement has evolved around the world, shifting from a focus on protection to one on prevention and finally to precaution. Nevertheless, the idea of the right of natural movement has become more widely accepted and has improved in various jurisdictions as time has progressed (Cano-Pecharroman & O'Donnell, 2024). Many indigenous and traditional societies have maintained their connection to the natural world for thousands of years, and they view it as an integral element of their existence (O'Donnell et al., 2020). The current ecological crisis caused by rapid climate change or the ineffectiveness of current environmental laws may be contributing factors to the acceleration of this idea's evolution. Regardless, it is part of a larger movement that is sweeping through civil society, particularly among young people, and is persistently advocating for a shift in paradigm in how people throughout the world view nature (Zelle et al., 2021). The human race has ceased to view nature as merely a resource; the idea of RoN elevates it to a position of inherent relevance (O'Donnell & Arstein-Kerslake, 2021). Where the natural world has the capacity to assert its rights in a legal setting, just like any other person or entity with an identity in the legal system. The idea behind this is that people should treat nature with the same reverence and care as any other living thing and even go so far as to act as a guardian for it (Tanasescu, 2016). Conversely, hydro-hegemony refers to a dominant power's control over water resources, often reinforced by asymmetry and control mechanisms (Hayat et al., 2022). India's dominance over transboundary rivers in South Asia, particularly with Bangladesh, reinforces its financial, political, and demographic authority (Obaidullah & Howlader, 2025). Consequently, natural resources like water are often treated as the centre of power, with tactics such as incentives, normative/legal treaties, or coercive actions highlighting the power dynamics in the region (Rawia, 2015).

THEORETICAL FRAMEWORK

Liberalism, being one of the major schools of international relations theory, believes that to minimize conflict, states can function together (Meiser, 2018). Liberalism suggests that international law can be effective through two major factors, which are international institutions and economic interdependence (Zangl, 2012). Using this economic interdependence, Kant transformed liberalism and neoliberalism where he viewed those democratic countries usually do not engage in wars since they are economically interdependent. Thus, neoliberalism looks into democratic peace through international institutions, which is one of the strongest contributions to International Relations theory (Meiser, 2018). Therefore, settling disputes is manageable following the theory of liberalism through diplomacy or legal means. Multitrack diplomacy covers all aspects of mediation, from meetings of high-level government officials (track 1) to meetings of federal officials facilitated by outside organizations (track 1.5), to informal discussions between intermediary officials from opposing groups or nations (track 2), to the involvement of grassroots organizations, communities, and private citizens (track 3) (Barua, 2018). This theory is highly relevant to analyze the prospect of NRCC in settling this transboundary issue.

Furthermore, on the legal aspect, equitable distribution of water is not always defined as an equal percentage of the resources involved or equal proportion of its usage and advantages; rather, it establishes reciprocal duties to prevent the deprivation of the distinct entitlements by other nations. The notion of equitable and adequate usage, with the purpose of delivering maximum benefit to riparian nations while limiting degradation, is addressed in article 5 of the Watercourse Convention, 1997. Moreover, the no harm principle is a widely established notion of customary international law (CIL) that holds governments accountable for encouraging, increasing, and regulating the likelihood of ecological distress to neighbouring states (Birnie, 2009). The Watercourse Convention, 1997 integrates this principle with article 7 specifying that a state is obligated to safeguard against serious injury to another state while operating an international waterway. If equitable utilization of the international watercourse has not been done, it may lead to harming the environment of the co-riparian states which leads to conflict among the riparian states.

In this equilibrium of international relations and international law, RoN movement has emerged granting the right of a legal person to the river itself nationally. This right has always been centrifuged towards indigenous groups and the communications between the state and those groups. Both India and Bangladesh have contributed to this new phenomenon nationally by declaring the river as a legal person and appointing its guardian. Now, the debate arises whether this RoN can contribute internationally and draw a fine line between multitrack diplomacy and legal obligations. In this chapter, the authors intend to scrutinize how much NRCC as an institution may contribute to multitrack diplomacy to resolve this water sharing dispute and generate environmental sustainability in South Asia within the framework of CIL, without specifically focusing on indigenous communities.

WATER SHARING GOVERNANCE MODEL IN INDIA AND BANGLADESH

Joint Governance Institution

The JRC, established in 1972 under the Indo-Bangladesh Friendship, Cooperation and Peace Treaty, aimed to maximize water resource benefits, but failed to ensure equitable rights for Bangladesh due to its expiration in 1997 (Islam, 2005). The 1977 accord was the initial agreement to acknowledge the need for shared water resources (SWRs) to share water. In order to share and exchange information regarding the availability of water, the Joint Committee of Experts (JCE) was founded in 1985, and during their sessions in early 1986, the JCE observed nine SWRs (Pandey, 2012). JCE was a failure as there was a disproportion between both countries' statistics given by the experts (Crow & Lindquist, 1990). After all these drawbacks, in 1996, both countries signed the Ganges Treaty to distribute water, focusing on 75% supply and disregarding environmental and climate change impacts, despite previous statistics (Najam, 2004). Despite the Treaty, the Farakka Barrage exploited up to 60% of the streams that are used for massive irrigation projects on the Indian side. According to the Bangladesh Water Development Board (BWDB), the arbitrary withdrawal of water is affecting the water level in the lower Ganges during the dry period (Baten & Titumir, 2016). A joint committee was mandated according to articles 4, 5, and 6 to regulate the treaty in terms of water sharing, and if they failed, the dispute would have been referred to the JRC.

Teesta Crisis

In 1983, Bangladesh and India signed an ad hoc agreement in terms of sharing the water of the Teesta, where Bangladesh was supposed to get a 36% share of the water. On the other hand, India would get 39%, and this was valid till 1985, but this was also not implemented according to that *ad hoc* agreement. In 1984, according to JRC, hydrological data showed that Bangladesh's share increased to 37.5% and India shared 42% of the water. Eventually, the governmental bodies of the two nations were satisfied with an agreement over the Teesta water by 2011 was approved to be authorized by the JRC, and that transitional accord provided India 42.5% and Bangladesh 37.5% allocation of Teesta water which was expected to remain for 15 years (Ujwala, 2017). The agreement was opposed by the CM of West Bengal, and Bangladesh failed to ensure its rights over the Teesta water (Samaranayake, 2018). Though many scholars opined that both countries have goodwill between each other, the outcome has been null since Bangladesh lacked a strong body of officials which would have been great for diplomatic relations regarding water sharing (Dey, 2014). As a result, both countries have tried to increase agriculture and afterwards failed due to mismanagement of the Teesta River water (Parven & Hasan, 2018). This historical water sharing road map between these two nations illustrates the times when cooperative water governance has fallen short.

INTERNATIONAL LEGAL FRAMEWORK ON TRANSBOUNDARY WATER DISPUTE

The comprehensive international environmental law on TWDs is codified in the Watercourse Convention 1997. However, the corpus of international law stemming from customary norms had been there along with the codified instruments. Thus, in the part the relevant judicial references are discussed earlier than the treaty obligations. The no-harm principle is a widely recognized doctrine of CIL which provides that countries have a duty to avoid, minimize, and manage the possibility of ecological suffering to neighbouring countries (Birnie, 2009). The *Trail Smelter Tribunal*, over 79 years ago, validated Transboundary Harm in International Environmental Law, imposing responsibility to prevent harm to another state's territory generated from the former state's territory (Cambridge University Press, 2017). Subsequently, the no-harm principle has been incorporated in the Watercourse Convention, 1997, where article 7 illustrates that a state is obligated to safeguard against serious injury to another state while operating an international waterway.

The pioneering case regarding the administration of transboundary water was *The Lake Lanoux arbitration* (France v Spain, 1957) where the tribunal affirmed the *Trail Smelter Arbitration's* no harm principle, which prohibits upstream nations from altering or redirecting transnational streams that adversely affect downstream nations (Tignino & Bréthaut, 2020). The case of *Gabcikovo-Nagymaros* (Hungary v Slovakia, 1997) is the prominent case where the duty of the state to prevent serious damage was upheld by applying equitable utilization (Rajsingh & Tripathi, 2020). The judiciary cited the advisory opinion on the *Legality of the Threat or Use of Nuclear Weapons (1996)*, which emphasized the ICJ's duty to ensure nations' operations respect the environment, a fundamental aspect of international environmental law. Additionally, in the *Pulp Mills Case* (Argentina v Uruguay, 2010), the notions of equitable utilization, environmental preservation, and sustainable development were acknowledged with an emphasis on protecting the environment not just inside the state itself but also for surrounding states. Similar reference was drawn in the *Corfu Channel Case* (United Kingdom v Albania, 1949), determining that a nation is accountable for conduct infringing international law that occurs within its boundaries and inflicts distress on the neighbouring nations. In the *North Sea Continental Shelf Case* (Germany v Netherlands, 1969), the ICJ decided that every country must follow this principle as this is a CIL. The most prominent case concentrating exclusively on international rivers following the *Lake Lanoux case* is the *Indus Waters Kishenganga arbitration*. The arbitration ruled that a nation ought to utilize natural assets in sustainable ways to minimize cross-border repercussions (Tignino & Bréthaut, 2020). In a subsequent dispute, as an upper riparian state, Bolivia was charged by Chile for violating customary law of equitable utilization and no-harm and disregarding the American Treaty on Pacific Settlement of 1948 through the enforcement of arbitrary initiatives, addressing the nature and the consumption of the resources of the Silala River (Dellapenna, 2023).

The notion of equitable and adequate usage, with the purpose of delivering maximum benefit to riparian nations while limiting degradation, is addressed in article 5 of the Watercourse Convention, 1997. The fundamental principle of equitable utilization of water first emerged in the Helsinki Rules of 1966. Article 4 covers the applications of the waters of international rivers, which affirms that every State possesses a legal entitlement to a reasonable and equitable allocation of the resources of an international river basin in terms of useful applications. Article 5 of the ILC Draft Articles of 1994 states that every nation enjoys the privilege to take part in water sharing in an adequate and equitable manner in transnational rivers. These principles took the form of a treaty law when the *UN Watercourse Convention 1997* was adopted in 1997.

Stockholm Declaration 1972 is the first international instrument that adopted this no harm principle. Principle 21 reflects that the State has the sovereign right to use and utilize natural resources in any manner inside its territory. However, this utilization of natural resources cannot be the cause of environmental harm to other States. Principle 2 of *the Rio Declaration on Environment and Development 1992*, Article 3 of *the Convention on Biological Diversity 1992*, and *the United Nations Framework Convention on Climate Change 1992* in recital 8 of its preamble has repeated the no harm principle precisely from Principle 21 of *the Stockholm Declaration*. All of these international instruments established and reaffirmed the no harm principle and equitable utilization strongly. The only watercourse convention of worldwide significance is the Watercourse Convention of 1997 (McCaffrey & Sinjela, 1998). Helsinki rules and various soft laws are derived from CIL. Table 2.1 shows both countries' position in the transboundary legal framework. Both countries are party to almost all the International Legal Framework (ILF) on Transboundary Water Governance except the most important Watercourse Convention 1997. India is not party to the Helsinki Rule 1966 as well (Table 2.1).

TABLE 2.1
Bangladesh and India's Membership to International Legal Framework on Transboundary Water Governance

Name of ILF on Transboundary Water Sharing	Relevant Provisions	Signatory
Sustainable Development Goal Target: 6.5 by 2030, implement integrated water resources management at all levels, including through transboundary cooperation as appropriate	**Goal 6:** Ensure availability and sustainable management of water	India: Yes Bangladesh: Yes
Helsinki Rule 1966	**Article 4:** Equitable sharing of basin water	India: No Bangladesh: Yes
Stockholm Declaration, 1972	**Principle 21:** No harm principle	India: Yes Bangladesh: Yes
Convention on Biological Diversity 1992	**Article 3:** No harm principle	India: Yes Bangladesh: Yes
United Nations Framework Convention on Climate Change, 1992	**Recital 8** of preamble No harm principle	India: Yes Bangladesh: Yes
Rio Declaration on Environment and Development, 1992	**Principle 2:** No harm principle	India: Yes Bangladesh: Yes
Convention on the Law of the Non-navigational Uses of International Watercourses, 1997	**Article 5:** Equitable utilization of water Article 7: No harm principle	India: No Bangladesh: No

Source: Compiled by authors.

RoN MOVEMENT AND EMERGENCE OF NRCC IN BANGLADESH

PROTECTION OF 'RoN' BY UNITED NATIONS

The United Nations provides some hope by recognising these rights in some manner, even though there is no formal law that protects these rights. Principle 1 of the Rio Declaration on Environment and Development asserts that individuals' health and productivity should be the primary goals of sustainable development that does not destroy nature. This idea has its roots in the original document. Following this approach to 'harmony with nature', which turned an undervalued concept into a framework and the United Nations General Assembly designated April 22nd as International Mother Earth Day in 2009, and nature began its journey to garner support for its legal personhood. The concept of sustainable development has been reaffirmed in international accords and resolutions passed by the General Assembly, which recognize the inherent value of natural resources (United Nations, 2015).

PROTECTION OF RIVERS WORLDWIDE

RoN as a philosophy is relatively new and has not yet fully developed. On the other hand, there have been outstanding court rulings safeguarding river rights, which are a logical extension of natural rights. These court rulings illuminate the path towards rivers' legal personality and usher in the long-sought revolutionary shift that leads to a more harmonious, environmentally conscious society. In order to have a better grasp of the global status and practice of river rights, this section will examine some domestic case studies precisely.

Ecuador And Bolivia

By a nationwide referendum in 2008, Ecuador became the pioneering state on the planet to acknowledge the RoN in its constitution (Revkin, 2008). The RoN was recognized in relation to indigenous groups' growing influence in politics. This included the RoN 'to exist, persist, maintain and regenerate its vital cycles, structures, functions and its processes in evolution, as well as the right of every individual and community to defend these rights' (Kauffman & Martin, 2023). In 2010, Bolivia promptly adopted 'Laws on the Rights of Mother Earth' to emulate this. The concept of 'collective public interest' allowed them to acknowledge nature, sometimes known as 'mother earth'. Bolivia used a systemic, rather than an individual, approach to protecting the environment, severing ties with human control and authority over it (O'Donnell, 2023).

New Zealand

In 2017, New Zealand bestowed legal personhood upon a river, honouring ancient principles of *kaitiakitanga*, which elevates humans to the role of custodians or guardians of natural resources, not proprietors (Sanders, 2018). When the *Te Urewera Act* was passed in 2016, the New Zealand government acknowledged *Te Urewera National Park* as a distinct legal body with its own set of responsibilities and rights. Instead of accurately emulating the revolutionary legislation of Bolivia or Ecuador, this followed the traditional notions of entrusting control to a generous and passionate board. Unlike the *Te Urewera* Act of 2014, which only permitted a specific entity, *Te Pou Tupoa*, the 2017 *Te Awa Tupua Act* acknowledged the *Whanganui River* as an indivisible and living whole by giving it legal personhood status (Tadak et al., 2023).

Australia

Australia's approach to river rights differs from New Zealand's, as it focuses on protecting rivers specifically, incorporating RoN. In order to effectively manage rivers, it combines modern understanding with traditional wisdom, recognizing the significance of indigenous participation in governance and protection (Tadak et al., 2023). Australia's federal system has separate water laws for each state, with Victoria in the South-East leading water regulation reforms (O'Donnell & Talbot-Jones, 2018). According to the Water Act of 1989, Victoria created the Victorian Environmental Water Holder (VEWH) in 2010 to make sure that environmental water management is independent and accountable and to be able to sue or be sued when problems emerge. While the river does not have any legal identity as a result, VEWH's water entitlements were required to be utilized for environmental purposes, thus ensuring that the rivers' and VEWH's interests were always aligned. As far as this river is concerned, VEWH can be considered a protector (O'Donnell & Macpherson, 2018). The Yarra River in Australia, an essential source of nutrition and earnings for Indigenous communities, was severely polluted by the 2017 Yarra River Protection Act, a move similar to New Zealand's self-governing rights, requiring effective environmental, social, and cultural integration. According to the Yarra River Protection (*Wilip-gin Birrarung murron*) Act 2017, the Birrarung Council was established with the purpose of safeguarding and preserving the river.

Colombia

The RoN and rivers are recognized by Colombia's courts, which have established a powerful connection framework known as 'biocultural rights'. Culture, ecosystems, biodiversity, and human and environmental reactions to change are all tracked by this connection (O'Donnell, 2023). It is claimed that the courts in Colombia did not recognize rivers and their ecosystems as distinct entities, which leads to the question of whether rivers should be directly granted legal personhood. A court case was brought up about the Atrato River, which is home to indigenous populations and biodiversity. The plaintiffs in the lawsuit claimed that the river's excessive mining pollution infringed on their fundamental rights to life, equality, and a healthy environment (Page & Pelizzon, 2022). The local administrative court dismissed the case, stating there was no violation of individual rights. However, the constitutional court ruled that pollution endangered rights to water, food, and culture, making water a necessary requirement for life. The breach of the river's rights and the constitutional environmental duties to maintain a healthy environment were acknowledged by the court in the Atrato case. The concept of rights was illustrated by the court as 'The land does not belong to man and, on the contrary, assumes that man is the one who belongs to the earth, like any other species'. In this instance, Colombia moved towards an eco-centric perspective, giving nature the same standing as people. Such a legislative representation of this eco-centric approach is significant because it forces the nation to make decisions that respect nature rather than just economic and utilitarian considerations. In order to establish a legal entity for guardianship of the river as an institutional remedy, the court referenced the New Zealand model, which includes representatives from all relevant parties (Page & Pelizzon, 2022).

In the Cauca River Case, the precedent set by the Atrato River case was cited by the court. The Supreme Court ruled, dismissing lower courts' concerns, that individual and community rights overlap and are nearly inseparable when it comes to fundamental environmental rights (O'Donnell, 2020). The court ruled that the Cauca River and the Atrato River are both legally subject and should be protected, conserved, maintained, and restored by the state so that future generations can have the same rights to dignity, water, security, food, and a healthy environment (Page & Pelizzon, 2022). The Magdalena River Case established an ecological constitution framework, which has been applied in other cases concerning nature rights. Within three months, the court established a guardian commission for the river and granted protection of communal rights after accepting a tutela action against a hydropower plant (Salgado, 2022).

An important ecological constitutional case based on the infrastructural requirements of the United Nations Committee of Economic, Social, and Cultural Rights and General Comment 15 of the CESCR is the Coello, Combeima, and Cocora Rivers Case. Following a pattern identical to that of the Atrato River case, the court determined a violation of community rights and imposed 26 orders, including the appointment of an expert advisory board and a guardian committee. Also included in this framework was the *La Plata River Case.*

India

Indian jurisprudence has recognized the legal RoN subjects, such as rivers like the Yamuna and Ganga, similar to Colombia's practice. This shift in environmental jurisprudence has made India a leading player in South Asia (O'Donnell & Talbot-Jones, 2018). Hinduism and religious devotion play a role in the Yamuna and Ganga cases, which could disrupt minority rights and secular democracy in India (Edirisinghe & Suchet-Pearson, 2024). A lawsuit was filed against the state of Uttarakhand in the Yamuna and Ganga Rivers Case, focusing on the government's inability to safeguard the rivers in question. The court halted mining on the Ganga River, established river control bodies, and ordered the eviction of unlawful squatters after hearing arguments. However, the court was unsatisfied with the Uttarakhand government's efforts, so it issued fresh instructions, sparking a flurry of debate (O'Donnell, 2018).

According to Hindu belief, the rivers Yamuna and Ganga have the ability to 'wash away all sins'. The Supreme Court has highlighted the rivers' status as autonomous entities in Hinduism. To ensure the continued existence of these rivers—and the honour and confidence placed in them—the court appointed three government officials to serve as human guardians (Islam & O'Donnell, 2020). In any case, in the middle of 2017, without providing any explanation, the Indian Supreme Court suspended both rulings. However, the Uttarakhand government requested this hold due to some uncertainties associated with this new knowledge of nature, rivers, and environmental law (Jackson et al., 2024). To illustrate the point, free-flowing cross-state rivers such as the Yamuna and the Ganga provide unique legal questions regarding who is liable in the case of flood damage and how this personhood is to be interpreted.

Comparison Of Legal Personhood Of Rivers As RoN Across The World

The aforementioned nations have all made efforts to incorporate the right to nature, but they have done so from their own unique vantage points. The river's legal personality has changed form and significance. The difference becomes more apparent when we compare the reasoning of the Indian Supreme Court with the decisions of the Colombian courts. As mentioned before, the Colombian court has taken a more expansive, eco-centric approach. In contrast, India is still grappling with jurisprudential questions and notions, and it leans towards applying classic human doctrines to environmental issues. New Zealand and Australian courts, on the other hand, have taken somewhat different tracks. In an effort to provide a unified view, Table 2.2 attempts to compare the types of rights granted by different jurisdictions.

TABLE 2.2
Comparative Legal Personhood of Rivers around the World

Attribution	Jurisdiction			
	New Zealand	Australia	Colombia	India
Legal personhood	✓	✓	✓	Yes (Initially)
Character of legal personality				
Contractual right	✓	✓	✓	✗
Property right	✓	✓	✓	✗
All human rights	✗	✗	✓	✗
Underlying rationale of the decision				
River values exclusive of human values	✓	✓	✓	✓
Legal representative/guardian				
Government interference	✗	✗	✗	✓
Funding sources	Dedicated funding	Mixed	✓	✗
Special protection by law	✓	✓	✓	✗

Source: Designed by authors.

EMERGENCE OF NRCC IN WATER GOVERNANCE IN BANGLADESH

Figure 2.2 suggests that Bangladesh water governance and management both nationally and internationally is highly complex from an institutional perspective. According to the Water Resources Planning Organization (WARPO), Bangladesh's water sector includes 35 public entities and bodies regulated by 13 distinct ministries, many of which have overlapping spheres of responsibility, creating challenges for effective water governance.

Existing Institutional Arrangement

The primary water administrator is the Ministry of Water Resources (MWR) responsible for controlling the availability and consumption (Islam & O'Donnell, 2020). On the other hand, the Department of Environment under the Ministry of Environment, Forest, and Climate Change (MEFCC) operates the function of primary water monitor for safeguarding the integrity of the water, with the right to institute legal proceedings against water emitters, usually industries (WWF and H&M, 2015). The WRPO and the BWDB are led by the MWR, the highest authority in the water sector which deals with policy framework, implementation, and monitoring (Islam & O'Donnell, 2020). The WRPO is the major institution engaged with the country's broad water-related strategy and administration, while the BWDB is in charge of designing and overseeing water management initiatives including flood control, drainage, and from agriculture to collecting rainwater, coastline safeguarding, and preventing erosion through collaboration with stakeholders (Alam, 2014).

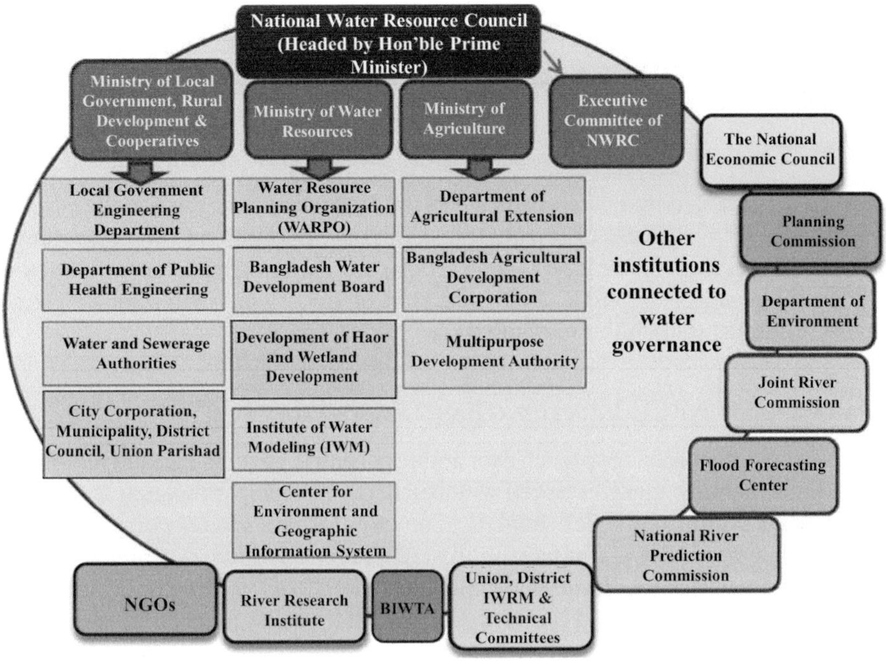

FIGURE 2.2 Water governance bodies and related institutions in Bangladesh. (*Source:* Arifuzzaman et al., 2019.)

The National Water Resources Council (NWRC), constituted under section 4 of the Bangladesh Water Act 2013 (BWA), holds authority for developing regulations and providing guidelines for sustainable development, effective utilization, effective allocation, safety, and preservation of the water supply through its Executive Committee. Additionally, the Bangladesh Inland Water Transport Authority (BIWTA), which reports to the Ministry of Shipping, is in charge of developing, maintaining, and managing navigable rivers and waterways (Islam & O'Donnell, 2020). The BIWTA focuses on river conservation projects such as reviving dead or dying waterways, channels, or streams, as well as providing new navigable channels and canals. In this complex web of institutional arrangements, coordination and collaboration among the water institutions is a perennial problem in Bangladesh (WWF and H&M, 2015). The NWRC is mandated with the power to exchange of international and regional cooperation (BWA, s. 7). The National Water Policy of Bangladesh (MoWR, 1999) articulates the need for endeavours like agreement on transboundary water sharing with co-riparian states (Salehin, 2010).

The JRC holds its position under the Department of Environment with the Flood Forecasting and Warning Centre, while NRCC emerged in 2013 through the NRCC Act 2013. Soon after the judgement of the Turag River case creation of NRCC was inevitable, and it was done without any further delay. The NRCC Act 2013 was introduced in Bangladesh with the goal to set up a body for stopping illegal occupation of rivers, environmental damage of water and contamination of waterways resulting from factories and other illegal structures, and maintaining multifaceted utilization of water for economic growth, such as regeneration of the natural flow of rivers, adequate regulations, and keeping them navigable (NRCC Act 2013, preamble).

FINDINGS

This study is divided into two broad categories. In the first part, the underlying challenges in legal frameworks of Bangladesh and India are identified which is followed by a SWOT matrix (Figure 2.3). In the second part, recommendations are discussed to show the prospects of NRCC in the discourse of TWD which can generate greater environmental sustainability in South Asia.

Challenges Of NRCC In Settling Bangladesh–India Twd

RoN favours an eccentric approach over anthropocentric governance, and the NRCC, as the implementing agent, is tasked with balancing the competing interests of many stakeholders within the state's borders. Therefore, it is a major worry as to how much involvement NRCC can have in the complex transboundary water issue with India, given the massive domestic threats that the NRCC is already dealing with. The assessment of associated barriers can potentially assist in locating its prospective role.

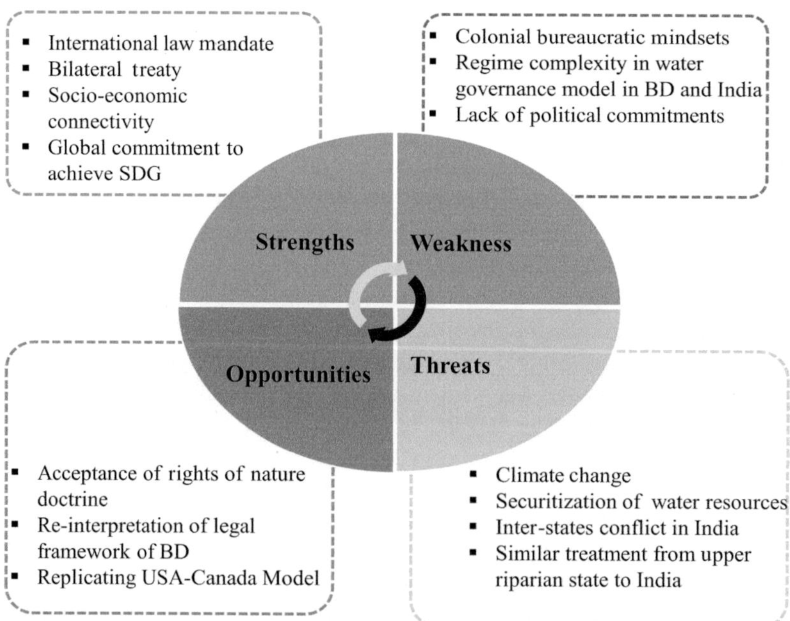

FIGURE 2.3 SWOT matrix of the transboundary water dispute between Bangladesh and India. (*Source:* Prepared by authors.)

Legal Barriers In Bangladesh

The preamble of the NRCC Act 2013 is the beacon for the prospective interventions of NRCC. However, the preamble does not identify any TWD issue that may be resolved by the NRCC. Furthermore, the law itself has cut down NRCC's potential to resolve any dispute within the country, let alone any transboundary issue. Section 12 of the NRCC Act 2013 articulated 13 functions for NRCC where the functions are formed in such a way as to make NRCC only a recommendation body (NRCC Act, 2013). This law has not given any mandate to NRCC for becoming the guardian of rivers in Bangladesh in transboundary disputes. Absence of legal mandate within the NRCC Act 2013 can be attributed to the High Court Division Judgement in Petition No. 13989/2016, and Appellate Division in Leave to Appeal No: 3039/2019 did not elaborate on any role of NRCC in the TWD issue, leaving a void here.

Bangladesh Delta Plan 2100 has specified six goals where goal 2 emphasizes enhancing water security and goal 3 highlights ensuring sustainable and integrated river systems. To achieve these goals, BDP 2100 has identified transboundary water management as one of the key areas where Bangladesh needs to develop strategies. For this purpose, BDP 2100 has labelled institutional mechanisms and governance issues as one of the setbacks and recommended institutional reforms to strengthen the core delta institutions (Bangladesh Delta Plan, 2100). However, this plan does not directly identify NRCC as a key institution for TWD. Bangladesh Climate Change Strategy and Action Plan 2009 (BCCSAP) prioritizes adaptation over mitigation, recognizing three natural calamities, including floods. The 10-year plan, based on six pillars, includes comprehensive disaster management, flood forecasting, and early warning systems. Three actions were taken to improve flood level forecasting within Bangladesh (BCCSAP, 2009). However, there was no mention of taking any steps with the neighbouring countries for data sharing of flood level forecast. The third pillar indicated infrastructure to be developed and adaptation against flood programme was initiated where three major factors for flood were identified, such as rainfall, river level rise, and outfall conditions, but failed to identify TWD as a reason for flood. The most prominent global development agenda, SDG in its Target 6.5 of SDG 6, calls for the implementation of integrated water resource management at all levels by 2030, including transboundary cooperation where required. The SDG tracker Bangladesh suggests JRC and the MWR will be responsible data source, leaving NRCC out of the tasks as well.[5]

Legal Barriers In Indian Jurisdiction

Compared to the National Water Policy of Bangladesh (MoWR, 1999), Indian Water Policy (MoWR, 2002) does not require a regional co-operation option (Jackson et al., 2024). In terms of water sharing in the river Teesta, West Bengal objected based on *the Constitution of India*, Article 253, which enunciates that the centre cannot take a decision arbitrarily without taking into consideration the social, political, and ecological impact of such a treaty in the catchment area. The economic dynamics relating to the Teesta River water sharing revolve around agriculture. West Bengal has objected that if any treaty is conducted, that will hurt the farmers of West Bengal (Ujwala, 2017). Agriculture is the lifeline of the people of the Teesta River basin where West Bengal is the biggest rice producing state in India. On the other hand, the Rangpur division is heavily dependent on agriculture, and without the Teesta River water, the agriculture of Rangpur will cease to exist (Waslekar & Futehally, 2013). For these reasons, negotiations through diplomatic tract have never been fruitful. This scenario reflects that hydro-hegemony has played a significant role in the TWD between India and Bangladesh (Obaidullah & Howlader, 2025). In fact, environmental sustainability has been compromised due to a lack of benefit sharing approach rather than political power asymmetry.

RECOMMENDATIONS

The prospects of NRCC in settling TWD between Bangladesh and India primarily lie with the policymakers of Bangladesh. The legal barriers discussed in the preceding part can be progressively overcome (Figure 2.4).

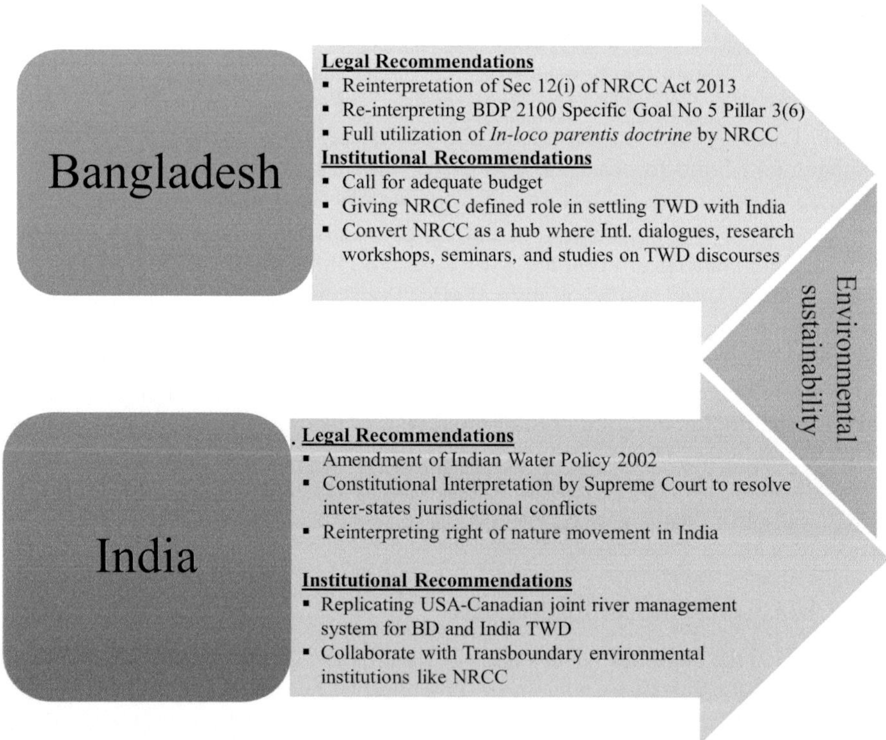

FIGURE 2.4 Legal & institutional recommendations for Bangladesh and India. (*Source:* Prepared by authors.)

Options In The Legal Mandates

Using the Complementarity Doctrine to resolve TWD, section 12(i) of the NRCC Act 2013 and the Preamble (the final two lines) should be interpreted to prevent a jurisdictional confusion between NRCC and JRC Bangladesh. The next help could be re-interpreting Bangladesh Delta Plan 2100 Specific Goal No 5 Pillar 3(6) 'Third party involvement (multilateral or bilateral development partner or country) to resolve transboundary water related issues'.

Thus, NRCC can take the lead to advocate and work as the driving coordination hub of the Bangladesh–India TWD for third parties, either bilateral or multilateral. It can advocate for the nascent consolidation of RoN in the existing legal system in both countries by arranging international dialogues, research workshops, seminars, and studies on TWD discourses frequently. NRCC can call for an adequate budget to do these activities from the Bangladesh government on need basis.

Additionally, the proposed amendment to the NRCC Act should include NRCC's complementary function in giving genuine meaning to the Apex Court's statement (in *loco parentis*) in settling TWD. This study also suggests that India should amend the Indian Water Policy 2002 and seek Constitutional Interpretation by the Supreme Court to resolve inter-state jurisdictional conflicts and RoN movement within India to generate environmental sustainability in South Asia.

Replicating USA–Canada Joint River Management System

It is imperative to abandon zealously securitized policies and underlying frameworks for the Hydropower Project in India as the upper riparian state which is opposed to international law consolidated within both customary and emerging RoN frameworks. To address conflicts over shared waters between the USA and Canada, the International Joint Commission (IJC) was formed to help both nations implement their obligations (Macfarlane, 2019). The treaty aimed to balance hydroelectric generation and navigational concerns by addressing disputes over water (Sedoff, 2014). The USA and Canada signed the Great Lakes Water Quality Agreement in 1972, which mandated reports from the IJC and a pledge to stop releasing permanent poisonous compounds into the environment (LeFeuvre, 1991). The Colombia River Treaty established collaborative advancement of water resources and control of flows in the upper Colombia River basin. The IJC was granted authority to decide when urgent conditions arise due to excessive or inadequate water in the Rainy Lake watershed (Friesen & Day, 1977).

These stories elaborate on how IJC has been successful in terms of water management between the USA and Canada. The potential of NRCC resides in its ability to supplement TWD settlement and wider basin management between India and Bangladesh if given a mandate alongside JRC.

CONCLUSION

Bangladesh and India share one of the world's largest and most complex transboundary river systems, the Ganges, Brahmaputra, and Meghna basins, which continue to be shaped by competing claims over water allocation. Despite the ecological interdependence of these rivers, transboundary water-sharing disputes (TWDs) persist, undermining efforts towards environmental sustainability. Addressing these challenges requires a robust community of trust-based, reciprocal actors—an institutional culture that remains largely absent across global watercourses, including in South Asia. As a lower riparian state, Bangladesh often bears a disproportionate burden in its water-sharing relationship with India, impeding both resource conservation and ecosystem protection. International legal frameworks rooted in customary law, alongside the growing global movement for the RoN, advocate for equitable and fair access to shared natural resources such as rivers. This movement has gained traction through the recognition of rivers as legal persons in various jurisdictions, signalling a shift towards eco-centric governance.

However, prevailing water diplomacy in South Asia remains dominated by interest-based paradigms. Despite the existence of bilateral mechanisms such as the JRC, the Ganga Water Sharing Treaty of 1996, and the Friendship Treaty between Bangladesh and India, these frameworks have yet to alleviate the lived hardships of affected communities. In this context, the NRCC emerges as a potential institutional actor capable of advancing environmental sustainability. Although NRCC has played a vital role in river restoration efforts within Bangladesh, its mandate has not been aligned with transboundary water governance. This chapter critically examines the legal landscape and finds that NRCC's role was notably absent in the landmark Turag River Case. Consequently, the NRCC Act of 2013 and related environmental legislation have sidelined its potential contribution to TWD resolution. Nonetheless, this study argues that through progressive interpretation of its founding legislation, NRCC can assume a more active role in transboundary water diplomacy. Drawing lessons from the USA–Canada International Joint Commission, the chapter proposes that Bangladesh and India could benefit from replicating such collaborative models to strengthen institutional cooperation and safeguard shared riverine ecosystems in South Asia.

NOTES

1. The Ganges and its tributaries the Madhumati, the Musa Khan, the Pashur, and the Kabodak are harmed badly, which also includes their tributaries, namely the Mathabhanga, the Kumar, the Baral, the Arial Khan, the Bhairab, the Gorai, and the Ichamati.
2. Eight out of the five transboundary rivers in the Ganga-Padma River basin and 12 out of the 18 transboundary rivers in the Brahmaputra basin, five dams and two barrages including the Gajaldoba Barrage across the Teesta were made by India. For more details, see Adel, M. M. (2001). Effect on water resources from upstream water diversion in the Ganges basin. *Journal of Environmental Quality, 30*(2), 356–368. Retrieved on 18 May 2024.
3. Bangladesh: Eastern Flash Floods 2024 Situation Report No. 02 as of 30 August 2024.
4. Human Rights and Peace for Bangladesh v Government of Bangladesh and others [2016] (Turag River Case) HCD (Writ Petition No. 13989 of 2016, judgement declared on 3 February 2019).
5. For more details, see SDG Tracker Bangladesh, https://sdg.gov.bd/page/indicator-wise/1/320/3/0#

REFERENCES

Adel, M. M. (2001). Effect on water resources from upstream water diversion in the Ganges basin. *Journal of Environmental Quality*, *30*(2), 356–368.

Adel, M. M. (2012). Downstream ecocide from upstream water piracy. *American Journal of Environmental Science*, *8*(5), 528–548.

Afroz, R., & Rahman, A. (2013). Transboundary river water for Ganges and Teesta rivers in Bangladesh: An assessment. *Global Science and Technology Journal*, *1*(1), 100–111.

Alam, M. M. (2014). An evaluation of water resources planning organization, an apex planning organization in water sector Bangladesh. *Asian Journal of Applied Science and Engineering*, *3*(1), 66–83.

Argentina v Uruguay, ICGJ 425 (2010).

Arif, K. H. (2015, July 28). Law opinion. *Teesta Water Rights and International Law*. Dhaka, Bangladesh: The Daily Star.

Arifuzzaman, M., Hannan, M. A., Rahman, M. A., & Rahman, M. R. (2019). Laws regulating water pollution in Bangladesh. *Journal of Sociology and Anthropology*, *3*(1), 15–24.

Bangladesh Climate Change Strategy and Action Plan. (2009). A publication by Ministry of Environment and Forests, Government of the People's Republic of Bangladesh, Ministry of Environment Forest and Climate Change. Accessed from https://iucn.org/content/bangladesh-climate-change-strategy-and-action-plan-2009.

Bangladesh Delta Plan, 2100. The Best Gift for the Future Generations by the Present Generation. General Economics Division (GED), Bangladesh Planning Commission, Ministry of Planning, Government of the People's Republic of Bangladesh. Accessed from https://policy.asiapacificenergy.org/sites/default/files/Bangladesh%20Delta%20Plan%202100%20%28Abridged%20Version%29%20%28EN%29.pdf.

Barua, A. (2018). Water diplomacy as an approach to regional cooperation in South Asia: A case from the Brahmaputra basin. *Journal of Hydrology*, *567*, 60–70.

Baten, M. A., & Titumir, R. A. (2016). Environmental challenges of trans-boundary water resources management: The case of Bangladesh. *Sustainable Water Resources Management*, 2, 13–27.

Begum, F. (2020). Transboundary water agreements in South Asia: Does the principle of equitable and reasonable utilization of water exist? *Bangladesh Journal of Law*, *18*(1), 179–194.

Birnie, P. (2009). *International Law and the Environment* (3rd ed.). Oxford University Press.

Bisht, M. (2012, April 25). Tipaimukh Dam: Some Myths, Some Facts. *IDSA Comment*.

Buocz, T., & Eisenberger, I. (2023). Demystifying legal personhood for non-human entities: A Kelsenian approach. *Oxford Journal of Legal Studies*, *43*(1), 32–52.

Cambridge University Press (2017). Trail smelter arbitral tribunal decision. *American Journal of International Law*, *35*(4), 684–736.

Cano-Pecharroman, L., & O'Donnell, E. (2024). Relational representation: Speaking with and not about Nature. *PLOS Water*, *3*(10), 1–25.

Crow, B., & Lindquist, A. (1990, February). Development of the Rivers Ganges and Brahmaputra: The Difficulty of Negotiating a New Line. *Discussion Paper*. Research Group The Open University.

Dellapenna, J. W. (2023). The dispute over the status and use of the Silala River (Chile v. Bolivia): The International Court of Justice again declines to apply international water law. *Wyoming Law Review*, *23*(2), 73–102.

Dey, B. (2014). Water management in the age of decolonization: An account of the development of the Ganges and Teesta water dispute in South Asia. *Journal of the Indian Law Institute*, *56*(3), 348–375.

Edirisinghe, A., & Suchet-Pearson, S. (2024). Nature as a sentient being: Can rivers be legal persons? *RECIEL*, *33*(2), 224–235.

France v Spain, RIAA 281 (1957).

Friesen, B., & Day, J. (1977). Hydroelectric power and science provisions of the 1950 Niagara Treaty. *Journal of American Water Resources Association, 13*(6), 1175–1190.

Germany v Netherlands, ICJ Rep 3 (1969).

Hanasz, P. (2014). Power flows: Hydro-hegemony and water conflicts in South Asia. *Security Challenges, 10*(3), 95–112.

Hayat, S., Gupta, J., Vegelin, C., & Jamali, H. (2022). A review of hydro-hegemony and transboundary water governance. *Water Policy, 24*(11), 1723–1740.

Higano, Y., & Islam, F. (2002). Rural poverty alleviation through large-scale irrigation planning: Problem and prospects of the Dalia Barrage Project, Bangladesh. *Bulletin of Research Institute of Economic Science, 32*, 371–384.

Hungary v Slovakia, ICJ Rep 7 (1997).

Islam, M., & O'Donnell, E. (2020). Legal rights for the Turag: Rivers as living entities in Bangladesh. *Asia Pacific Journal of Environmental Law, 23*(2), 160–177.

Islam, M. N. (2001). The 1997 Watercourse Convention and its relevance to Bangladesh-India water sharing issue. *Bangladesh Journal of Law, 5*(1&2), 1–34.

Islam, M. N. (2005). Water resources: Limitations in the Indo-Bangladesh legal regime. *The Dhaka University Studies, 16*(2), 89–102.

Jackson, S., O'Donnell, E., Godden, L., & Langton, M. (2024). Ontological collisions in the northern territory's aboriginal water rights policy. *Ocenia, 93*(3), 259–281.

Kauffman, C. M., & Martin, P. L. (2023). How Ecuador's courts are giving form and force to rights of nature norms. *Transnational Environmental Law, 12*(2), 366–395.

LeFeuvre, D. A. (1991). The Great Lakes water quality agreement: An exercise in co-operation and commitment. *Canadian Water Resources Journal, 16*(3), 261–265.

Macfarlane, D. (2019). Negotiating Niagara falls: US-Canada environmental and energy diplomacy. *Diplomatic History, 43*(5), 916–943.

Macpherson, E. J. (2019). *Indigenous Water Rights in Law and Regulation: Lessons from Comparative Experience*. Cambridge University Press.

McCaffrey, S., & Sinjela, M. (1998). The 1997 United Nations Convention on International Watercourses. *The American Journal of International Law, 92*(1), 97–107.

Meiser, J. W. (2018). Introducing Liberalism in International Relations Theory. *E- International Relations*. https://www.e-ir.info/2018/02/18/introducing-liberalism-in-international-relations-theory/

Mianabadi, H., Mostert, E., & Gie, N. v. (2015). Trans-boundary River basin management: Factors influencing the success or failure of international agreements. In K. W. Hipel, L. Fang, J. Cullmann, M. Bristow, K. W. Hipel, L. Fang, J. Cullmann, & M. Bristow (Eds.), *Conflict Resolution in Water Resources and Environmental Management*. Berlin Heidelberg: Springer. doi:http://dx.doi.org/10.1007/978-3-319-14215-9_7

Naher, M. K., Rahman, A. M., & Chowdhury, M. O. (2021). Policy recommendations towards water sharing geopolitics between India and Bangladesh. *International Journal of Humanities Social Sciences and Education, 8*(7), 77–85.

Najam, A. (2004). The environmental challenge to human security in South Asia. In R. Thakur, & O. Wiggen (Eds.), *South Asia in the World: Problem Solving Perspective on Security, Sustainable Development and Good Governance* (pp. 225–247). Tokyo: United Nations University Press.

National Water Policy. (1999). Ministry of Water Resources, Government of the People's Republic of Bangladesh. Accessed from https://www.ircwash.org/sites/default/files/822-BD-15935.pdf.

National Water Policy. (2002). Government of India, Ministry of Water Resources. Accessed from https://www.nwm.gov.in/national-water-policy-2002.

O'Donnell, E. (2018). At the intersection of the sacred and the legal: Rights for nature in Uttarakhand, India. *Journal of Environmental Law, 30*(1), 135–144.

O'Donnell, E. (2020). Rivers as living beings: Rights in law, but no rights to water? *Griffith Law Review*, *29*(4), 643–668.

O'Donnell, E. (2023). Repairing our relationship with rivers: Water law and legal personhood. In R. Larson, & V. Perez (Eds.), *Research Agenda for Water Law*. Edward Elgar Publishing Ltd.

O'Donnell, E., & Arstein-Kerslake, A. (2021). Recognising personhood: The evolving relationship between the legal person and the state. *Griffith Law Review*, *30*(3), 339–347.

O'Donnell, E., & Macpherson, E. (2018). Voice, power and legitimacy: The role of the legal person in river. *Australasian Journal of Water Resources*, *23*(1), 35–44.

O'Donnell, E., Poelina, A., Pelizzon, A., & Clark, C. (2020). Stop burying the Lede: The essential role of indigenous law(s) in creating rights of nature. *Transnational Environmental Law*, *9*(3), 403–427.

O'Donnell, E., & Talbot-Jones, J. (2018). Creating legal rights for rivers: Lessons from Australia, New Zealand and India. *Ecology and Society*, *23*(1).

Obaidullah, M., & Howlader, M. R. (2025). Assessing geopolitical and socio-economic consequences of India-Bangladesh water. *Discover Global Society*, *3*(1), 1–25

Page, J., & Pelizzon, A. (2022). Of rivers, law and justice in the Anthropocene. *The Geographical Journal*, *190*(2) 1–11.

Pandey, P. (2012). Revisiting the politics of the Ganga Water Dispute between India and Bangladesh. *India Quaterly*, *68*(3), 267–281.

Parven, A., & Hasan, S. (2018). Trans-boundary water conflicts between Bangladesh and India: Water governance practice for conflict resolution. *International Journal of Agricultural Research, Innovation and Technology*, *8*(1), 79–84.

Pecharroman, L. C. (2018). Rights of nature: Rivers that can stand in court. *Resources*, *7*(1), 1–14.

Putzer, A., Lambooy, T., Jeurissen, R., & Kim, E. (2022). Putting the rights of nature on the map. A quantitative analysis of rights of nature initiatives across the world. *Journal of Maps*, *18*(1), 89–96.

Rahman, K. S., Islam, Z., Navera, U. K., & Ludwig, F. (2019). A critical review of the Ganges Water Sharing arrangement. *Water Policy*, *21*(2), 259–276.

Rajsingh, S., & Tripathi, N. (2020). Equitable and reasonable utilization of water resources: A critical analysis in context of Indo-Bangladesh Ganges Water Sharing. *BiLD Law Journal*, *5*(2), 103–118.

Rawia, T. (2015). *Revisiting Hydro-hegemony from a Benefit Sharing Perspective: The Case of Grand Ethiopian Renaissance Dam*. Bonn: German Development Institute.

Revkin, A. (2008, September 28). Ecuador Constitution Grants Rights to Nature. *The Opinion Pages*. The New York Times.

Salehin, M. (2010). Opportunities for transboundary water sharing in the Ganges, the Brahmaputra and the Meghna basin. *Water: Policy and Performance for Sustainable Development* (pp. 29–43). New Delhi: Oxford University Press.

Salgado, J. (2022). Causes and consequences of recent degradation of the Magdalena River basin, Colombia. *Limnology and Oceanography Letters*, *7*(6), 451–465.

Samaranayake, N. (2018). Bangladesh: The strongest advocate of basin-wide management. In J. Wuthnow, N. Samaranayake, & S. Limaye (Eds.), *Raging Waters: China, India, Bangladesh, and Brahmaputra River Politics*. Virginia: Marine Corps University Press.

Sanders, K. (2018). 'Beyond human ownership'? Property, power, and legal personality for nature in Aotearoa New Zealand. *Journal of Environmental Law*, *30*(1), 207–234.

Sedoff, A. (2014). Sustainable power and scenic beauty: The Niagara River Water Diversion Treaty and its relevance today. *Energy Policy*, *66*, 526–536.

Tadak, M., Clapcott, J., Holmes, R., & McNeil, C. (2023). Transforming freshwater politics through metaphors: Struggles over ecosystem health, legal personhood, and invasive species in Aotearoa New Zealand. *People and Nature*, *5*(2), 496–507.

Tanasescu, M. (2016). *Environment, Political Representation and the Challenge of Rights.* Springer Nature.

Tignino, M., & Bréthaut, C. (2020). The role of international case law in implementing the obligation not to cause significant harm. *International Environmental Agreements: Politics, Law and Economics*, *20*, 631–648.

Ujwala, P. (2017, May 27). *Teesta River Dispute between India and Bangladesh Explained.* Retrieved from ClearIAS. https://www.clearias.com/teesta-river-dispute/

United Kingdom v Albania, ICJ Rep 4 (1949).

United Nations. (2015). *Future We Want – Outcome document.* Retrieved from Sustainable Development Knowledge Platform. https://sustainabledevelopment.un.org/futurewe-want.html

UN [UNWC]. (1997). *Convention on the Law of the Non-Navigational Uses of International Watercourses* 189 UNTS 137.

Warner, J. (2004). *Plugging the GAP Working with Buzan: The Ilisu Dam as a Security Issue. University of London.* London: SOAS Water Issues Study Group, School of Oriental and African Studies/King's College London.

Waslekar, S., & Futehally, I. (2013). *Rivers of Peace: Restructuring India Bangladesh Relation.* Mumbai: Strategic Foresight Group.

WWF and H&M (2015). *Water Governance in Bangladesh: Challenges and Opportunities around Policy, Institutional Function and Implementation for a Sustainable Water Future.* Switzerland: World Wide Fund for Nature.

Zangl, B. (2012). Between law and politics: Explaining international dispute settlement behavior. *European Journal of International Relations*, *18*(2), 369–401.

Zelle, A., Wilson, G., Adam, R., & Greene, H. (2021). *Earth Law: Emerging Ecocentric Law—A Guide for Practitioners.* New York: Aspen Publishing.

3 Riparian Neighbours
Exploring India–China Water Disputes and Pathways to Resolution

Neeraj Singh Manhas and Genevieve Donnellon-May

INTRODUCTION

The geopolitical landscape of South Asia is intricately shaped by its vast transboundary river systems, which traverse national borders and connect the livelihoods of millions across multiple countries (Figure 3.1). Among these, the shared river basins between India and China, the Ganges–Brahmaputra–Meghna, the Indus, and the Irrawaddy stand out due to their immense scale, the volume of freshwater they transport, and their critical role in sustaining agriculture, drinking water supplies,

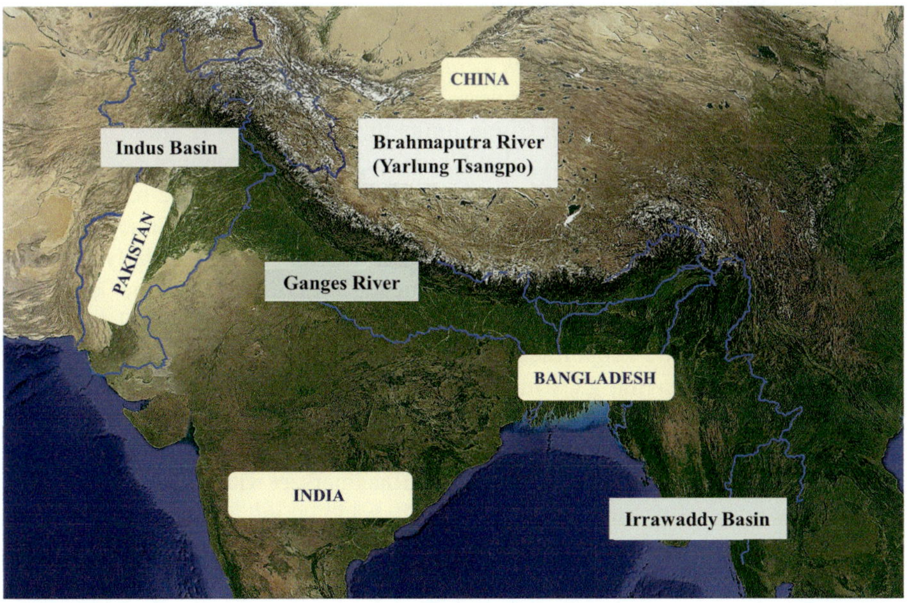

FIGURE 3.1 Geopolitical landscape of South Asia shaped by transboundary rivers linking nations and livelihoods. (*Source:* Prepared by authors.)

DOI: 10.1201/9781003599692-3

and hydroelectric power generation for some of the world's most populous regions. Originating primarily in the Tibetan Plateau, often referred to as the "Third Pole" due to its vast freshwater reserves and glaciers, these rivers flow from China into India, linking the two nations as riparian neighbours. However, the shared reliance on these vital water resources has not translated into robust cooperation. Instead, the absence of a comprehensive bilateral treaty or regulatory mechanism has created a precarious situation, marked by intermittent cooperation and conflict, further complicated by China's upstream control and perceived lack of transparency in sharing hydrological data.

The India–China water relationship is deeply embedded within the broader context of their geopolitical rivalry, which spans territorial disputes, military standoffs, and competing regional ambitions. The Brahmaputra River (known as the Yarlung Tsangpo in Tibet), in particular, has emerged as a flashpoint due to China's ambitious dam-building projects, which have raised significant concerns in India about potential disruptions to downstream water flows. Such disruptions could lead to severe droughts or catastrophic flooding, threatening the livelihoods of millions in India's northeastern states and Bangladesh (Ramachandran, 2015). The 2017 Doklam standoff, a 73-day military confrontation over a disputed border area, further strained bilateral relations, exacerbating mistrust and hindering efforts to establish cooperative water-sharing mechanisms. China's subsequent suspension of hydrological data-sharing for the Brahmaputra, citing technical reasons, underscored the fragility of existing arrangements and highlighted the strategic use of water as a geopolitical tool (Sinha, 2018).

Recent literature on India–China transboundary water relations provides valuable insights into the geopolitical, environmental, and legal dimensions of these disputes. Scholars like Wouters and Chen (2013) emphasize the need for legal frameworks grounded in international water law principles, such as equitable utilization and no-harm, to govern transboundary rivers. However, they note that the absence of a binding treaty between India and China limits the applicability of these principles. Similarly, Chellaney (2011) argues that China's upstream dominance and its reluctance to engage in multilateral water governance frameworks exacerbate tensions, framing water as a potential trigger for conflict in Asia. Studies by Bandyopadhyay and Ghosh (2016) highlight the environmental challenges, particularly climate-induced glacial melting in the Tibetan Plateau, which threatens the long-term sustainability of these river systems. These works underscore the interplay of environmental stressors and geopolitical rivalries but often focus on either the legal or ecological aspects, leaving gaps in integrating these dimensions with the broader India–China strategic competition.

A notable gap in current research lies in the limited examination of how transboundary water disputes between India and China intersect with broader dimensions of their strategic rivalry, namely competition for regional influence, military posturing along the Line of Actual Control (LAC), and expanding economic ambitions across South Asia. While scholars such as Samaranayake (2019) have explored the strategic implications of China's Belt and Road Initiative (BRI) in relation to water infrastructure, few studies have explicitly linked these developments to the dynamics of water-sharing. Furthermore, there is a scarcity of research on trust-building mechanisms such as joint monitoring systems, data-sharing protocols, or third-party mediation that could help de-escalate water-related tensions. Equally overlooked is the potential for sub-regional cooperation involving other riparian states, including Bangladesh and Myanmar, which are directly impacted by the India–China water equation. Given the transboundary nature of these river systems, a broader regional lens is essential for fostering sustainable and equitable water governance.

The implications of India–China water disputes extend well beyond bilateral concerns, posing significant risks to regional stability in South Asia. These rivers support the livelihoods of over a billion people, making their management central to food security, energy generation, and economic development. China's upstream position affords it asymmetric control over water flows, a reality that India views as a strategic vulnerability, especially in the context of recurring border tensions and China's expanding footprint in the Indo-Pacific. The lack of transparency in hydrological data-sharing, notably during the Doklam standoff, has deepened India's apprehensions regarding China's intentions, amplifying fears of potential "water wars" in an already volatile geopolitical landscape (Wirsing & Jasparro, 2012). Climate change compounds these challenges by accelerating glacial melt, disrupting seasonal flows, and increasing the frequency of extreme weather events such as floods and droughts. Together, these environmental and geopolitical pressures underscore the urgent need to embed water diplomacy within a broader framework of India—China strategic relations—one that prioritizes cooperation, resilience, and regional sustainability.

This study aims to unpack the complex interplay of geopolitical, legal, and environmental factors that shape transboundary water relations between India and China. Building upon existing scholarship, it bridges the strategic dimensions of bilateral rivalry with the ecological and legal challenges inherent in managing shared river systems. Its core contribution lies in adopting a holistic lens, treating water disputes not as isolated technical issues, but as integral to the broader geopolitical contest between the two nations. By examining the potential for cooperative mechanisms such as joint hydrological data-sharing protocols and inclusive regional governance frameworks, the study proposes actionable strategies to mitigate conflict and promote sustainable water management. It also expands the analytical scope by considering the role of sub-regional actors and international legal norms, offering a fresh perspective on a topic often confined to bilateral analysis. To structure this inquiry, the research is guided by the following three central questions:

1. How do strategic rivalries between India and China shape their transboundary water disputes?
2. What environmental and legal barriers hinder effective cooperation over shared river basins?
3. Which institutional mechanisms, ranging from bilateral treaties to regional platforms, can support sustainable and equitable water governance?

These questions illuminate the multifaceted nature of the issue and provide a framework for understanding how geopolitics, environmental pressures, and legal regimes converge in India–China water relations. The significance of this study lies in its timely engagement with a critical yet underexplored dimension of South Asian geopolitics. As tensions over shared rivers increasingly influence regional stability and global water governance, this research fills a vital gap by integrating strategic, ecological, and legal perspectives. Its novelty stems from a focus on trust-building and regional cooperation, essential ingredients for transforming water from a source of friction into a foundation for collaboration. Through this approach, the study offers practical insights for policymakers, scholars, and stakeholders committed to securing the long-term sustainability of South Asia's transboundary river systems.

OVERVIEW OF INDIA–CHINA TRANSBOUNDARY RIVER RELATIONS

The transboundary river relations between India and China are influenced by the region's unique geographic, geopolitical, and hydrological characteristics (Table 3.1). As two of the world's most populous and rapidly developing nations, both countries rely heavily on shared water resources to meet their agricultural, industrial, and domestic needs. The key rivers crossing their borders include the Ganges–Brahmaputra–Meghna system, the Indus, and the Irrawaddy. Among these, the Brahmaputra stands out as the most significant and contentious, owing to its strategic importance and the vast scale of its basin.

A core feature of India–China water relations is the stark asymmetry between the two nations in terms of riparian position and relative power. China, as the upstream state, exercises considerable hydrological and strategic control over the headwaters of several major rivers, particularly the Brahmaputra and Sutlej. This control enables Beijing to undertake unilateral development projects, such as hydropower dams, without requiring consent from downstream states (Biba, 2018; Hanasz, 2017). Furthermore, the economic and military gap between the two countries reinforces

TABLE 3.1
Summary of Major Transboundary Rivers between India and China

River Basin	Origin	Flow Through	Strategic Importance
Brahmaputra (Yarlung Tsangpo)	Tibet (China)	China, India (Arunachal Pradesh, Assam), Bangladesh	High: hydropower, agriculture, ecology, and regional stability
Ganges	India and Nepal	India, Bangladesh	Moderate (between India and Bangladesh): integrated into the wider Brahmaputra system
Indus	Tibet (China)	China, India, Pakistan	Moderate: upstream tributaries in Chinese territory affect water availability
Irrawaddy	Eastern Himalayas (Yunnan)	China, Myanmar	Low: less politically contentious, but ecologically significant

Source: Created by the authors.

this power imbalance. China's larger economy and stronger state capacity enable it to execute infrastructure and water engineering projects at a scale and speed unmatched by India, further complicating cooperative efforts (Biswas and Tortajada, 2016). This asymmetry shapes negotiation strategies: China typically frames its projects as internal developments under its sovereign jurisdiction, while India, being more vulnerable to upstream alterations, prioritizes transparency, data-sharing, and risk mitigation.

BRAHMAPUTRA RIVER (YARLUNG TSANGPO)

The Brahmaputra River, known in China as the Yarlung Tsangpo, is one of South Asia's most important transboundary rivers. Originating in the Tibet Autonomous Region (TAR), it flows eastward along the northern slopes of the Himalayas before turning southward through a deep gorge near Namcha Barwa into India's Arunachal Pradesh. It then traverses the floodplains of Assam and finally enters Bangladesh, where it merges with the Ganges and Meghna rivers to form the world's largest delta. The Brahmaputra sustains millions through agriculture, fisheries, and hydropower, and its ecological importance is matched by its role in regional food and water security. China's control over the river's upper reaches allows it to construct hydropower projects such as the Zangmu Dam and plan further development including controversial diversion schemes which could impact downstream water availability and seasonal flow variability (Figure 3.2). India has voiced concerns over these activities, particularly in the absence of legally binding agreements or basin-wide coordination

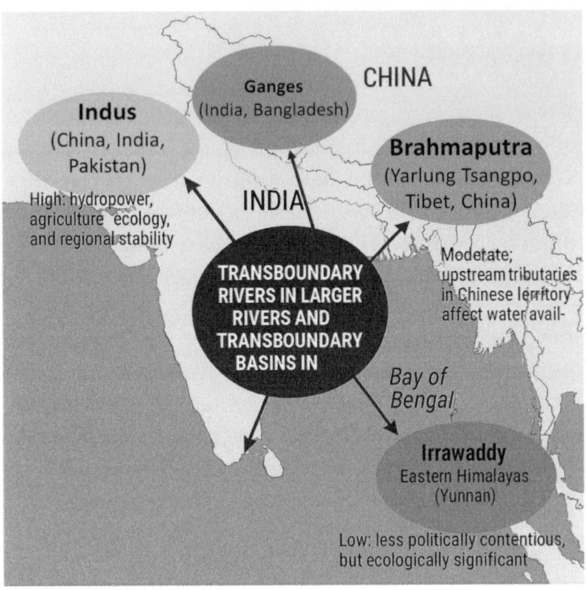

FIGURE 3.2 Major transboundary rivers in South Asia, highlighting their shared nature, regional significance, and role in shaping interstate relations. (*Source:* Prepared by authors.)

mechanisms. In response, India has pursued diplomatic engagement and infrastructure development in its own territory, both as a countermeasure and as a means of asserting riparian rights.

GANGES RIVER

Although the Ganges primarily flows through India and Bangladesh, its integration into the larger hydrological network of the region links it indirectly to China. Some of its headwaters and tributaries originate in Tibetan territory. While not as politically charged as the Brahmaputra, the Ganges remains crucial to India's internal water security. It supports a dense population reliant on its waters for irrigation, drinking water, and cultural practices. This makes India particularly sensitive to any upstream developments that could affect the region's already precarious water balance.

THE INDUS BASIN

The Indus River also originates in Tibet, and although it flows predominantly through India and Pakistan, its upper tributaries, including the Shyok and Sutlej, are partially under Chinese control. China's activities in these upper reaches, such as potential dam construction or river regulation, are closely monitored by India, especially given the already fraught hydro-politics of the India–Pakistan Indus Waters Treaty (IWT). While not the central axis of India–China water tensions, the Indus basin represents a broader strategic concern for India regarding upstream Chinese control and potential trilateral flashpoints.

THE IRRAWADDY BASIN

The Irrawaddy River, though primarily associated with Myanmar, has its origins in the eastern Himalayas near the China–India border. Its headwaters are fed by rivers that flow from China's Yunnan province, linking it to the broader transboundary water system of the region. While less politically contentious, the Irrawaddy's hydrology highlights the interconnectedness of regional water systems. China's upstream influence in Yunnan province affects the river's ecological flow, with implications for downstream ecosystem services and livelihoods in Myanmar (Figure 3.2). Its case illustrates the potential for transboundary rivers to become politically significant depending on how upstream interventions evolve. As such, while not a major point of dispute, the Irrawaddy's headwaters remain important in discussions about regional environmental health, water-sharing agreements, and sustainable management of cross-border water systems.

HISTORICAL CONTEXT

The history of India–China transboundary river relations is a complex interplay of cooperation, tension, and strategic manoeuvring (Table 3.2). Unlike other regions, where riparian states have established comprehensive treaties to govern shared water

TABLE 3.2
Timeline: Key Milestones in India–China Water Relations

Year	Event
1954	Limited hydrological data-sharing initiated during the early Nehru–Zhou Enlai cooperation
1962	Sino-Indian border war halts early cooperation; deepens mistrust
2002	First MoU signed for hydrological data-sharing on the Brahmaputra during the flood season
2004	Second MoU signed for Sutlej River flood-season data exchange
2006	Establishment of Expert-Level Mechanism (ELM) for dialogue on transboundary waters
2010	China confirms construction of the Zangmu Dam, the first major dam on the Yarlung Tsangpo
2013	Third MoU signed; expands frequency and duration of data-sharing on the Brahmaputra
2017	China withholds hydrological data following the Doklam standoff, citing flood damage
2020	India voices concern over reports of China's planned 60 MW "super dam" near the Great Bend, Tibet
2021	Chinese authorities confirm feasibility studies for new dams near the Arunachal border

Source: Created by the authors.

resources, India and China have largely relied on ad hoc arrangements and informal agreements. This lack of formalized frameworks has shaped the way both countries approach issues related to water management, with the absence of structured agreements leaving room for both collaboration and contention.

The earliest formal interactions between India and China concerning transboundary rivers date back to the 1950s and 1960s, a period characterized by relative cooperation despite the broader geopolitical tensions of the Cold War and the ongoing Sino-Indian border disputes. During these years, the focus was primarily on the exchange of basic information and data related to river flows and flood control, especially within the Brahmaputra basin. This era of cooperation was marked by a pragmatic approach, where both nations recognized the importance of sharing vital hydrological data, particularly in the context of managing floods and ensuring the stability of the region's water resources. However, this cooperation occurred against the backdrop of growing tensions, laying the foundation for a more complicated relationship in the decades that followed.

The late 20th century marked the intensification of tensions between India and China over water-related issues, particularly as both nations accelerated their infrastructure development projects. China's ambitious construction of hydropower dams and other hydrological projects in Tibet, especially along the Yarlung Tsangpo (Brahmaputra), became a major source of concern for India (see also Donnellon-May, 2022). These concerns were compounded by China's opacity in its water management policies and its reluctance to enter into binding agreements on water sharing. This lack of transparency and cooperation only served to deepen India's anxieties, as it feared the potential for upstream water manipulations that could severely affect the availability of water for millions of people downstream in India and Bangladesh. The relationship between China and India has long been marked by complexity, volatility, and structural tensions. As the world's two most populous nations, they share a history fraught with territorial disputes and competing strategic interests. The Sino-Indian War of 1962 set the stage for decades of acrimony, marked by sporadic military confrontations and protracted standoffs along their Himalayan boundaries.

Beyond the territorial disputes, the Sino-Indian relationship has also been marred by historical factors, including India's long-standing concerns over China's growing ties with Pakistan. The Sino-Pakistani alliance has been a source of significant unease for India, particularly as it relates to regional security and the future of the disputed Kashmir and Jammu regions. These enduring tensions, coupled with the competition for influence in the broader Asian and global geopolitical landscape, have made cooperation on issues such as water management difficult to achieve, despite the urgent need for dialogue and collaboration. In response, India raised the issue at various diplomatic forums, advocating for stronger mechanisms for data-sharing and joint river basin management. However, progress was slow, hindered by the deep mistrust between the two nations and exacerbated by broader geopolitical tensions. These obstacles prevented the establishment of any meaningful bilateral treaties on water management, leaving both countries in a state of uncertainty and uncoordinated action.

As the 21st century progressed, the complexities of India–China water relations grew more pronounced, particularly in the context of both nations' rapid economic and military expansion. One of the most significant sources of tension has been China's ongoing dam-building activities, notably the construction of the Zangmu Dam on the Yarlung Tsangpo (Brahmaputra). These projects have the potential to alter the flow of the Brahmaputra River, which could significantly impact water availability for both northeastern India and Bangladesh. Such changes pose not only immediate economic risks threatening agriculture, livelihoods, and regional stability, but also long-term environmental consequences, further underscoring the urgency for effective and cooperative management of their shared water resources.

The challenges surrounding water management have been further complicated by geopolitical tensions. Bilateral relations between India and China have soured as India has increasingly aligned itself with the United States, particularly through initiatives like the Quad and Indo-Pacific military coordination. From China's perspective, India's growing ties with the United States represent a strategic attempt at containment, intensifying mistrust. Meanwhile, India perceives China's expanding technological and data presence overseas, particularly in South Asia, as a growing threat to its own national and economic security.

The complexities of the broader geopolitical rivalry between the two countries came to a head in 2020 with the deadly Galwan River Valley clash—the first such confrontation in five decades. This violent incident, in which dozens of soldiers from both sides lost their lives, further highlighted the deeply rooted strategic mistrust and the risks associated with unresolved territorial and water disputes. The clash was a stark reminder that the India–China relationship, particularly concerning shared water resources, remains fraught with potential for conflict unless managed with care and foresight. The geopolitical and geoeconomic rivalry between India and China has significantly complicated their water-sharing arrangements and broader water relations. The Doklam standoff in 2017, though primarily a border dispute, underscored the deeper tensions between the two nations, including those concerning shared water resources. In the aftermath of this incident, India has grown increasingly cautious about China's intentions regarding transboundary rivers, fearing that water could be wielded as a strategic tool in future conflicts.

Although a comprehensive treaty remains elusive, India and China have established some bilateral mechanisms to address their shared water resources. One of the most notable of these is the Memorandum of Understanding (MoU) signed in 2002, which was designed to facilitate the exchange of hydrological data during the monsoon season. This MoU has been periodically renewed and expanded, with efforts to include more detailed data-sharing, particularly concerning the Brahmaputra and Sutlej rivers. While there is no formal water-sharing agreement, these MoUs provide some space for cooperation. But they are inadequate to meet the complexities of transboundary water management. For example, the MoU on the Brahmaputra River was signed in 2002 and renewed in 2008, 2013, and 2018, and the MoU for the Sutlej River was signed in 2004 and renewed in 2010 and 2015 (see also Ministry of Jal Shakti, n.d.). Under these arrangements, China shares with India real-time data on water levels, discharge, and rainfall from hydrological stations on the two rivers during the flood season (June–October). India utilizes this data to oversee its water resources and reduce flooding dangers, which is important to the agriculture and livelihoods of those impacted.

Both MoUs have, however, lapsed in 2023, and while China has continued to provide data on the Sutlej River, both sides are currently in the process of renewing these accords (PTI, 2020). However, MoUs are not legally binding and do not formalize a dispute resolution mechanism. As a result, many divisive issues related to collective water management remain unaddressed. Besides the three MoUs, India and China have the Expert-Level Mechanism (ELM), which has met regularly since 2006 to discuss transboundary water issues. While it has indeed provided a platform for dialogue, the ELM has done little to address any of the significant disputes plaguing the two nations (Manhas, 2023).

KEY HYDRO-DIPLOMATIC MECHANISMS

Despite high-level mistrust, India and China have signed several MoUs and initiated bilateral dialogue forums (Table 3.3). However, these frameworks remain limited in scope, lacking dispute resolution mechanisms, enforcement provisions, or basin-wide coordination.

These efforts reflect incremental trust-building, but fall short of creating a binding legal regime akin to the IWT between India and Pakistan. While these agreements have proven useful in facilitating the exchange of hydrological information, they are limited in scope and often criticized for their lack of enforceability and comprehensiveness. The MoUs primarily focus on data exchange rather than joint river basin management or conflict resolution, which leaves many critical issues unresolved. As a result, while they provide a foundation for cooperation, they fall short of addressing the full spectrum of challenges related to the management of shared rivers between India and China. These gaps underscore the need for more robust and binding agreements that go beyond information sharing to include

TABLE 3.3
Summary of Key Bilateral Hydro-Diplomatic Mechanisms between India and China

Mechanism/Agreement	Year	Focus Area	Limitations
MoU on Brahmaputra	2002	Flood-season data-sharing	Seasonal, non-binding, no dispute resolution clause
MoU on Sutlej	2004	Flood warning and data-sharing	Narrow focus; no provisions for water use restrictions
Expert-Level Mechanism	2006	Institutional dialogue on rivers	Consultative only; no enforcement capacity
Updated MoU on Brahmaputra	2013	Expanded duration of data exchange	Still excludes hydropower transparency

Source: Created by the authors.

proactive collaboration on water governance, dispute resolution, and long-term sustainability.

Beyond their bilateral engagements, both India and China have also participated in various multilateral forums aimed at addressing transboundary water management. These include initiatives within the South Asian Association for Regional Cooperation (SAARC) and the Shanghai Cooperation Organization (SCO). However, the effectiveness of these multilateral efforts has been limited, as water disputes are frequently overshadowed by broader geopolitical issues, such as security concerns and regional rivalries, which dominate the agendas of these forums. The role of international law in India–China water relations has similarly been minimal. Unlike other transboundary river disputes, which are governed by international conventions such as the United Nations Watercourses Convention (UNWC), India and China have been hesitant to subject their water relations to global legal frameworks. Both nations prefer to manage their shared water resources through bilateral or regional mechanisms, reflecting their reluctance to bring sensitive issues into international legal arenas.

As of today, the transboundary river relations between India and China remain complex and fragile. While there has been some progress in terms of data-sharing and dialogue, the persistent challenges of trust, transparency, and strategic rivalry continue to hinder more substantial cooperation. The potential for conflict over water resources—especially in the context of the broader geopolitical tensions between the two nations—remains a significant concern, casting a long shadow over efforts to achieve lasting collaboration on shared water management. The stark asymmetry in India–China transboundary water relations, where China occupies the upper riparian position and India the lower, fundamentally shapes negotiation dynamics (Table 3.4). As the upstream state, China controls the sources and timing of river flows, allowing it to pursue large-scale hydropower and diversion projects with minimal downstream consultation (Biba, 2018). This hydrological leverage is bolstered by China's superior economic resources and military capability, which together afford Beijing greater strategic autonomy in water governance.

TABLE 3.4
Summary of Asymmetries Shaping India–China Water Negotiations

Asymmetry Type	India	China
Riparian Position	Lower riparian—reliant on upstream flows from China	Upper riparian—controls the headwaters of major transboundary rivers
Geopolitical Leverage	Seeks transparency, data-sharing, and joint management mechanisms	Uses upstream position to maintain strategic autonomy
Economic Power	Developing economy with high water stress in certain regions	Larger, more resource-capable economy with extensive dam-building plans
Military Power	Regional military power with a defensive posture	Superior military strength enabling assertive geopolitical behaviour
Domestic Politics	Nationalist pressures and regional water scarcity shape cautious diplomacy	National sovereignty and regional development (e.g. Tibet) prioritized
Negotiation Approach	Advocates for binding agreements and institutionalized cooperation	Prefers non-binding MoUs and avoids formal treaties

Source: Created by the authors.

India, by contrast, confronts both geographic vulnerability and limited leverage: its dependency on consistent monsoon-fed flows makes it especially sensitive to upstream alterations (Ho, 2017). Consequently, India's negotiation strategy emphasizes institutional safeguards—such as data-sharing MoUs and the ELM—seeking transparency and predictability. China's preference for non-binding, bilateral arrangements further entrenches this power imbalance, compelling India to adopt reactive measures and domestic infrastructure development to mitigate perceived upstream risks.

Domestic political considerations intensify these asymmetries. In India, chronic water scarcity in northeastern states and the politicization of water security drive a nationalist narrative that casts China's upstream dams as existential threats (Singh, 2022). State and central leaders face electoral pressure to demonstrate strength on water issues, limiting their willingness to compromise on sensitive hydropolitical matters (Vij et al., 2020). In China, internal imperatives such as securing water for its arid north and bolstering economic development in western provinces prioritize upstream control over transboundary cooperation. Nationalist discourse reinforces the notion that China must guard its sovereign right to develop Tibetan rivers, further reducing political incentives to engage in binding international agreements. Together, these domestic pressures on both sides harden negotiating positions and curtail the political space for collaborative, basin-wide water governance.

CURRENT CHALLENGES AND ISSUES

India–China transboundary river relations are shaped by geographic, geopolitical, and hydrological complexities. As two populous and industrializing nations, both depend heavily on shared rivers such as the Brahmaputra (Yarlung Tsangpo), Indus, and Irrawaddy to meet growing water demands. Among these, the Brahmaputra is the most strategically significant due to its trans-Himalayan flow, economic value, and ecological importance. A defining feature of these relations is the asymmetry in riparian position and power. China, as the upstream state, holds hydrological dominance over rivers like the Brahmaputra and Sutlej. This positional advantage allows Beijing to undertake unilateral hydropower projects without consulting downstream countries (Biba, 2018; Hanasz, 2017). China's superior economic and infrastructural capabilities further enable it to implement large-scale water infrastructure swiftly, while India seeks greater transparency and risk mitigation (Tortajada & Biswas, 2016).

The commissioning of the Zangmu Dam in 2015 marked a significant shift in Sino-Indian hydro-politics. Built in the TAR, the 510 MW dam was presented by China as essential for regional development (Reuters, 2015). However, it triggered an alarm in India about potential upstream alterations to water flow into Assam and Arunachal Pradesh. Despite Chinese claims that the dam would not significantly impact downstream flow (The Paper, 2020), the lack of transparent data exacerbated India's anxieties (Zhang, 2016). This project forms part of a broader strategy of hydropower development in TAR, serving strategic and economic objectives (State Council of the PRC, 2021; Tibet DRC, 2021; Xinhua News Agency, 2024). India views these efforts as consolidating China's upstream dominance, stoking fears over water security, regional flooding, and agricultural sustainability (Kashyap, 2014; Keerthana, 2021; Padmanabhan, 2014; Shaw, 2020).

Chinese narratives have sought to assuage Indian concerns. PowerChina, for example, described its Yarlung Tsangpo project as a "historic opportunity" (The Paper, 2020), while Chinese academics dismissed "water war" fears (Zhang, 2016). Hydrological estimates suggest that China contributes just 15–25% of the Brahmaputra's total flow, with the majority originating within India and Bhutan (Jyoti Deka, 2021). From Beijing's viewpoint, this limits the scope of disruption to India's water supply. Yet concerns persist. India remains wary of China's reluctance to provide transparent data or detailed project information, which deepens mistrust and fuels the "water wars" narrative (Donnellon-May, 2022). China's refusal to share comprehensive hydrological data creates an asymmetry of information, reinforcing India's perception of strategic vulnerability.

Formal cooperation remains limited. While China shares hydrological data during the monsoon season, there is no joint river basin management mechanism. Beijing prefers case-by-case bilateral engagement and has not signed any binding international river treaties (Ho, 2014, 2016). India, meanwhile, has pushed for treaty-based agreements to regulate water sharing and increase transparency (Ho, 2016), but China's aversion to multilateralism has left few avenues for structured cooperation. These disputes reflect broader geopolitical dynamics. India is particularly concerned about potential diversions of the Yarlung Tsangpo before it enters Indian territory. Although China insists that such plans will not harm downstream regions and that it remains open to dialogue, the absence of a binding agreement leaves India exposed. A significant alteration in river flow could severely impact agriculture and livelihoods in northeast India (The Economic Times, 2021). Environmental degradation is another pressing issue. The Brahmaputra basin is ecologically vital, supporting biodiversity and communities in northeast India. Dam construction can disrupt sediment transport, reduce soil fertility, and increase riverbank erosion, undermining agriculture and displacing vulnerable populations (Monga, 2022; Pradhan et al., 2021). Climate change compounds these challenges. Melting glaciers, erratic monsoons, and extreme weather events threaten the long-term stability of river systems.

Hydrological data exchange remains sporadic. Although several MoUs exist, the data shared is often seasonal and inadequate for long-term forecasting. Moreover, data-sharing is vulnerable to political tensions, as demonstrated when China suspended information exchange during the 2017 Doklam standoff, citing technical issues (Gupta, 2018; Khadka, 2017). While data resumed in 2018 (India Today, 2021), the incident highlighted India's dependence on Chinese goodwill. China controls the Nugesha, Yangcun, and Nuxia (Brahmaputra) and Tsada (Sutlej) stations, key data sources (Ministry of Jal Shakti, n.d.). Without transparency, trust, and formalized cooperation, the risk of miscalculation and long-term environmental degradation will only grow. Effective management will require sustained political will, shared norms, and depoliticized governance to safeguard the region's water security and ecological resilience.

SINO-INDIAN TERRITORIAL DISPUTES

The broader geopolitical rivalry between India and China deeply influences their transboundary water relations, reinforcing a cycle of mistrust and limited transparency. Both nations perceive each other as strategic competitors, and this adversarial lens extends to their management of shared rivers. India, as the downstream country, is particularly concerned that China could exploit its upstream position on rivers like the Brahmaputra to manipulate water flows for strategic advantage, especially during times of political or military tension. These anxieties are compounded by China's increasing hydrological infrastructure in Tibet and its refusal to enter binding water-sharing agreements.

A critical factor exacerbating these fears is the unresolved territorial dispute between the two countries, particularly over Arunachal Pradesh (called South Tibet by China). This Eastern Himalayan region, through which the Yarlung Tsangpo flows before becoming the Brahmaputra in India, holds immense strategic and symbolic value for both sides. Encompassing approximately 90,000 km^2, the territory lies at the intersection of sovereignty, resource control, and historical memory—especially given China's military success there in the 1962 war. For China, relinquishing its claim would challenge its broader narrative of sovereignty over Tibet (Pak, 2016), while for India conceding the territory is politically unacceptable. These territorial tensions feed into a wider pattern of strategic defensiveness that hampers bilateral water cooperation. The unresolved border and competing claims not only fuel diplomatic hostility but also undermine trust in technical and scientific collaboration over shared rivers. As a result, India and China often prioritize sovereignty and national security over cooperative river governance, making transparent dialogue and joint solutions elusive.

LEGAL AND REGULATORY FRAMEWORKS

The absence of a comprehensive legal framework governing India–China transboundary rivers presents a major challenge to the sustainable and cooperative management of shared water resources. Unlike many international river systems, there is no bilateral treaty or multilateral legal mechanism that binds the two countries to defined obligations, leaving room for unilateral actions and heightened tensions. Both nations have, at times, accused each other of hydro-hegemonic behaviour. China, as the upstream country controlling key rivers originating in the Tibetan Plateau, has refused to sign water-sharing agreements with its downstream neighbours, including India. Although it claims to follow principles of international water law—such as equitable and reasonable use—it has not ratified the 1997 UNWC (Wouters & Chen, 2011). India, similarly, is not a party to the UNWC and has invoked Chinese hegemony narratives to justify its own hydropower development projects (Zhang, 2016). This mutual suspicion limits the scope for cooperative water governance. Some informal cooperation mechanisms do exist, such as hydrological data-sharing agreements, but these have proven unreliable. Notably, during the 2017 Doklam standoff, China withheld critical hydrological data, undermining trust and illustrating the fragility of non-binding arrangements. These events highlight the limitations of informal cooperation in the absence of a binding legal architecture.

Most international transboundary rivers are governed by treaties that define water rights, mandate data-sharing, and provide dispute resolution mechanisms. India and China, however, have historically resisted involving third parties or international legal institutions in their water disputes, preferring bilateral negotiations. This reluctance further complicates efforts to institutionalize cooperative river basin management. The absence of a formal treaty continues to hinder effective cooperation and contributes to a precarious environment where transboundary water disputes could escalate, especially amid broader geopolitical tensions. Establishing a robust legal framework remains essential to promoting trust, transparency, and sustainable water governance between the two powers.

CLIMATE CHANGE IMPACTS

Climate change adds a critical dimension to the already strained transboundary water relations between India and China, complicating efforts to achieve environmental sustainability in South Asia. The rivers originating from the Tibetan Plateau, especially the Brahmaputra, are acutely vulnerable to climatic shifts, as they depend heavily on glacial meltwater. Accelerated glacial retreat due to rising global temperatures is altering the hydrological patterns of these rivers, leading to both short-term and long-term challenges. In the near term, increased glacial melt may lead to heightened river flows, particularly during monsoon seasons, raising the risk of flooding in downstream regions (Prakash, 2020). Over time, however, diminishing glaciers will reduce the meltwater supply, especially in the dry season, threatening water availability for agriculture, drinking, and energy in both countries. This shift could deepen existing water insecurity and amplify the likelihood of inter-state tensions over access and allocation of transboundary resources. The unpredictability of climate-induced hydrological change presents additional governance challenges. Climate change is intensifying the frequency and severity of extreme weather events floods, droughts, and storms that further strain transboundary water systems. The Brahmaputra basin, for example, regularly experiences seasonal floods that are becoming more destructive due to erratic rainfall and glacial melt. Likewise, prolonged droughts are placing increasing pressure on already water-stressed regions. Despite these shared vulnerabilities, there is no coordinated mechanism for joint flood management or early warning systems. The lack of institutionalized disaster response frameworks leaves both populations and infrastructure exposed to escalating environmental threats.

Beyond environmental impacts, climate change intersects with national political and economic agendas. In China, the construction of hydropower dams in Tibet is part of a dual strategy: fulfilling domestic energy demands and reinforcing geopolitical leverage (Manhas & Yadav, 2024). Though coal still dominates China's energy mix, hydropower is central to its plans to peak carbon emissions by 2030 and achieve carbon neutrality by 2060 (State Council, 2021, 2024). India, meanwhile, faces immense pressure to secure water for its agrarian economy and growing population, particularly in drought-prone regions (WaterAid, 2024). These internal pressures compound transboundary tensions, as both states prioritize domestic stability and development over regional cooperation. Climate change acts as a threat multiplier in India–China water relations. The lack of collaborative frameworks, combined with geopolitical mistrust and national development imperatives, creates a volatile environment that undermines environmental sustainability. Addressing these challenges requires not only technical adaptation strategies but also renewed political will for cooperation. Without joint climate-resilient governance of shared rivers, the region risks deepening ecological and geopolitical instability in the coming decades.

DOMESTIC PRESSURES

Domestic pressures in both India and China significantly shape their approaches to shared transboundary rivers, adding complexity to water-related tensions. In India, the Brahmaputra basin supports a rapidly growing population heavily reliant on agriculture (European Foundation for South Asian Studies, 2022). This region is not only economically vital but also politically sensitive, marked by long-standing insurgencies, ethnic tensions, and territorial disputes (Pak, 2016). Water scarcity exacerbates these challenges, heightening the Indian government's responsibility to ensure a reliable water supply to the northeastern states. Perceptions of upstream water diversion or dam construction by China often provoke fears of adverse impacts on local agriculture, water access, and broader regional stability.

In China, internal pressures are driven primarily by water scarcity and energy security. Despite being home to major river systems, China suffers from acute water shortages, particularly in its northern and western provinces. Rapid industrialization, urban expansion, agricultural demand, and pollution further strain the available water supply (Cheng et al., 2009). This situation has elevated the strategic importance of water-rich regions like Tibet, where several transboundary rivers—including the Brahmaputra (Yarlung Tsangpo)—originate. Ensuring water access and harnessing hydropower from these rivers has become a priority for Beijing's development agenda.

Hydropower is now China's second-largest energy source, critical to achieving national energy goals and reducing reliance on fossil fuels. The government has committed to peaking carbon emissions by 2030 and achieving carbon neutrality by 2060, making the expansion of renewable energy infrastructure, particularly in Tibet, central to its plans (State Council of the People's Republic of China, 2021; Tibet Autonomous Region Development and Reform Commission, 2021). These efforts are framed not only as tools of green development but also as mechanisms for stimulating economic growth in Tibet and enhancing political stability in this strategically sensitive region (Tibet Autonomous Region Party Committee Office, 2024). However, this development-focused approach often overlooks the downstream implications for countries like India and Bangladesh, which depend on the same river systems for agriculture, energy generation, and domestic water use. India, in particular, sees such unilateral upstream projects as undermining regional environmental security and violating norms of equitable water-sharing.

Regional geopolitics further complicate these domestic drivers. India's water-related diplomacy with Bangladesh, for instance, is closely tied to its concerns over Chinese upstream behaviour. Cooperation—or the lack thereof—on Brahmaputra water-sharing can influence India's broader strategic posture in South Asia. Meanwhile, China's BRI, which involves major infrastructure and water-related investments across South and Southeast Asia, raises alarms in New Delhi about Beijing's expanding regional influence (Zhang & Zhang, 2021). India perceives the BRI as a potential tool for Beijing to gain strategic leverage over countries reliant on shared water systems. As a result, both countries increasingly view water not only as a domestic necessity but also as a strategic asset. Water is securitized, seen as both a resource to be protected and a potential instrument of statecraft. This strategic framing undermines prospects for environmental sustainability and joint river basin management. Rather than fostering cooperation on data-sharing, ecosystem preservation, or disaster preparedness, domestic imperatives and geopolitical competition encourage unilateralism and mistrust.

Domestic socio-economic and political pressures combined with strategic ambitions have turned transboundary rivers into contested arenas (Table 3.5). Without institutional mechanisms that bridge national interests and environmental sustainability, these rivers risk becoming flashpoints in the broader India–China rivalry. Effective, cooperative governance will require both countries to recalibrate domestic priorities in ways that recognize the transboundary and ecological nature of shared water systems.

COMPARATIVE CASE STUDIES

Examining successful examples of transboundary water cooperation globally provides valuable lessons for managing the complex water dynamics between India and China. These cases illustrate how countries facing geopolitical tensions, power asymmetries, and upstream–downstream challenges can achieve sustainable and equitable water governance. A prominent example is the Nile River Basin,

TABLE 3.5
Domestic Pressures on India and China in Transboundary Water Management

Aspect	India	China
Population & Economy	Rapidly growing population in the Brahmaputra basin, largely dependent on agriculture (European Foundation for South Asian Studies, 2022)	Industrializing rapidly, with significant water demands from agriculture, industry, and urbanization (Cheng et al., 2009)
Political Context	Politically sensitive northeastern region with insurgencies, ethnic tensions, and territorial disputes (Pak, 2016)	Tibet viewed as a strategic region; development linked to political stability and national security (Tibet Autonomous Region Party Committee Office, 2024)
Water Scarcity	Water scarcity heightens vulnerability, intensifying concerns over upstream activities by China	Severe water shortages in northern and western China; managing transboundary rivers critical to addressing scarcity
Energy Security	Significant pressure to secure water for agriculture and economic development	Hydropower is the second-largest energy source; focus on expanding renewable energy to meet emissions targets (State Council of the People's Republic of China, 2021)
Environmental Concerns	Fear of altered river flows impacting agriculture and water security downstream	Hydropower and water infrastructure development in Tibet prioritized despite potential downstream impacts
Geopolitical Dynamics	Water-sharing relations with Bangladesh influence broader regional diplomacy	BRI investments in South and Southeast Asia raise concerns over Chinese leverage in regional water politics (Zhang & Zhang, 2021)
Strategic View of Water	Water seen as a critical resource and potential geopolitical leverage in regional power dynamics	Water securitized as a strategic asset and tool for national development and regional influence

Source: Created by the authors.

where 11 countries share critical water resources. Historically contentious due to its importance for agriculture, drinking water, and hydropower, the basin saw a major shift with the establishment of the Nile Basin Initiative (NBI) in 1999 (Nile Basin Initiative, n.d.). Despite significant power imbalances—particularly between downstream Egypt, with entrenched historical water rights, and upstream Ethiopia, which is developing the Grand Ethiopian Renaissance Dam—the NBI has fostered dialogue, data-sharing, and technical cooperation. These mechanisms have helped reduce tensions and build trust among riparian states, enabling joint management despite their divergent interests.

In Europe, the Danube River Basin offers another instructive case. Flowing through ten countries, the Danube is one of the world's most international rivers.

The International Commission for the Protection of the Danube River (ICPDR), established under the 1994 Danube River Protection Convention, provides a legally binding, comprehensive framework for cooperation (Danube River Protection Convention, n.d.). The ICPDR emphasizes sustainable water use, pollution control, and disaster management, supported by transparent institutions and robust legal mechanisms. This example underscores the importance of binding agreements and inclusive institutional structures in ensuring long-term cooperative water governance. The Mekong River Commission (MRC), formed in 1995, also offers parallels, especially due to the geographical and political dynamics involved. The Mekong flows through six countries, including China, which is an upstream state and a dialogue partner but not a full member of the MRC (Mekong River Commission, n.d.). Despite China's limited formal participation, the MRC has facilitated cooperation among the lower Mekong countries through data-sharing and joint initiatives, albeit with challenges (Uprety, 2014). This case highlights how dialogue and technical collaboration can persist even without full upstream involvement.

These international experiences offer important lessons for India–China water relations. First, as seen with the NBI and MRC, establishing a formal institution to manage shared river basins could significantly improve cooperation. Such an institution might focus on transparent data-sharing, joint scientific research, and coordinated disaster response, reducing uncertainties related to China's upstream activities and ensuring more equitable water distribution downstream in India. Second, the importance of strong legal and institutional frameworks, exemplified by the Danube and Nile cases, cannot be overstated. While India and China face historical tensions and power imbalances that complicate binding treaty negotiations, even a formalized non-binding framework could create a platform for regular cooperation. This could include dispute resolution mechanisms, standardized data-sharing protocols, and ongoing dialogue addressing emerging water challenges. Third, trust-building is vital. Long-standing mistrust between India and China has often stalled cooperation, but regular, transparent communication and joint projects such as flood forecasting, disaster preparedness, and sustainable water management research could foster confidence and demonstrate mutual benefits. These initiatives could serve as stepping stones towards a more positive bilateral relationship on water issues. Finally, preventive diplomacy is key to avoiding escalation of water conflicts. Many successful cases have employed proactive diplomacy to manage tensions before crises arise. By establishing robust communication channels and cooperation mechanisms, India and China could shift from reactive crisis management to anticipatory conflict prevention, thereby promoting regional stability and environmental sustainability.

Thus, comparative case studies show that even amid geopolitical rivalry and asymmetrical power, cooperative water governance is achievable through institutions, legal frameworks, trust-building, and preventive diplomacy (Table 3.6). Applying these lessons to South Asia's transboundary waters could help transform India–China water relations into a foundation for regional sustainability rather than a source of conflict.

TABLE 3.6
Summarizing the Comparative Case Studies and Lessons Relevant to India–China Transboundary Water Cooperation

River Basin	Countries Involved	Key Institution/ Mechanism	Main Challenges	Outcomes and Lessons
Nile River Basin	11 countries (e.g. Egypt, Ethiopia, Sudan)	Nile Basin Initiative (NBI), 1999	Historical water rights disputes, power asymmetries	Fostered dialogue, data-sharing, and technical cooperation. Shows that joint management is possible despite tensions
Danube River Basin	10 countries in Europe	International Commission for the Protection of the Danube River (ICPDR), 1994	Pollution, disaster risk, and water usage	Legally binding framework with strong institutions. Demonstrates the value of binding agreements and inclusive structures
Mekong River Basin	6 countries (including China)	Mekong River Commission (MRC), 1995	China's limited formal participation, upstream dominance	Enabled lower Mekong cooperation via data-sharing and joint projects. Shows partial cooperation is possible even without full upstream commitment

Source: Created by the authors.

PATHWAYS TO RESOLUTION

Resolving India–China transboundary water disputes requires a multifaceted and forward-looking strategy that addresses both immediate challenges and long-term governance deficits. The absence of a binding legal framework between the two countries has exacerbated mistrust, reduced transparency, and impeded cooperative efforts. To mitigate these challenges, a structured approach that encompasses legal, technical, and diplomatic tools is necessary. This section outlines a five-pronged framework: (1) Confidence-Building Measures (CBMs), (2) Early Warning and Data-Sharing Mechanisms, (3) Multilateralizing Bilateral Issues, (4) Eco-Diplomacy and Joint Research Initiatives, and (5) Lessons from Global Case Studies.

Confidence-Building Measures

A critical first step towards sustainable cooperation is the establishment of robust CBMs. Trust remains a scarce commodity in the India–China relationship, particularly in the realm of transboundary water governance. Historical tensions, power asymmetries, and a lack of communication mechanisms have contributed to a cycle of suspicion and reaction. To break this cycle, both countries should pursue joint projects that deliver immediate, visible benefits such as flood forecasting systems, early warning platforms, or drought mitigation plans. These initiatives can serve as neutral, apolitical platforms for cooperation and reduce misperceptions about each side's intentions. Equally important is the enhancement of institutional dialogues. Government-to-government channels, particularly those involving military or border management authorities, should incorporate water cooperation as a strategic confidence-building topic. Complementing this, track 2 dialogues involving think tanks, academics, and civil society can help broaden stakeholder engagement and foster a shared understanding of ecological interdependence between the two countries. These interactions can also act as diplomatic buffers during periods of geopolitical tension, maintaining continuity in dialogue even when official communication stalls.

Public diplomacy initiatives such as educational exchanges, water diplomacy forums, and media campaigns can further encourage domestic constituencies supportive of bilateral cooperation. Over time, these measures may reduce the (geo) political risk for leaders engaging in sustained river basin diplomacy, helping to insulate water cooperation from broader geopolitical rivalries.

Multilateralizing Bilateral Issues

While water-sharing issues between India and China are primarily bilateral, the experience of other regional river basins demonstrates that multilateral frameworks can depoliticize transboundary water governance and create institutional incentives for cooperation. China has historically resisted multilateral engagement in river governance, especially in South Asia, due to sovereignty concerns. Nonetheless, selective multilateralism, such as observer status, or even joint initiatives within forums, such as the MRC or the South Asia Cooperative Environment Programme, could offer a middle ground to help address concerns.

Notably, China's growing presence in the Mekong Basin, where it has partially engaged in multilateral hydrological data-sharing and basin management programmes, provides valuable precedents. Lessons can be drawn from these frameworks—including transparency protocols, benefit-sharing models, and joint scientific assessments, which, in turn, could inform the development of a semi-formal, multi-stakeholder platform for the Brahmaputra River. Involving third parties such as Bhutan and Bangladesh would also reflect the transboundary nature of the river and distribute the burden of cooperation across a wider coalition of stakeholders. India and China could also collaborate within broader regional organizations, such as the SCO or the BRICS New Development Bank, to fund and coordinate infrastructure and environmental resilience projects in shared basins. These platforms could serve as "soft" venues for dialogue, where politically sensitive issues are discussed through the lens of development, sustainability, or climate adaptation.

Eco-Diplomacy and Joint Research Initiatives

Another viable path for fostering cooperation is through eco-diplomacy, where shared environmental challenges become the basis for collaborative engagement. Climate change presents an existential risk to water security in both India and China, particularly in the Himalayan region, which serves as a critical source of glacial run-off for major rivers. Joint research initiatives on glacial melt, snowpack variation, and watershed degradation can build scientific consensus and encourage coordinated adaptation strategies. Bilateral environmental task forces or working groups composed of hydrologists, climatologists, geographers, and legal scholars could support integrated river basin management strategies. These bodies could jointly assess the environmental impacts of proposed infrastructure projects (such as hydropower dams or water-diversion projects) and recommend mitigation strategies that benefit both countries. Establishing shared monitoring stations in ecologically sensitive areas, such as the Yarlung Tsangpo–Brahmaputra corridor, would further institutionalize cooperation. Academic institutions and think tanks can play a pivotal role in generating impartial, data-driven analysis that informs diplomatic efforts. Universities in both countries could establish joint water research centres, host annual symposia on Himalayan hydrology, or develop student exchange programmes focused on river basin management. These collaborations promote long-term capacity building and deepen epistemic communities that are more resilient to political disruptions.

Case Studies and Legal Precedents

The absence of a legally binding treaty between India and China on water-sharing is a major obstacle to sustainable cooperation. By contrast, other transboundary water regimes such as the IWT between India and Pakistan offer instructive lessons. Despite recurrent hostilities, the IWT has largely endured, thanks to its clear dispute resolution mechanisms, detailed water allocation protocols, and an independent commission for implementation oversight. Similarly, the Mekong Cooperation Mechanism illustrates how a basin-wide institutional approach can help manage asymmetries between upstream and downstream states. While not perfect, the Mekong model's emphasis on data-sharing, joint infrastructure planning, and stakeholder consultations provides a blueprint for possible adaptations in the Brahmaputra Basin.

Furthermore, India and China should consider initiating negotiations towards a legally binding treaty based on the principles enshrined in the 1997 UN Convention on the Law of the Non-Navigational Uses of International Watercourses. Although neither country is currently a signatory, the Convention provides a robust normative foundation for equitable and reasonable water use, no-harm principles, and procedural transparency. Adopting select provisions such as prior notification of planned measures and compulsory dispute resolution mechanisms could fill the regulatory void that currently defines India–China water relations.

Synthesis and Forward-Looking Recommendations

To move from reactive conflict management to proactive cooperation, both India and China must acknowledge the transboundary and ecological nature of their shared river systems. The existing reliance on informal MoUs and narrowly scoped technical exchanges is insufficient to manage the growing complexity of water governance under conditions of climate stress, geopolitical competition, and increasing demand. In the next decade, the two countries should prioritize the formalization of a treaty-based framework for river cooperation that includes enforceable dispute resolution clauses, basin-wide environmental assessments, and multi-stakeholder governance structures. Establishing a bilateral water commission with the mandate to oversee data exchange, infrastructure coordination, and joint emergency responses would mark a significant step forward. These measures should be complemented by broader institutional reforms, including the integration of water diplomacy into foreign policy dialogues and the alignment of water governance with national climate adaptation plans.

At the same time, India and China must explore ways to embed water cooperation into regional development narratives. Infrastructure corridors, energy grids, and agricultural value chains that depend on river flows can serve as anchors for cooperative engagement, provided they are governed by principles of transparency, equity, and sustainability. Ultimately, transboundary water governance between India and China is not just a technical or legal challenge; it is a strategic imperative. A cooperative framework rooted in shared ecological responsibility, mutual benefit, and international best practices can serve not only to manage current tensions but also to build the foundations of long-term regional stability.

CONCLUSION

India–China transboundary water relations remain burdened by persistent challenges that have hindered meaningful cooperation. At the heart of these tensions is the absence of a formal, binding legal framework to govern the management of shared river systems. This institutional void has enabled unilateral actions, most notably China's dam construction on rivers such as the Brahmaputra, which have heightened downstream concerns in India. The situation is further exacerbated by a lack of transparency in hydrological data-sharing, perpetuating mistrust and limiting opportunities for coordinated response. Power asymmetries between the two nations compound the difficulty of negotiating equitable solutions, while existing legal and regulatory mechanisms across the region remain inadequate to address the evolving complexities of transboundary water governance. In response to these structural gaps, several pathways for resolution emerge. Chief among them is the establishment of formal treaties that mandate regular data exchange, joint decision-making, and dispute resolution protocols. Such agreements would foster trust, enhance predictability, and lay the groundwork for more effective basin-wide management.

Transparency in hydrological data, especially in the context of extreme weather events, is increasingly vital for disaster preparedness and long-term planning. CBMs, including collaborative water management initiatives and sustained diplomatic dialogue, could further institutionalize cooperation and reduce the risk of escalation. Climate change adds urgency to these efforts. The Himalayan region, which serves as the source for many of Asia's major rivers, is experiencing accelerated glacial melt, erratic monsoon behaviour, and shifting precipitation patterns. These trends threaten water availability, agricultural productivity, and disaster resilience in both India and China. Shared vulnerabilities present a compelling rationale for collaboration. Joint climate monitoring, cooperative adaptation strategies, and integrated watershed management could help both countries build resilience while reducing the potential for conflict. Ultimately, the sustainable and equitable management of transboundary water resources is not only essential for bilateral relations but also a strategic imperative for regional stability and climate resilience. A cooperative framework grounded in transparency, legal clarity, and ecological foresight will be critical to meeting the growing water demands of both populations in the decades ahead.

REFERENCES

Bandyopadhyay, J., & Ghosh, N. (2016). Holistic engineering and hydro-diplomacy in the Ganges-Brahmaputra-Meghna Basin. *Economic and Political Weekly, 51*(44–45), 50–60.

Biba, S. (2018). *China's Hydro-Politics in the Mekong: Conflict and Cooperation in Light of Securitization Theory*. Routledge.

Biswas, A. K., & Tortajada, C. (2015). Water security, climate change and sustainable development: an introduction. In *Water Security, Climate Change and Sustainable Development* (pp. 1–5). Springer Singapore.

Chellaney, B. (2011). *Water: Asia's New Battleground*. Georgetown University Press.

Cheng, H., Hu, Y., & Zhao, J. (2009). Meeting China's water shortage crisis: Current practices and challenges. *Environmental Science & Technology, 43*(2), 240–244. https://doi.org/10.1021/es801934a

Danube River Protection Convention. (n.d.). *Convention*. International Commission for the Protection of the Danube River. https://www.icpdr.org/about-icpdr/framework/convention

Donnellon-May, G. (2022, December 9). China's super hydropower dam and fears of Sino-Indian water wars. The Diplomat. https://thediplomat.com/2022/12/chinas-super-hydropower-dam-and-fears-of-sino-indian-water-wars/

European Foundation For South Asian Studies. (2022). *Transboundary River Management in Bangladesh and India | Bilateral Cooperation and Environmental Securitization*. https://www.efsas.org/publications/study-papers/transboundary-river-management-in-bangladesh-and-india/

Gupta, J. (2018, May 17). China resumes sharing Brahmaputra water flow data with India. Dialogue Earth. https://dialogue.earth/en/water/china-resumes-sharing-brahmaputra-water-flow-data-with-india/

Hanasz, P. (2017). Muddy waters: International actors and transboundary water cooperation in the Ganges-Brahmaputra problemshed. *Water Alternatives, 10*(2), 459–474.

Ho, S. (2014). River Politics: China's policies in the Mekong and the Brahmaputra in comparative perspective. *Journal of Contemporary China, 23*(85), 1–20. https://www.tandfonline.com/doi/abs/10.1080/10670564.2013.809974

Ho, S. (2016). 'Big brother, little brothers': Comparing China's and India's transboundary river policies. *Water Policy, 18*(S1), 32–49. https://iwaponline.com/wp/article-abstract/18/S1/32/20262

Ho, S. (2017). Introduction to 'transboundary river cooperation: actors, strategies and impact'. *Water International, 42*(2), 97–104.

India Today. (2021, June 1). *Annual data sharing by China on Brahmaputra, Sutlej rivers begins*. https://www.indiatoday.in/india/story/annual-data-sharing-by-china-on-brahmaputra-sutlej-rivers-begins-1809723-2021-06-01

Jyoti Deka, B. (2021). Hydro-politics between India and China: The 'Brahma-Hypothesis'and securing the Brahmaputra. *Asian Affairs, 52*(2), 327–343. https://doi.org/10.1080/03068374.2021.1914449

Kashyap, S. G. (2014, September 18). *Convey Assam's concern to China over construction of dams on Brahmaputra: Gogoi to Modi*. The Indian Express. https://indianexpress.com/article/india/india-others/convey-assams-concern-to-china-over-construction-of-dams-on-brahmaputra-gogoi-to-modi/#sthash.ak1Zz7V5.dpuf

Keerthana, R. (2021, December 3). *Why is China's new dam a concern for India?* The Hindu. https://www.thehindu.com/children/why-is-chinas-new-dam-a-concern-for-india/article61734775.ece

Khadka, N. (2017, September 18). *China and India water 'dispute' after border stand-off.* British Broadcasting Corporation News. https://www.bbc.com/news/world-asia-41303082

Mahapatra, S. K., & Ratha, K. C. (2016). Brahmaputra river: A bone of contention between India and China. *Water Utility Journal*, 13, 91–99.

Manhas, N. S. (2023, November 21). A fragile lifeline: India and China must collaborate on water. Lowy Institute. https://www.lowyinstitute.org/the-interpreter/fragile-lifeline-india-china-must-collaborate-water

Manhas, N. S., & Yadav, G. H. (2024). Beyond the border: Exploring the complex dynamics of water tensions between India and China. *Discover Global Society*, *2*, 57. https://doi.org/10.1007/s44282-024-00089-x

Mekong River Commission. (n.d.). About. Mekong River Commission. https://www.mrcmekong.org/

Ministry of Jal Shakti, Department of Water Resources, River Development and Ganga Rejuvenation. (n.d.). *India-China Cooperation*. Ministry of Jal Shakti. https://www.mowr.gov.in/india-china-cooperation

Monga, B. (2022). *Impact of Chinese Activities on Brahmaputra River.* https://nilebasin.org/our-goals

Nile Basin Initiative. (n.d.). Our goals. Nile Basin Initiative. https://nilebasin.org/our-goals

Padmanabhan, S. (2014). *Next China-India War: World's First Water War-2029.* Manas Publications.

Pak, J. H. (2016). China, India, and war over water. *Parameters*, *46*(2), 25–37. https://doi.org/10.55540/0031-1723.2920

Pradhan, N. S., Das, P. J., Gupta, N., & Shrestha, A. B. (2021). Sustainable management options for healthy rivers in South Asia: The case of Brahmaputra. *Sustainability*, *13*(3), 1087. https://doi.org/10.3390/su13031087

Prakash, A. (2020). *Retreating Glaciers and Water Flows in the Himalayas: Implications for Governance.* https://www.orfonline.org/research/retreating-glaciers-and-water-flows-in-the-himalayas-implications-for-governance

PTI. (2020). China initiates process of sharing hydrological data for Sutlej, 10 days before scheduled date. Deccan Herald. https://www.deccanherald.com/india/china-initiates-process-of-sharing-hydrological-data-for-sutlej-10-days-before-scheduled-date-840678.html

Ramachandran, S. (2015, April 13). Water wars: China, India and the great dam rush. The Diplomat. https://thediplomat.com/2015/04/water-wars-china-india-and-the-great-dam-rush/

Reuters. (2015, October 13). China completes Tibet's biggest hydropower plant. Reuters. https://www.reuters.com/article/markets/commodities/china-completes-tibets-biggest-hydropower-plant-idUSL3N12D3NO/

Samaranayake, N. (2019). *China's Engagement with Smaller South Asian Countries.* United States Institute of Peace.

Shaw, S. (2020, December 18). *War over water: The new flashpoint in the India-China stand-off.* Byline Times India. https://bylinetimes.com/2020/12/18/war-over-water-the-new-flashpoint-in-the-india-china-stand-off/

Sinha, U. K. (2018). *Indus Water Treaty and Beyond: Sharing Water Resources.* Routledge.

Singh, A. (2022). Better water and land allocation for long-term agricultural sustainability. *Water Resources Management*, *36*(10), 3505–3522.

State Council of the People's Republic of China. (2021, March 13). *中华人民共和国国民经济和社会发展第十四个五年规划和2035年远景目标纲要*. State Council of the People's Republic of China. https://www.gov.cn/xinwen/2021-03/13/content_5592681.htm

State Council of the People's Republic of China. (2024, September 19). 能源供给保障有力节能降碳成效显著——新中国75年经济社会发展成就系列报告之十三. State Council of the People's Republic of China. https://www.gov.cn/lianbo/bumen/202409/content_6975422.htm

The Economic Times. (2021, February 7). *China's hydropower strategy: Threats, challenges and responses*. https://economictimes.indiatimes.com/news/defence/chinas-hydropower-strategy-threats-challenges-and-responses/articleshow/80732494.cms?utm_source=contentofinterest&utm_medium=text&utm_campaign=cppst

The Paper. (2020, November 28). 重磅 | 雅鲁藏布江下游水电开发决策敲定，规模近6000万千瓦，相当于"再造3个三峡". *The Paper*. https://www.thepaper.cn/newsDetail_forward_10183087

Tibet Autonomous Region Development and Reform Commission. (2021, March 29). 西藏自治区国民经济和社会发展第十四个五年规划和二〇三五年远景目标纲要. Tibet Autonomous Region Development and Reform Commission. https://drc.xizang.gov.cn/xwzx/daod/202103/t20210329_197641.html

Tibet Autonomous Region Party Committee Office. (2024, January 8). 政府工作报告（摘登）*2024* 年工作安排. Tibet Autonomous Region Party Committee Office. http://jxt.xizang.gov.cn/xzweb/detail?id=6244&channelId=8&categoryId=60

Uprety, K. (2014). *Challenges, Lessons, and Prospects for Operationalizing Regional Projects in Asia*. https://openknowledge.worldbank.org/server/api/core/bitstreams/a6cc2709-a9df-524c-be7b-7806f68d93ad/content

Vij, S., Warner, J., & Barua, A. (2020). Power in water diplomacy. *Water International*, *45*(4), 249–253.

WaterAid. (2024). *Water Scarcity – Causes, Issues & Problems of Water Scarcity in India*. https://www.wateraid.org/in/blog/water-scarcity

Wirsing, R. G., & Jasparro, C. (2012). *Spotlight on Sino-Indian Water Relations*. Asia-Pacific Center for Security Studies.

Wouters, P., & Chen, H. (2011). China's 'soft-path' to transboundary water cooperation examined in the light of two UN global water conventions – exploring the 'Chinese way'. *Journal of Water Law*, *22*(6), 229–247. http://www.lawtext.com/pdfs/sampleArticles/Wouters229-247.pdf

Xinhua News Agency. (2024, December 25). 雅鲁藏布江下游水电工程已获核准. Xinhua News Agency. http://www.news.cn/politics/20241225/b61f14cad5c046a0a209bf36fa99d69b/c.html

Zhang, H. (2016). Sino-Indian water disputes: The coming water wars? *Wiley Interdisciplinary Reviews: Water*, *3*(2), 155–166. https://doi.org/10.1002/wat2.1123

Zhang, L., & Zhang, H. (2021). Water diplomacy and China's bid for soft power in the Mekong. *China Review*, *21*(4), 39–75. https://hdl.handle.net/10356/160280

4 From Colonial Land Regimes to Contemporary Water Politics

Tracing Hydro-Social Transformations in the Bengal Delta

Debojyoti Das

INTRODUCTION

The theme of water, rain, rivers, and the monsoon as governing systems of life, and as paradigms for research in their own right, has acquired global relevance today. Worster's classic Rivers of Empire (1988) exemplifies this approach, arguing that the social space of any region is profoundly shaped by the relationship between humans and their environment. In this sense, the triadic relationship between land, water, and society has drawn increasing scholarly attention, particularly in light of contemporary challenges: projected sea-level rise, the retreat of polar ice, shifting monsoon patterns, transboundary river disputes, and large-scale infrastructural interventions that endanger the fragile ecologies of the Bay of Bengal littoral (Amrith, 2013a, Amrith 2013b 2018). Many of these issues can be traced to the colonial-era separation of land from water, a project rooted in late 18th-century science. This was premised on a Cartesian principle of control, supported by notions of progress, improvement, modernisation, and the so-called civilisation of the Oriental 'other'. At its core lay the division of nature from culture and the appropriation of land as an economic resource (D'souza, 2014). Such epistemological shifts not only reshaped landscapes but also set in motion enduring patterns of ecological disruption and contestation, whose consequences remain evident in present-day environmental and governance challenges.

This thinking continued to gain momentum in the decolonised period in the name of 'national development' through publicly funded dams and embankment projects initiated by the premier Jawaharlal Nehru, who conferred full faith on technology and scientific experts. However, this scientific vigour of authoritarian high

DOI: 10.1201/9781003599692-4

modernism as James Scott puts so eloquently in his work *Seeing Like a State* has come under intense scrutiny in recent decades (Klingensmith, 2007; Mehta, 2005) as river water disputes and the submergence of land by big dams have raised important questions surrounding social justice, equity and challenged the very foundation of techno managerial understand of nature, culture, and natural resource management. New research looked into the cultural geographies of water and land through the prism of hybridity, as a third space (Appadurai and Breckenridge, 2009), such as the delta braided landscape with temporary islands, oxbow lakes, streams, and rivers that constantly change their material state (Lahiri-Dutt, 2014a, Lahiri-Dutt, 2014b). A notable development in this regard has been the increasing recognition that we need to understand water as a social object, or as Linton and Budds put it: 'it is not just society's relationship with water that is at stake, but the social nature of water itself' (Linton and Budds, 2014: 171). This work draws attention to how water and land should be seen as part of a complex ecological and social system and how they change meaning and subjectivities in a particular cultural context. Similarly, my research based on the Sundarbans delta will showcase how the riverine waterways have shaped people's livelihood and the cultural geography of the littoral.

I endeavour to complicate this dominant epistemic narrative of global processes by looking at local anthropogenic interventions such as colonial land reclamation programme that has had far-reaching impact on community life and local ecology dominated by water. Delta ecology presents a unique formless space and finds resonance in local customs and literature. For example, centuries before the term Sundarban came to define mangrove forest as the masterwork of the unique delta ecology (a nomenclature that developed from colonial science – botany, geology, and oceanography), the littoral was known as *atharo bathir desh*, meaning land of 18 tides (Sarkar, 2010). Here, the importance of water over land was a predominant feature of Bengal's social life. Inopportunely, deltas have been the subject of neglect in South Asia's Environmental History discourse due to a methodological bias in favour of the land (forest, agriculture, and urban landscape) or the sea-oceanic world (maritime environment, sea routes, trading guilds, and port networks).[1] Likewise, in earlier studies of the environmental history of South Asia, the delta was considered as a unique geographical region, and there was less consideration on how the functioning of the delta was impacted by colonial processes of land reclamation, local anthropogenic intervention, legislation that privileged land as a resource and water as waste, and hazard for human settlement.

Though deltas remain anonymous and discrete geographical features within 'area studies', they have historically shaped connected worlds between land and water through tributaries and backwaters that unite maritime and terrestrial sites, enabling circulation, mobility, and spatial histories. This chapter examines how a land-centric understanding of the delta environment since the late 19th century in Bengal led to catastrophic anthropogenic interventions at the local level and embodies the current environmental dispensation of population at risk to hydro hazards. Therefore, I will diachronically examine the cause and consequences of the current crisis by historicising hazards, unlike the instrumental and technocratic discourses of state planning units and disaster mitigation policies. To put it differently, state planning works with a particular instrumental view of the environment. Here, I affirm my understanding of *hydro-sociality* following Linton and Budds' (2014) definition, namely as 'a relational-dialectical approach to conceptualize the hydro-social cycle as a socio-natural process by which water and society make and remake each other over space and time'. This framing deliberately moves away from viewing water and society as distinct, bounded entities, a perspective institutionalised by colonial cartographers, hydrologists, and revenue officials, who historically ascribed economic value to water and land while marginalising their social, cultural, and political dimensions. Instead, this chapter foregrounds how water actively shapes society and functions as a vital agent of connection across cultures, identities, and geographies. The historiography and ethnography presented here centre on this entanglement, emphasising how littoral communities in the Bengal delta experience and interpret nature's serendipity through their fluid, evolving relationships with water.

THE RESEARCH SO FAR

The Bay of Bengal delta is like the edge of a knife. They are environmentally vulnerable, densely populated, and at the fault line of struggles between India and China for regional dominance (Amrith, 2013a). But was it the same a few centuries ago? In this chapter, I envisage re-examining the historical connections that shaped vibrant littoral histories of communities inhabiting the bay in light of the Cartesian politics of dividing the land from its water resources. Economic historian Tirthankar Roy, in his essay 'Where is Bengal' published in *Past and Present* (2011), makes a blunt point that Bengal became a vibrant space for trade and commerce in the late 18th century with the establishment of the British Empire and in doing so rejects the idea of 'Sonar Bangla' golden province before the advent of colonialism. This chapter aims to understand how this transformation took place and at whose cost. In the great transformation that took place in the Bengal delta between 18th and 20th centuries, the colonials and local intermediaries played a catalytic role in transforming the littoral geography into an agrarian frontier. Though the process was initiated in a small way with the penetration of Islam from the 12th century, major intervention through embankment and land reclamation programme began in the 19th century, giving shape to the land-centric discourse on revenue settlement that undercut the importance of water in the local economy (D'souza, 2015).

The cultural context in which riverine communities evolved in the Bengal delta can perhaps be best understood through their food habits and objects of material culture. Rivers, ponds, canals, and river islands have long been integral to Bengal's physical and social landscape, with communities depending equally upon land and water across the delta. However, from the late 18th century, when colonial policies sought to expand the revenue base of the Bengal Presidency, large-scale land reclamation in the Sundarbans began to gather momentum. This process was far from straightforward and spanned several decades, often bringing the colonial state into conflict with local Zamindars. The efforts of Claude Russell (Collector General, 1770–1790) and Tilman Hinckell (Settlement Officer, 1780–1785) are particularly noteworthy, as both were tasked with adjudicating land boundary disputes in order to expand the agrarian frontier. Hinckell even established local courts, known as Cutcherries of Reference, to resolve conflicts between Zamindars and the distribution of khas land to tenants and landless labourers who cleared the land for settled agriculture (Hunter, 1875). It soon became evident that the revenue administration of the Lower Bengal delta could not be sustained without firm colonial policies on tenure development. Throughout the 19th century, land tenure legislation facilitated the systematic reclamation of deltaic swamps and forests, yet these new settlements remained acutely vulnerable to coastal flooding and cyclones in the Sundarbans.

BRINGING DELTAS BACK INTO ENVIRONMENTAL HISTORY

In the last decade, several scholarly works by Iftekar Iqbal (2012), Rohan D'Souza (2015), Dutt (2007, , 2013, 2014), and Debjani Bhattacharya (2018) have looked into colonial law and land use regime that have transformed the fluvial landscape in the Bay of Bengal and the adjoining settlement, including the important townships such as Kolkata, Burdwan, Chittagong, and Dhaka. In this context, the Bengal Alluvion and Diluvion Act 1847 takes centre stage as it played a crucial role in classifying land that temporarily emerged and disappeared to the delta with the creation of braided streams and islands (Lahiri-Dutt, 2014b). The law gave teeth to the segregation of productive alluvium land from wasteland and classified permanent from temporary landmass. The legal control of rivers went hand-in-hand with engineering controls. Once the idea of land as being in need of protection from unpredictable waterways was entrenched, the fluid world of waterscape became invisible. Thereafter, these shifting waterways continued to be presented as a hydrological hazard that needed technocratic intervention by those experts who had studied how to manage rivers through engineering measures.

These recent works can be grouped within the broad field of environmental humanities and history that poignantly speak of the problem of technocratic and bureaucratic interventions that generated land as an object of 'capital' and extractive economic resource in the Bengal Presidency separating it from its life breath 'water'-that is mixed in freshwater systems in the floodplain and delta archipelago. The colonial attempts to flood proof the Ganges and its tributaries in the plains emerged out of a 'property thinking' (D'souza, 2014; Whitehead, 2010) and revenue generation aims of the British administration. In the littoral of Kolkata, the delta environment was shaped by land grants that led to the slow reclamation of marshlands and mangrove swamps for human settlement (Roy, 2012). Within two centuries, the landscape of the Bengal delta was transformed with speculative investments that led to the growth of the *Bhadralok* gentry, who made windfall profits through tenants who worked their fields as sharecroppers and bonded labourers (Bose, 1993).

While most of the work has focused on the entire delta and explored national and Imperial archives, my sources are based on informal and private collections kept in peripheral places such as Gosaba – one of the important entry points to the Indian part of the Sundarban delta settled by Sir Daniel Hamilton during 1920–1945. During my year-long fieldwork in the Sundarbans conducted in two phases (2012–2013 and 2015–2016), I investigated the private archival collection of Sir Daniel Hamilton that has a good number of court cases, petitions, land-related litigations that became a dominant scene within the welfare co-operative Hamilton Zamindari. The documentation of court cases in the collection made me realise the centrality of land as a commodity that could be fragmented and used to generate 'wealth' through taxes and appropriation of labour. The watery landscape featured insignificantly in the proceedings with the exception of embankment expenditure tables and petitions from aggrieved parties on privately open ponds that were periodically poisoned by rival landowners. The centrality of land in the lives of the people was embedded in the colonial politics of land grants that were given to expand the revenue base of the Bengal Presidency. As additional land grants were allotted, it attracted many landless peasants from other parts of India, including East Bengal (present-day Bangladesh), who came to make a living by tilling the land that was reclaimed from the tidal marshes and swamps vital to the local biodiversity and floodplain ecology. The creation of the new settlement was premised on the idea of protecting reclaimed land from annual flooding and salinity. Sir Daniel Hamilton issued paper currency in 1931 that contained an important message for the people of Gosaba to protect embankments and their land. The paper currency was launched with a promise to give 100 percent dividend to all peasants and free them from the clutches of money-lenders (Figure 4.1).

The concept of wasteland not only took a utilitarian turn as discussed by scholars like Gidwani (2008) and Whitehead (2012) but also came to be widely discussed within the colonial administration, based on the physical and relational nature of the land to its ecology (Hoop and Arora 2017). For example, in the Bengal delta, there were interchanges with terms like uncultivated land, not yet cultivated land, and forest land (Jungle) based on the soil quality, fertility, and land use cover. Anthropologist K. Sivaramarishnan, in his book *Modern Forest* (1999), also outlines how the 'Jangal mahals' in Bengal province were periodically cultivated by slash-and-burn farmers and kept fallow for a long period of time to be used once the land regained its vegetation and fertility for swidden agriculture. But this did not imply that they were not used by farmers and forest goers. In fact, the jungle was regularly visited for extracting resources, and merchants lived and navigated through these landscapes following the tides and periodically escaped the vagaries of weather by being mobile traders and seafarers. By clearing land of its forest and settling landless peasants in the delta,

Rs. 1 SIR DANIEL MACKINNON HAMILTON Rs. 1

promises to pay the Bearer, on demand, at the Co-operative Bhundar, in exchange for value received, One Rupee's Worth of rice, cloth, oil or other goods.

No. AW 50

Gosaba

Manager. Zamindar.

FIGURE 4.1 One-rupee note issued by Sir Daniel Mackinnon Hamilton in his Zamindari, Gosaba estate. (*Source:* Collected by author.)

the colonial administration made a strategic compromise with the local estuarine geography by trying to manipulate the river flow through a series of embankments along the Ganges and its tributaries and by protecting islands with levees that were not suitable for human settlement. This was achieved through the grant of long leases to prospective Zamindars and speculative investors who could make windfall profits out of land development and taxation. While Sir Daniel Hamilton acquired three islands in the Sundarbans to experiment with his philanthropic vision, his managers were involved in appropriating the resources and exploiting the tenants who brought land at the centre of dispute between the Zamindari and the people who tilled the land.

PRE-COLONIAL REPRESENTATION OF NATURE AND CULTURE

I begin this section by providing a contextual background of how the watery landscape of the Sundarbans was imagined in the pre-colonial text and scriptures. Much of the folk and cultural history of the delta since the Medieval period has been framed with water, rivers, and tributaries as ephemeral geographical features that constantly modified with the flow of sediments and water as the rivers shifted its course and tributaries meandered across the floodplains to join the sea, creating braided streams and channels of small islands that became part of the Bird's footed delta. As has been recorded in the local ballads and poems composed by Bengali writers, these landscapes were never permanent. Also, the worship of goddesses such as *Manasa* and the popularity of *Manasa Mangal* suggest the importance of water over land as snakes, crocodiles, and other amphibious creatures dominate these mythical stories in old Bengali literature. The shifting course of the Ganges also testifies to this fact.

In the past decades, a number of archaeological excavations have shown that there was human settlement before the arrival of the British in the Sundarban. These were always threatened by the rising tides, cyclones, tsunami waves, and the constant denudation processes that created the delta. Krishnokali Mondol, a resident of Baruipur and an active excavator of settlement sites in the Sundarban, observed in his interview with the author in 2013 that the collected artefacts in his museum date back millennia to the Gupta, Mauriya, and Sunga Dynesty (c.185-73BEC), and indeed it was difficult for him to date their age as they were found in different layers of the delta's subsoil. Such evidence corroborates with other recent excavations done by the French archaeologist in Bangladesh, who have confirmed the existence of human settlements in the delta through excavations much before the arrival of European powers (Van Schendel, 2009). Two historians of lower Bengal, Satish Chandra Mitra and Kalidas Dutta, have proposed the idea that the Sundarbans had a grand civilizational history that was periodically tempered by the estuarine regime of saline sea and fresh river water in Bengal (Dutta, 1989; Mitra, 1914).

In the ancient Bengali literature, this region was identified as *rosatol* (land saturated by saline water) and was also known as the lowland. The East Bengali ballads written on the region from the Middle Ages, 12 centuries onward, highlight cultural attributes that link Bengal's maritime history with riverine trade and commerce that were taking place with the neighbouring region and the presence of sea pirates from Spain, Portugal, and the Arakan region of lower Burma that shaped 'connected histories' (Subrahmanyam 2004). The popular *Mangal Kavya* narrates the story of Chand Saudagar, a local merchant who travels through the Lower Bengal delta from the *Radh* part of Bengal (Bardhaman), following the river course of the *adi ganga* – the dried out ancient course of the Ganges.[2] In the story, we see allegorical narratives of emerging and vanishing islands that indicate the amorphous nature of terrestrial landscape. The changing course of the Ganges also exemplifies the fragility of the delta floodplain where geomorphic processes incessantly keep the settlements vulnerable to natural calamities and hydro hazards. Similarly, in a later Islamic text composed by Banayuddin in 1287, titled *Banabibi Jahauranama*, the centrality of the forest and the rivers is highlighted by the story of Dukhe, who is protected by the forest goddess *Banabibi* from the king of the jungle tiger. In this old Bengali *punthi* literature, people's everyday struggle with nature is featured with the centrality of maritime commerce in wax, honey, coral products that don't grow over land but are sourced from the littoral forests and shores. The deities worshipped in the delta, like Manasa, the serpent goddess, represent the centrality of water, rivers, and forests in people's everyday life and ritual culture. Similarly, in *Bananbibis Jahaurnama* these days retold to tourist and visitors through popular open air theatres performances local know as (*pala*) performance, the centrality of river, water, boats, and jungle goers take precedence over sedentary agrarian activities. Dukhe, the protagonist of the play, as the story goes, was left in the forest by the greedy merchant in exchange for honey. *Banabibi*, the forest goddess, protected Dukhe from being killed by Tiger personified as Dakhin Rai, the tiger god, is later brought back to his mother by a crocodile, a figurative representation of the amphibious nature of the delta water space.

Rivers, backwaters, and seas have therefore played a crucial role in shaping connected histories in the littoral Bay of Bengal since medieval times. The discovery of artefacts dating back to the Mauriya, Sunga, Gupta, Pala, and Sen periods in places like Chandraketugarh, Harinarayanpur, Deoulpota, Atghetha, Sita Kundu, Karanjali, Roydighi, and Tota has made historians and archaeologists of Bengal develop a narrative of grand civilizational history that flourished in the delta, but given the archaeological evidence of sedimentation and denudation well established permanent development of towns and as civilizational hubs never developed in the region. Instead, the discovery of a number of artefacts, particularly, Greco-Roman inscriptions, pottery, idols, and numismatic coins, points to the commerce and trade that connected the inland waterways with the sea and backwaters through the Indian Ocean trade routes and entry ports.[3] In fact, as maritime historian Rila Mukherjee rightly observes, the region's waterways presented itself as a poly-focal landscape that saw the circulation and trade in commodities across and beyond the region. One of my interlocutors, Khitish Bishal, curator and trustee of the Matla Ancholik Sangroyshala, a regional museum located in Canning, holds a large collection of artefacts, both archaeological and archival, showcasing how life and society in the Sundarbans have evolved over centuries through maritime and transoceanic networks and trade exchanges. The Greco-Roman artifices and numismatic inscriptions present important evidence of vibrant maritime trade that flourished between 12 century BC and 16 century AD in the region, showcasing the connected history with the surrounding littoral region (Figure 4.2).

FIGURE 4.2 Khitish Vishal displaying his museum terra cotta collection. (*Source:* Collected by author.)

The expansion of the British administration in the delta during the later 18th century changed this cross-cultural tie for good and led to the evolution of a new land-use regime based on the economic importance of land as a revenue resource. Colonial attempts to tame nature in South Asia began with the expansion of railways that deforested land and degraded hills. Railway tracts across the great plains also necessitated the development of embankments, culverts, flood gates, and dams (D'souza, 2007, 2008). In other parts of South Asia, canal irrigation was promoted to improve the productivity of the land. In the Sundarbans, likewise, land grants and reclamation of marshlands began on a mammoth scale in the late 19th century. In the next section, I turn to describe how the delta communities developed hierarchies on caste and class lines with the expansion of the agrarian frontier and the creation of new tenures that tilled the land with Zamindars as the top of the exploitative cycle who established tenure rights and revenue through the extraction of tax.

CASTE AND THE CONSTRUCTION OF LITTORAL ETHNICITY

Rivers, salt marshes, and the mangrove forest have played a great levelling field for the growth of both laminal religiosity, syncretic beliefs and the absence of a rigid caste hierarchy in the Sundarbans. The region is dominated by the caste population –

untouchables and tribes who came as migrant labourers from central parts of India. Even during partition, most of the people who came to inhabit the Sundarban delta were lower caste people who could not find space among the *Bhadralok* (upper caste *babus*) living in Kolkata. The Sundarbans were inhabited by fishers, salt makers, forest goers, and sailors – professions attached to water and wetlands looked down by the upper caste in Bengal. With the opening up of land to outside speculators and Zamindars in the 19th century, the caste and class division sharpened with the creation of the new landless tenants who worked as sharecroppers for the landed gentry, the urban Zamindars. In the areas where land was leased out to greedy Bengali Zamindars Sir Daniel Hamilton a Scotsman envisioned an idealistic masterplan across three islands to establish a co-operative movement free of moneylenders and middlemen (Bandopadhyay and Matilal, 2003). However, Hamilton's utopian project was vandalised by the *bhadrolok* managers of the estate who were recruited from Calcutta and started exploiting the local tenants in Hamilton's absence and defeated the ideals of co-operative development.

It is noteworthy to highlight that the division and hierarchies emerged with the development of land and the establishment of private property rights in the delta. Annu Jalias, in her work *Forest of Tigers*, notes that in the forest people worship *bonobibi*[4] and go with a pure heart and co-operate to collect honey and wax. The hostile environment of the forest swamps does not discriminate between people. All were treated equally. In the island settlements, the contest for resources continues to grow and create inequalities as unequal distribution of land and resources creates hierarchies and disputes over property. While water acts as a great equaliser, the creation of new cultivable land has created social boundaries and division within society. The fluid bond of co-operation broke. Jalias further argues that caste differences are more apparent in the agriculture-dependent communities of Bengal. To put it in her words, 'land eventually is seen by islanders as hierarchizing and corrupting' (Jalias, 2010, p. 42). Among fisherfolk and forest goers, there is no social stratification as the tiger, unlike humans, does not differentiate its prey. Inside the forest, caste groups work collectively and without prejudices towards one another. Also, because forest goers and fishers do not own land or have a tenure relationship, unlike peasant societies, egalitarianism is generally the rule. This relationship also extends to the religious sphere as *bonobibi* – the forest goddess is worshipped both by Hindu and Muslim forest goers (Roy, 1983).

My positionality during fieldwork was shaped by a complex intersection of familiarity and distance as I was a researcher who had entered the field from abroad. As a native anthropologist, my identity closely aligned with my interlocutors as they had migrated from East Bengal during and after India's partition. My cultural background as a Sylheti closely resonated with their belongingness as refugees from Khulna and Barisal. I could share my empathy with them, myself having a family genealogy, being born into a migrant family. However, this also carried ethical responsibility to examine things objectively. I straddled between a thin line, speaking the native language yet being distant from that place, as I was brought up in a different part of India, in an urban cosmopolitan environment, away from the rhythms of the littoral Bengal delta environment. Thus, my fieldwork entailed navigating the thin line between cultural proximity and analytical distance, allowing me to reflect on how memory, displacement, and belonging operate across generations and geographies. The research was grounded in a historical and anthropological exploration of Gosaba, which was developed through an analysis of the Hamilton Zamindari records and further enriched by ethnographic fieldwork. This encompassed participant observation, informal conversations, group discussions, and everyday interactions, such as leisurely walks to the village's weekly market and visits to farmlands and freshwater ponds alongside my informants, as well as attending village festivals and open-air theatre performances (*jatra*). All these events are related to aspects of life in the littoral environment that are closely connected to land and water.

My interlocutor, Mr Mukuno Gayne, explained how caste discrimination is experienced by villagers when they venture out of the delta. Mukundo frequently visits the city to participate in *Kabita sammelon* (poetry recitation conclave) and conferences. On one occasion, he was invited by the Ramakrishna Mission monks in Gauria. On his arrival, Mukunda was offered food separately which he soon realised was because he belonged to the lower caste. He was shattered by this experience. Mukunda also narrated how he was treated by the Forest Guard officials in the Sajnakhali Forest resort. The attitudes of the caste Hindus who migrated to the region created a stigmatised identity of communities that live in the delta and survive on the forest. This is because the *Bhadralok* babus who come from Kolkata to work as school teachers, medical staff, and forest officials look down upon the peasants and fishers as *chotolok* (in Bengali, meaning little people). The term has both class and caste dimensions embedded in it.[5] The process of exploitation and differentiation between the *bhadrolok*, the Zamindars, and the *chotolok*, the tenant landless peasantry, was reinforced with the land reclamation programme and the entry of outsiders in the region. Colonial legislation and the classification of caste groups further accentuated difference and exploitation. Nicholas Dirks, in his work *Caste of Mind*, observes that the caste system in India was reinforced by the colonial state. In the Sundarbans, the local ecology was significantly transformed by the process of land reclamation, which also affected the social stratum of communities residing in the region as a new land tenure system emerged with the Zamindari system.

All of the above means that during the colonial period, the very occupation of fishing or activities related to fishing came with many negative connotations. Again, we can see this negativity reflected in popular culture accounts of people-environment relations in the Sundarbans. In the Bengali novel, *Titas the Name of a River* (1956), Advaitya Malla Barman presents a graphic and tender narrative of the life of the fishing community in the Bengal delta. These fishing communities are socially ostracised and economically backward. Movement over water and fish trade with *araddars* (fish traders) shape their relation with the outside world. Inside the forest and in the sea, the social hierarchies are replaced by cooperation as there is a constant threat from tigers, crocodiles, rising tidal waves, and tropical disturbance. In a similar way, Manik Bandopadhyay's novel *Padma River Boatman* (1934) chronicles the struggle of the marginalised fisher folk in Bengal and glorifies their relation with water and the bank that shaped their livelihood. The fishermen have developed a strong relationship with their local environment that is periodically disrupted by cyclones and flash floods caused by the state's failure to regularly repair and maintain embankments. Samesh Basu's *Ganga* (1957), published at a later period, talks about the life of fishers along the Ganges River bank. The novel, unlike the former, is located in West Bengal and chronicles the struggle faced by fishers who mostly engage in inland fishing and are vulnerable to the vagaries of monsoonal floods. All these vernacular novels were to a great extent influenced by Marxist class struggle, and they highlight an important point that established the fact that 'the lives of the fishing peoples are shaped by their relations to water and to the broader natural environment in which they live' (Pokrant et al., 1998: 127). This is strikingly illustrated by Barman, who is an insider narrating his own *Malo* community in an ethnographically rich description.

As long as the land was under water, the Malos moved on it; it was theirs. The moment it has floated above water, water; it has become the farmers. They will plant seeds here; they will harvest and take home the crop. This right will always stay theirs; no one will be able to steal it. This right of theirs is staked in solid reality, rooted in soil. And the Malos' right was in the flowing water, in its formless, groundless ever-moving fluidity. That right never found the solid touch of real ground; it never had a stable support, a hard foothold. Hence, they are the floaters. No matter how much they, the fishermen, befriend the trees and the homes on the ground, they remain floating like vapour. No matter how hard they cling to the bosom of the earth, the soil of the earth is forever pushing them away: 'No space for you, no place for you!' as long as there is water in the river, only then they float on the water. When the water dries, they too evaporate and disappear.[6]

The contemporary period also demonstrates the continuities of the caste-class relationship and state official perception of people living in the region. Amitesh Mukhopadhyay's 2016 book *Living with Disaster* highlights how the apathy of the state triggered the 2009 Aila cyclone and massive destruction of livelihoods in the delta as the storm inundated the islands with saline water. Soon after the cyclone, the Planning Commission Chairman Mr Montek Singh Ahluwalia visited the islands and the government announced Rs 2000 crores as a financial package for relief, rehabilitation, and, most importantly, the reconstruction of concrete embankments that would protect the islands from future calamities. During my fieldwork in Gosaba, I interviewed numerous villagers on the issue of embankment reconstruction and cyclone shelters. An undivided disappointment was expressed by villages as they did not see major embankment work being undertaken by the government departments, with the exception of a few earth diggers and public works department labourers who were involved in repairing embankment breaches.

LAND, WATER, AND LITIGATION

In this section, I described through analysis of court cases how land became a bone of contention between emerging peasantry and their intermediaries, the Zamindars, who reclaimed the marshes by exploiting the labour of the landless peasants. From Henckell to Hamilton, the wastelands witnessed many changes. The uninhabited landscape of Gosaba and Khulna, for example, underwent a total transformation under the stewardship of these benevolent administrators and philanthropists.[7] A cursory glance at the statistics prepared by John F. Richard and Elizabeth P. Flint shows the transformation in land use. Between 1880 and 1950, a period of land reclamation and crop intensification in the delta, the wetlands reduced from 1289 ha to

867 ha while the net cultivated and settled built-up area increased from 1375 ha to 2047 ha (Sarkar, 2010).

Parallel to the land reclamation project, the development of a Port in Canning further changed the morphology of the delta environment, although the project subsequently failed to have commercial success. The railhead from Kolkata was developed and extended towards the delta. This ambitious project undertaken by Viceroy Lord Canning was in itself a major game changer for the deltaic ecology that was connected by ephemeral meandering rivers and tributaries that connected the sea through bird-footed delta formation. The establishment of railways up to Caning not only expanded the suburban commuter belt of Calcutta but also pushed a large settlement to grow in and around Canning, which is today a bustling town. The construction of the railway lines spanned many rivers, and hundreds of culverts, embankments, and sluice gates were developed to keep the railways running during the rainy season. This disrupted the river regime of the delta. A classic example is the high sedimentation on the mouth of the Matla where the port was conceived. The necessity for creating Port Canning emerged from a mercantile capitalist thinking and the fury of gales and rough weather that capsized many ships in Calcutta. Henry Paddington, a weather expert in his series of papers published in the Royal Asian Society Journals, developed the 'Law of Storm' to help merchant ships evade strong gales and safely trade in the Bay of Bengal waters (Paddington, 1849, 1853).

As the Calcutta port became busy and shallow in the late 18th century, with increasing damage to merchant investment and Vessel, the need arose to develop an alternative port. This debate was rekindled by the devastating cyclone of 1842, which damaged the ships anchored at Calcutta port. In 1853, the Bengal Chamber of Commerce represented to the government the necessity of establishing a port on the Matla which would be connected to Calcutta by railways or canals. The railhead option was favoured over canal construction primarily because of the cost involved, despite the challenges and dangers of tempering the local ecology of the delta waterscape. The rail link was seen as a major expansion for speculative investment and a hinterland link to carry the bulk of the food grain for export. The creation of the port also necessitated the development of townships and a wet dock. In connection with the Port Canning Scheme, a company was created called the Port Canning Land Investment, Reclamation and Docks Company Limited for the purchase and reclamation of wasteland bordering the swamps of the Matla River as it joins the Bay of Bengal (O'Melley, 1914). This speculative investment on land was claimed to bring major profit to investors. The township and a proposed rice mill quickly developed, but it failed to attract major commercial ships that led to the abandonment of the project within a decade of its establishment in 1871 and the assets of the Port Canning Dock Company Limited were liquidated through the creation of the Port Canning Land company limited that proposed to generate revenue through agricultural expansion on reclaimed land. The failure was triggered by deep sedimentation at the head of the Matla River once the dock was stablished that interrupted the flow of sediments into the sea.

Failed colonial capitalist commercial interest and the vagaries of speculative investments in the delta had permanently put marginal landless labourers at the forefront of environmental risk posed by river bank erosion, denudation on the land reclaimed, and the periodic annual cyclones and tropical disturbances that threatened to disrupt crops and vulnerable livelihoods in the newly settled parts of the delta.[8] Before the land reclamation programme began on a rapid scale, the economy was based on non-permanent and seasonal activity, such as salt trade, trade in dry fish, crab catching, wax and honey collection, and fishing that did not require large-scale settlement of villages and townships in the transforming delta islands due to sedimentation. One of the failures of the Port Canning investment lay in stabilising the built environment as it was constantly denuded by erosion. The company had heavily invested in building retention walls and the dock infrastructure that were continuously eroded by the saline environment. This led to overspent expenses that could not be recovered leading to the non-viability of the port.

Following the failure of the Canning port project, the colonial land grant project did not deter both local and outside speculators from investing in land that could make a profit through the expansion of paddy culture. One of the visionary investors was Sir Daniel Hamilton. His intention to lease three islands in the Sundarbans was different from his contemporaries. His shipping company had brought him an enormous fortune and he championed the idea of a co-operative movement in India through his speeches and writings. To bring his idea into practice, he bought 90,000 acres of land in the Sundarbans and started to promote settled cultivation through free land grants to people who did not own land. Slowly and steadily, scores of people joined the estate who came from various social backgrounds and religions. In 1909, it was estimated that there were 900 people in Gosaba. To help the process of reclamation, land was offered free of charge and several facilities were established for safe drinking water and a co-operative model farm, bank, school, and health centre, all artefacts of a civilising mission. Hamilton also appointed an estate Manager who was educated and belonged to a Brahmin family from Kolkata.

The most famous was the case of Father Mazerik who filed a case in the Boshirhat sub judicial magistrate court the grievances of his follower peasants who were exploited and oppressed by Sudanshu Bushan Mazumdar, the estate manager. Mazerik also antagonised the Baptist missionaries in the area who sided with the Manager of the trust during the court trials. Finally, Mezerik was recalled by the Archbishop of Calcutta but the dissensions did not die. The struggle over land continued for decades and had a lasting legacy during the mid-19th century. I spotted an anonymous letter written to the Amrit Bazar Patrika, where the peasants petitioned about the exploitation of the Zamindaris.[9] The estate managers controverted such charges and, in a booklet published by the estate, they strongly disclaimed such discontent as reactionary and instigated by Marxist revolutionaries. The Hamilton Zamindari had employed six Nayabs under the manager who looked after the estate and reported on encroachment. These *latials* (henchmen) were migrants from Uttar Pradesh, and because of their strong, built bodies, they were employed by the Zamindari. One of the *latials'* grandsons, who I interviewed during fieldwork, observed that his father Jang Bahadur Rai was the head *latial* who used to protect the Mazumdars from peasant protest and revolutionary marches and often came to Sudanshu Bushan Mazumdar's rescue.

The seeds of discontent were carried forward after Father Mazerik's departure from the Island by local revolutionaries like Jogen Maity, who was arrested and released on several charges brought by the estate to till the land without paying revenue to the Zamindari. Land-related encroachment, poisoning of waterbodies, tax evasion, and false penalties on farmers who could not pay rent because of failed crops and the cessation of cultivable land by the estate became a dominant feature of everyday life. In fact, the court cases became so heavy that the ideal co-operative model failed to survive. Corruption among Jamini Rice Mill officials led to the failure of the self-sufficient village's economic production. The village merchants' abuse of weighing scales led to the *handor andolan* in the estate. The co-operative movement, therefore, received a setback.[10] In all these, the old hierarchies of peasant societies were established, superimposed on Daniel Hamilton's ideal co-operative movement. As new immigrants started to populate the region in the wake of Indian Independence and the Redcliff award, the Muslim landowners in West Bengal exchanged their land with their Hindu counterparts in East Bengal, the present-day Bangladesh. The Hindus who migrated to Sundarbans and established their tenure came in conflict with Hamilton's *latihals*. My host in Dayapur village, Mukunda Gayen's family, migrated from East Bengal in 1946, just before partition, and exchanged land with another Muslim family, Isap Pir. Mukundas uncle, Murari Gayne, during his lifetime, fought with the Zamindar's *latihals* in Dayapur village which bordered the Hamilton estate.

The root of these problems related to land adjudication lies in the colonial project of dividing Bengal's wasteland through land reclamation and the artificial division between swamps and forest, land and water bodies in the Delta. Kuntala Lahiti-Dutt writes poignantly on this issue that: 'The water/lands of the Bengal delta are hybrid environments not only because they are natural and lived-in waters, but because colonial history constructed land as more productive, and rivers as challenging. The portrayal of rivers as uncivil (and hence in need of control) in riverine Bengal led to colonial and postcolonial engineering and legal controls, through land- and water-management policies and the revenue system' (Lahiri-Dutt and Samanta, 2013: 522). The problem of land demarcation and appropriation by neighbours went on to become a major point of contention with the Indian state after India's independence, when the land ceiling act was passed in the late 1950s. After Hamilton's death, the manager of the estate Sudanshu Bushan Majumdar and his family took control over the estate as trustees and ran the affairs of the estate. In the 1970s, when the left-led Communist Party of Indian (Marxist) coalition came to power in Bengal, they decided to redistribute the trust land to the people, and over a period of long-drawn lawsuit, it was finally decided to turn it into a charitable public trust with the Deputy Commissioner of South 24 Parganas as its Chairman.

From these examples, we can see how land has become the centre of the local economy and ownership when it was part of the commons before the immigration of landless peasants in the region. The shift of local economic practices from forest and water to land has imposed social hierarchies and economic differences based on the ownership and use of land. Today, land is a commoditised asset in the Sundarbans with the popularisation of tourism and the declaration of Sundarban as a World Heritage Site for the conservation of the Bengal tiger. Land is partitioned for sale to outside investors and tourist lodge operators. In the last decade, many unfinished resorts and guest houses have sprung up in the delta through investments made by chit fund and private companies such as Sahara and Gulshan group of investors (Murilidharan, 2013). Even the forested part of the delta is reserved with core and buffer territories where access is permitted through a boat licence issued by the Forest Department. The local fishers, woodcutters, and honey collectors work under strict scrutiny and checks imposed by forest patrol parties who encourage people to stay away from the core area of the forest. Alternatively, livelihood regeneration projects and public work man-days in building and repairing earthen embankments and roads are seen as vacations that will keep community away from the delta waters and its aquatic resources.

The socio-political scene of the 19th century also begs the question: How have the legacies of colonial land-water management influenced contemporary hydro-social arrangements (e.g. in land reclamation or legal conflicts)? One of the reasons for rural dissonances in West Bengal post 1947 was the unequal distribution of land established by the Zamindari system in Bengal. The Naxalbari movement of the late 1960s was a direct fallout of the discontent and growing hierarchy caused by the unjust land tenure system introduced by British lawmakers and revenue collectors who gave primacy to land reclamation and undermined the role of art and craft, weavers, and the handloom cottage industry in shaping Bengal's cultural geography. For example, katha embroidery work that was so popular until the late 19th Century carried motifs of aquatic life and the esturian environment. Today, they have become things of the past as museum artefacts. Kacia Paprocki, in her book *Threatening Dystopias* (2024), argues that colonialism, capitalism, and local agrarian struggles have shaped the Bay of Bengal's agrarian landscape more than the impact of Climate Change and predicted sea level rise.

Colonial legislation like the Permanent Settlement Act of 1793 and the Bengal Alluvium and Diluvium Act sought to establish permanent tenurial land rights so that the littoral population could be mapped and settled. The sole intention was to augment the revenue base of the Bengal Presidency through 'land reclamation' and 'permanent entitlement'. These policies came into conflict with local ecological relations, where the shifting alluvium in *char* lands (floodplain islands) led to frequent disputes among landowners, as the boundaries of these lands constantly changed with fresh siltation and flooding. In the postcolonial period, the rise of commercial aquaculture, particularly shrimp farming in the 1980s, has led to further dispossession of marginal farmers in Bengal. Soil salinisation has increased with the expansion of shrimp cultivation, damaging lands previously used by small paddy farmers. Capital-intensive commercial farming, now conducted inland on formerly cultivable paddy fields consolidated by large landholders, has transformed the rural landscape, producing a dystopian vision of agrarian ruination.

While climate change and sea-level rise are often blamed for rural dispossession in Bengal, the real causes lie in colonial land reclamation policies and capitalist land consolidation. The Zamindari system and the Permanent Settlement in the Sundarbans created a patron-client relationship between landowners and tillers in Bengal, generating decades of legal disputes and land conflicts in a region where land was once abundant and not permanently owned due to the vagaries of cyclones and the risks of flooding. The part-land, part-water landscape where communities once adapted to the river's rhythm through fishing, wax and honey collection in the backwaters and creeks has been disrupted by permanent agriculture, land litigation, peasant movements, and contemporary commercial aquaculture. The current development buzzword is the 'Blue Economy'. But will this become yet another top-down intervention that disrupts the lives of small fishers and seafarers who are already struggling with declining fish catches and transborder maritime competition in the Bay of Bengal-Indian Ocean rim between India and China for regional dominance?

CURRENT PREDICAMENTS AND WATER FUTURE IN THE SUNDARBANS

The contemporary problems that feature largely in media coverage of the Sundarban delta focus on two major issues. Firstly, the loss of habitat for endangered tigers and secondly the hydro hazard periodically faced by communities in the form of cyclone disasters. Often, natural disasters that strike the region are directly correlated with El Niño and climate change events. Major cyclones have left terrible scars in the landscape, the most recent one being Cyclone Aila that not only led to the destruction of standing crops and mud huts but also led to widespread salinisation of the soil. While climate change can be linked to rising sea levels, increasing intensity and frequency of deadly cyclones, local level changes, and impact on environment are made adverse by human intervention. Poet Mukunda Gayen and his activist friend Supobitro Pradhan who runs a printing press in Gosaba and regularly writes in local newspapers during my fieldwork conversations observed that the islands were not suitable for human habitation because of the saline marshlands. It is only through embankments and earthworks that low-lying islands have been made suitable for cultivation and human habitation during the course of the late 18th and 19th centuries. The Aila cyclone of 2009 inundated vast parts of the arable land because of poor embankment management and corruption at the local administrative level. Amitesh Mukhopadhyay, in his latest book *Living with Disaste*rs (2016), shows how the lack of embankment protection and corruption at the village level heightened the scale of Aila devastation in the Sundarban delta. It was a man-made disaster. In many parts of the Sundarban delta, small rivers and rivulets have been blocked to connect roads and to block the saline water ingress during high tide. This has resulted in drying up of streams, increased sedimentation, and shallowing of the river-bed and backwaters that raise the level of water during cyclones and produce embankment breaches and flash floods.

In the 1980, the West Bengal government's Agriculture Department started to promote rice cultivation in a big way by introducing new and improved varieties. In the Hamilton estate records, I discovered files dating back to the 1980s that hold maps and plans for establishing an agriculture research station in Gosaba by the government's Agricultural Research Institute on land owned by the Hamilton Charitable Trust. These infrastructures needed the creation of new embankments to protect the experimental farm. Similarly, as the population of these islands grew, new jetty *ghats* emerged and concrete embankments followed to protect valuable land and road infrastructure that connected the village settlements. New culverts were laid, and small streams and water bodies that worked as flushers of rainwater joining the main river were blocked in order to develop connectivity between settlements in Gosaba and the surrounding island settlements.

During the Aila cyclone in 2007, these critical infrastructures were tested, and at many places embankment breaches were recorded after the cyclone-led flooding had destroyed the channels. One of the local residents of Gosaba, Shaktipodo Nath, a retired schoolmaster, observed that he has been watching the ebb and flow of the river since his childhood. The construction of embankments by engineers and the Public Works Department (PWD) does not take into consideration the tacit knowledge of the local communities who have lived knowledge of natural disasters expressed through community memory and everyday knowledge of breaches and which embankments have survived the furies of cyclones. The civil engineers build embankments based on their technical training and the village *Pradhan* colludes with contractors and district administrators to siphon off cash and often weaken the construction and delay the repairs. Shaktipodo has composed his experience through poems and short plays that reflect his concern for the people of Gosaba. Khitish Vishal, my other interlocutor, a fine arts painter by training, showed me his water colour canvas that he developed on natural calamities and the different kinds of boats that were used in the Sundarbans before the advent of *bhotboti*, a diesel-powered country boat. Small boats and steamers still are the main means of transport connecting delta settlements through *haat bazaar* (weekly village market). However, the landscape outside the core area of the Sundarban is fast changing as two major bridges have been built to connect Sundarban with Calcutta and Canning town, the gateway to the littoral region. These new physical infrastructures over the river Matla have further reduced the transportable capacity of the river as it was since the creation of the port and the wet dock in the 18th century and have led to erosion of the delta islands in other places. Today, alongside floods and cyclones, erosion and loss of productive built-up settlements are big problems in the villages. Simultaneously, in the last decade, the government has built metalled concrete roads connecting villages and helipad infrastructure in Glosaba on Hamilton Trust land that has made the state's presence more legible in the fragile delta landscape.

The genesis of the problem lies in the colonial imagination of the landscape as a revenue reserve and a paddy frontier. The changes in land use from watery wastelands to reclaimed paddy fields have transformed the economy of the region at a cost. The impact of large ecological changes is felt by vulnerable lower caste and marginalised communities who now live on the edges of these islands susceptible to regular embankment breaches, coastal erosion, denudation resulting from shifting river course, and devastation from tropical cyclones. In the aftermath of the shattering Aila cyclone in 2009, thousands of farmers migrated to cities like Kolkata, Delhi, Mumbai, and Chennai for work as the land was salinised by seawater and crops failed. The influx of partition refugees during the 1950s and 1960s was mostly composed of fishers who depended on marine fishing as they had migrated from Khulna in present-day Bangladesh. However, their water-based economy has been restricted by the Forest Department, who tightly guard the forest and does not allow free fishing. This has put pressure on land where the opportunity to grow more food lies only in multi cropping and mixed cropping systems. This has changed the agricultural scene in the village. The cultivation of crops like chillies, betel leaf, and melons dominated the landscape before the Aila cyclone of 2009. The decline in the trade of minor forest products such as honey and wax and freshwater fish has forced many fishers to change their profession and many are forced to seek alternative livelihood.

The policies pursued by the colonial government have had a deep impact on the local economy. While it was premised on the idea of converting wastelands into cultivable land and to clear the deltaic backwaters of its bandits – a civilising mission – the spurt of population that came through immigrant labour force from different parts of Eastern India and partition refugees from East Bengal has made the region one of the densely populated parts of the Indo-Gangetic River basin (Mallick, 1999). With the threat of coastal flooding and the deadliest tropical cyclone, the threat of tiger poaching and loss of mangrove forest have proved to be a major catastrophe for the region and besieged the confrontation of man and nature. In light of these historical transformations in the delta, we get a diachronic view of the littoral region that has been shaped by countervailing forces of law, legislation, state interference, and human endeavour to civilise the landscape. The importance of human agency, labour of love, and the dedication of luminaries like Sir Daniel Hamilton and Tillman Hinckell played a significant role in shaping settlements that transformed water-based culture and economy to a land-based property regime.

The impact of these anthropogenic interventions has not been positive in light of the claims made by the architects of these programmes – civilising the amorphous deltaic wilderness through settled economy and co-operative colonies. Old hierarchies of caste, class, and ethnicity resurface within the cooperative and free land grants were taken over by the Zamindars and their local henchman through distress sale and loan repayment deeds. The creation of a classless co-operative society in Sundarban dissolved over time as Daniel Hamilton left Gosaba to settle in Balmacara, Scotland. Land had become the contention of disputes and litigation and is also the cause of concern for the state Forest Department and environmental NGOs who see the World Heritage Site as the only home of the endangered Bengal Tiger encroached by villages and settlements around the reserve. The role of the river Ganges and its tributaries is central to the formation of the Bengal delta. The morphology of the islands is constantly undergoing change due to the geomorphic process of alluvial deposition and transportation. The construction of physical infrastructures like embankments and bunds for the protection of private property (such as hotels and guest houses) and farmlands has created blockages to the free flow of sediments and the resultant erosion of islands and rising water levels during high tides. Revenue generating policy of colonial land reforms and the incentive to make a profit out of reclaiming wasteland has gone hand in hand with the contemporary dream to promote tourism in the islands. The land-centric thinking of economic development and improvement promoted by colonial administration was ingrained in the Co-operative movement initiated by Sir Daniel Hamilton in Gosaba who had witnessed this with his own eyes during the Scottish land reform movement.

In the post-independence period, land again became the symbol of the people's struggle in West Bengal, with the Communist Party snatching power in Bengal with a peasant movement. Land redistribution and agricultural intensification programmes became central to bridge rural backwardness and bring about development

(Rogaly et al., 1995). The declaration of the Sundarbans as a wildlife reserve, and subsequent UNESCO declaration of World Heritage Site, has further ingrained the contest over terrestrial space between humans and tigers. The surrounding villages in the Sundarban delta are now presented as buffer territories that are in direct conflict with the tiger habitat. Two centuries ago, when land was free of its inhabitants, the delta amphibious space was only periodically disturbed by salt makers, sea pirates, and forest goers who followed the ethics of the forest that *bonobibi* had imbibed through her rituals – not to be greedy and not to have more than what people need for their everyday survival. Today, with the globalisation of the local economy, a number of marine activities have further threatened the ecology of the water space. In the late 1980s, prawn hatchling became a major economic activity for the villagers as it was fetching good money. However, over a decade, the global demand for prawns declined as the process of shrimp farming did not meet international regulations and prawn exports from the Bay of Bengal littoral waters were banned by major Asian and European importers (Pokrant et al., 1998). Following the decline of shrimp, the demand for crab in China has fuelled illicit crab catching and overfishing of crab meat in the delta creeks and backwaters. Often the local community are blamed by government officials and conservation NGO staff as greedy and destroyers of the delta environment. These narratives undermine the overarching livelihood and survival challenges of marginalised communities who have come to inhabit the delta not by their own will but because of historical processes of colonisation, forced migration, and kinship relations that attracted new immigrants once the Sundarban began to evolve as a settled landscape.

NOTES

1. The environmental history literature of South Asia developed out of a reaction to colonial and post-independence forest appropriation and the building of big dams and river valley projects by the Indian government. See, for example, Guha and Gadgil's work: *The Unequal Woods and the Fissured Land*. The interrelationship of water and land in amorphous landscapes such as deltas and lagoons has entered very late in the environmental discourse of South Asia, with the seminar contemporary works of Kunatala Lahiri-Dutt (2014), Debjani Bhattacharjee (2018), William Van Schendel (1993), Iftekhar Iqbal (2010), and Sunil Amrith (2018).
2. David L. Curley, in his immensely important book, *Poetry and History: Bengali Mangal–Kavya and Social Change in Precolonial Bengal* (2008), defined *Mangal Kavya* as a genre of "narrative poetry, composed, as far as we know from surviving texts, during the period of the fifteenth through the eighteenth centuries."
3. During my fieldwork in the Sundarbans (2012–2015 and 2016–2017), I discovered a variety of community owned depositories that were dedicated to the preservation of the delta's archaeological history. I digitised the collection of *puthis* kept in the Sundarban Ancholik Sangroyshala in Baruipur with the support of the British Library Endangered Archives Project EAP 759 and Queen's University, Belfast. As part of the project, we digitised and restored 18 palm leaf and handwritten manuscripts and catalogued 180 manuscripts for digital recording of the collection.
4. Forest deity worshipped by people who enter the jungle to collect honey and wax. It is a belief among forest goers in Sundarban that Binobibi protects goers from tiger attack.
5. During my fieldwork I interviewed several poets, writers and literary figures in the Sundarbans who represented the Dalit community. They had similar experience of discrimination in their lifetime from town people who represented the *Bhadrolok* (in

Bengali meaning refined, English educated, gentle people). This concept has been widely discussed by Joya Chatterjee in her book *Spoils of Partition* and Annu Jalias in her book *Forest of Tigers*.

6. Barman, *Titas Ekti Nadir Naam*, p. 261.
7. Both Tillman Hinckell and Sir Daniel Hamilton had grand ideas to bring about improvement among the landless peasantry in Bengal by empowering the community through co-operatives and land grants. Nonetheless both these luminaries had to work within the parameters set by colonial mercantile and revenue interests. For Himalton, the Bhadrolok Trustees of the Estate usurped both his good work and vision by indulging in corrupt practices. In the Hamilton Estate Archives in Gosaba, I discovered anonymous petitions sent to the *Amrit Bazaar Patrika* complaining about the excesses of Hamilton's appointed Zamindars and *Latihals* (henchmen) in the estate while simultaneously applauding the philosophy and tenet laid down by Daniel Hamilton. Peasants were not against his philosophy but the officials who managed the estate. In fact, Hamilton worked very closely with Indian Nationalists such as Rabindranath Tagore and Mahatma Gandhi and planned schemes like the Gosaba-Bolpur-Shantiniketan project, but this scheme did not see the light of day.
8. Khitish Bishal fondly recollected minute details of the Canning Port town project during my interview. He had collected bricks that can still be found in the abandoned port site and has made it part of his museum collection to memorialise the long journey of humans' tryst with destiny in the ever-changing Sundarbans delta environment. More significantly, he had the original copy of one of the project documents in his archive that he fondly cherished as a rare and endangered document in his depository. People in Canning Town are nostalgic about their past, and the Canning Port project is symbolised by other colonial imprints such as the Bungalow that was constructed for Lord Canning and way the town came to bear his name and the broken legacy of port infrastructure.
9. Sir Daniel Hamilton's private paper collection, Gosaba.
10. I examined the correspondence of these movements and that of a communist revolutionary alerted in the letters written by Sudanshu Bushan Majumdar to Sir Daniel Hamilton in Balmacra, Scotland. These have also appeared in the pamphlet titled Estate Farming in India written by the estate Manger S. B. Majumdar in 1938.

REFERENCES

Amrith, Sunil S., 2013a. 'The Bay of Bengal: In Peril from Climate Change'. *New York Times*.

Amrith, Sunil S., 2013b. *Crossing the Bay of Bengal: Furies and Fortunes of Nature*, Cambridge: Harvard University Press.

Amrith, Sunil S., 2018. *The Unruly Waters: How the Mountain Rivers and Monsoons Have Shaped South Asian History*, London: Penguin.

Appadurai, A. and C. A. Breckenridge, 2009. Foreword. In Anuradha Mathur and Dilip Da Cunha (eds.), *Soak: Mumbai in an Estuary*, New Delhi: Rupa & Co., pp. 1–3.

Bandopadhyay, Alapan and Anup Matilal (eds), 2003. *The Philosophers Stone: Speeches and Writings of Sir Daniel Hamilton*, Gosaba, Kolkata: Sir Daniel Hamilton Estate Trust.

Bandopadhyaya, M., 1934. *Padma River Boatman*, translated by Barbara Painter and Yann Lovelock (St Lucia, University of Queensland Press, 1973). The Bengali original, *Padma Nadir Majhi*, was first published in Calcutta by Granthalaya in 1934.

Barman, A. M., 1956. *A River Called Titash*, translated by Kalpana Bardhan (New Delhi, Penguin, 1992). The original Bengali novel, *Titash Ekti Nadir Nam*, was first published in Calcutta by Puthighar in 1956.

Basu, Samresh, 1957. *Ganga*, Kolkata: Mousami Publication.

Bhattacharjee, Debjani, 2018. *Empire and Ecology in the Bengal Delta: The Making of Calcutta*, Cambridge: Cambridge University Press.

Bhattacharyya, D., 2018. Discipline and drain: Settling the moving Bengal delta. *Global Environment*, 11(2), 236–257.

Bose, S., 1993. *Peasant Labour and Colonial Capital: Rural Bengal Since 1770*, Cambridge: Cambridge University Press.

Curley, David L., 2008. Poetry and History: Bengali Maṅgal-kābya and Social Change in Precolonial Bengal. A Collection of Open Access Books and Monographs. 5. https://cedar.wwu.edu/cedarbooks/5

D'Souza, R. 2007. *Drowned and the Dammed: Colonial Capitalism and Flood Control in Eastern India*. New Delhi: Oxford University Press.

D'Souza, R., 2008. Framing India's hydraulic crises: The politics of the modern large dam. *Monthly Review*, 60(3), 112.

D'souza, R., 2014. 'Mischievous rivers and evil shoals: The English East India Company and the colonial resource regime'. In Vinita Damodaran, Anna Winterbottom and Alan Lester (eds.), *East India Companies and the Natural World 1600-1850*, London: Palgrave Macmillan, pp. 128–146.

D'souza, R., 2015. 'Drainage, River Erosion and Chaurs: An environmental history of land in Colonial Eastern India', Nehru Memorial Museum and Library, Working Paper Series. http://www.nehrumemorial.nic.in/images/pdf/occasional/Rohan_D_Souza_Ecology_and_History_23_December_2015.pdf, downloaded on 15/01/2019.

Dutta, Kalidas, 1989. *Dakhin Chabbis Pargana Ateet*, 2 vols., Baruipur: Deep publcation.

Eaton, R. 1990. Human Settlement and Colonization in the Sundarbans, 1200–1750. Agriculture and Human Values 7(2), 12-31.

Gidwani, V. (2008). The subaltern moment in Hegel's dialectic. *Environment and Planning A*, 40(11), 2578–2587.

Hoop, Evelyn de and Saurabh Arora, 2017. 'Material meanings: 'Waste' as a performative category of land in colonial India', *Journal of Historical Geography*, 55: 82–92.

Hunter, W. W., 1875. *A Statistical Account of Bengal*, Vol. 1, District of the 24 Parganas and the Sundarban, London: Trubner & Co.

Iftekhar Iqbal. 2010. The Bengal Delta: Ecology, State and Social Change, 1840–1943. Cambridge, Cambridge University Press.

Iqbal, I. (2010). *The Bengal Delta: Ecology, state and social change*, 1840–1943. Springer. https://doi.org/10.1057/9780230289819.

Iqbal, I., 2012. *The Bengal Delta: Ecology, State and Social Change*, London: Palgrave Macmillan.

Jalias, Annu, 2010. *Forest of Tigers: People, Politics and Environment in the Sundarbans*, New Delhi: Routledge.

Joya Chatterjee. 2011.The *Spoils of Partition: Bengal and India, 1947–1967*, Cambridge University Press, Cambridge.

Klingensmith, D., 2007. *One Valley and a Thousand Dams: Nationalism, and Development*, Oxford: Oxford University Press.

Lahiri-Dutt, Kuntala, 2014a. 'Beyond the water-land binary in geography: Water/lands of Bengal re-visioning hybridity', *ACME: International Journal of Critical Geography*, 13 (3): 505–529.

Lahiri-Dutt, Kuntala, 2014b. 'Chars: Islands that float within rivers', *The Sima International Journal of Research into Island Cultures*, 8 (2): 22–38.

Lahiri-Dutt, K & Samanta, G 2007, 'Like the Drifting Grains of Sand: Vulnerability, Security and Adjustment by Communities in the Charlands of the Damodar River, India', *South Asia-Journal of South Asia Studies*, vol. 30, no. 2, pp. 327–349.

Lahiri-Dutt, Kuntala and Gopa Samanta, 2013. *Dancing with the River: People and Life on the Chars of South Asia*, New Haven: Yale University Press.

Linton, Jamie and Jessica Budds, 2014. 'The hydrological cycle: Defining and mobilizing a rational dialectical approach to water', *Geoforum*, 57: 170–180.

Mallick, Ross, 1999. 'Refugee settlement in forest reserve: West Bengal policy reversal and the Maricjjhanpi massacre', *Journal of Asian Studies*, 58 (1): 108–125.
Mehta, Leyla, 2005. *The Politics and Poetics of Water: Naturalising Scarcity in Western India*, Delhi: Orient Longman.
Mitra, Satish Chnadra, 1914. *Jassor Khulnar Itihas*, 2 vols. Calcutta: Granthagara.
Mukhopadhyay, A. (2016). *Living with Disasters*. Cambridge University Press.
Murilidharan, S., 2013. How the Big Chit Fund Chits, *The Hindu Business Line*, 23 March, https://www.thehindubusinessline.com/opinion/columns/s-murlidharan/how-large-chit-funds-cheat/article20605895.ece1, downloaded on 21/07/2019.
Nightingale, A. J. (2024). Threatening dystopias. *Contemporary South Asia*, 32(4), 593–595.
O'Melley, L. S. S., 1914. Bengal district Gazetteer, Twenty-Four Parganas, Calcutta, 1914, pp. 223–225. Papers relation to the formation of Port Canning on the Matla River, op. cot., p. 52. Report on the Committee of the Matla Association established in February, 1859 with the object of promoting the progress of the Mutlah as an auxiliary port, Calcutta, 1861, p. 3.
Paddington, Henry, 1849. 'A seventeenth memoir on the law of storm in India; being storm of China seas from 1842–1847, and some of the northern Pacific Ocean from 1797', *Journal of Asiatic Society of Bengal*, 18: 1–45.
Paddington, Henry, 1853. 'A twenty first memoir of the law of storm in the Indian and China Seas', *Journal of the Asiatic Society of Bengal*, 21: 283–328.
Pokrant, Bob, Peter Reeves and John McGuire, 1998. 'The novelist's image of South Asian fishers: Exploring the work of Manik Bandopadhyaya, Advaita Malla Barman and Thakazhi Sivasankara Pillai', *South Asia: Journal of South Asian Studies*, 21 (1): 123–138.
Rogaly, B., Harriss-White, B., & Bose, S. (1995). Sonar Bangla? Agricultural growth and agrarian change in West Bengal and Bangladesh. *Economic and Political Weekly*, 30(29): 1862–1868.
Roy, A., 1983. *The Islamic Syncretic Tradition in Bengal*, Princeton NJ: Princeton University Press.
Roy, Tirthankar, 2011. 'Where is Bengal? Situating an Indian region in the early modern world economy', *Past and Present*, 21 (3): 115–146.
Roy, Tirthankar, 2012. *Natural Disasters and Indian History: Oxford India Short Introductions*. Oxford University Press.
Sarkar, Sutapa Catterjee, 2010. *The Sundarbans: Folk Deities, Monsters and Mortals*, Delhi: Orient Blackswan.
van Schendel, W. (1993). 'Easy come, easy go: Smugglers on the Ganges', *Journal of Contemporary Asia*, 23(2), 189–213.
van Schendel, W. 2009. *The History of Bangladesh*, Cambridge: Cambridge University Press.
Sivaramakrishnan, K. (1996). *Forests, Politics, and Governance in Bengal*, 1794–1994. Yale University.
Subrahmanyam, Sanjay, 2004. *Exploration in Connected History: From the Tagus to the Ganges*, New Delhi: Oxford University Press.
Whitehead, J., 2010. 'John Locke and the governance of India's landscape: The category of wasteland in colonial revenue and forest legislation', *Economic and Political Weekly*, 45 (50): 83–93.
Whitehead, J. (2012). John Locke, accumulation by dispossession and the governance of colonial India. *Journal of Contemporary Asia*, 42(1), 1–21.
Willem Van Schendel. 1991. *Three Deltas: Accumulation and Poverty in Rural Burma, Bengal and South India*. New Delhi: Sage Publications.
Worster, D., 1988. *Rivers of Empire: Water, Aridity and the American West*, London: Oxford University Press.

5 Challenges of Stakeholder Engagement in Integrated Transboundary River Basin Management

Kausik Ghosh

INTRODUCTION

Integrated River Basin Management (IRBM), also referred to as basin-wide management, has evolved as a comprehensive approach to managing river basins. IRBM encompasses multiple aspects of river basin management, including physical, chemical, geological, hydrological, geographical, natural and socio-economic, and political (Bhowmik, 1998). It involves the coordinated development and management of water, land, and related resources across sectors within a river basin to maximize economic and social benefits equitably while ensuring the sustainability of essential ecosystems (Agarwal et al., 2000). Despite widespread international support for IRBM, its meaning and practical application remain debated, with implementation challenges often arising, particularly in transboundary river basin contexts. Integration of multiple aspects of the river basin and water-centric approach complicated the implementation of IRBM principles for transboundary river basin management. As a result, it lost its popularity in managing transboundary rivers and resolving the water-sharing issues between riparian states. These challenges are particularly acute in South Asian transboundary river basins, where prolonged mistrust led to water-sharing disputes. In South Asia, the complex administrative boundaries, geographic contiguity, and power asymmetries among riparian nations have fuelled unilateral actions, resulting in intense water-sharing tensions. Bilateral relations in transboundary river basins have also suffered due to a long history of mistrust, dam-induced river control, and regional hostility, exacerbating tensions. The lack of an effective basin-wide management strategy and the failure to enforce river management through water-sharing treaties have further strained the region's water resources over the decades. A similar case has been witnessed in the case of the Tista/Teesta River water-sharing between India and Bangladesh, where decades of negotiations over water-sharing agreements have yet to yield a resolution, and both nations have failed to agree on a water-sharing framework and a basin-wide management plan. This long negotiation process has undermined a basin-wide multilateral management approach while focusing only on the 50-50 equitable water share

 DOI: 10.1201/9781003599692-5

proposition (Ghosh, 2021). As a result, it raises concerns regarding the proposed treaty's sustainability as the water demand has increased multifold compared to the available water resources in the basin.

There is a growing and urgent need to adopt a comprehensive, basin-wide water management strategy on which the IRBM principles are grounded to address escalating water scarcity challenges. Riparian communities residing along transboundary river basins are among the most severely affected, particularly those in downstream regions. These communities often accuse upstream regions of constructing water infrastructure projects to maximize their own benefits, thereby exacerbating water scarcity downstream. At the same time, upstream communities are equally affected as they face growing water shortages driven by increasing demands across sectors, unequal water distribution, and the intensifying risks associated with climate change. This complex scenario suggests that both upstream and downstream communities are adversely impacted by global freshwater scarcity, albeit in different ways. Despite bearing the brunt of these challenges, riparian communities are frequently excluded from key decision-making processes related to water-resource allocation and basin-scale water governance. Their marginalization within IRBM frameworks calls for deeper empirical investigation and policy attention to ensure more inclusive and equitable water management practices.

As a result, a growing body of literature advocates for a fundamental shift in river basin management, emphasizing the need for more integrated, participatory, and multi-scalar planning approaches (de Souza et al., 2025; Hüesker & Moss, 2015; Warner et al., 2014). This perspective increasingly frames river basins as complex socio-ecological systems with interconnected environmental and social dynamics (Hein et al., 2021). In this context, stakeholder participation emerges as a critical mechanism for fostering inclusive management, enabling the incorporation of diverse perspectives, values, and priorities of those inhabiting the basin (Carr, 2015; Karambelkar & Gerlak, 2020). Achieving such inclusive outcomes requires the genuine integration of multiple forms of knowledge and lived experiences, grounded in the principles of equity, inclusiveness, and social justice (Reed et al., 2018). Significantly, participatory processes must begin with a transparent and informative engagement phase, initiated by policymakers. Such transparency is essential for building trust and fostering multilateral dialogue in transboundary river basins. Ultimately, these processes contribute to the development of shared visions and the formulation of effective, integrated policies for sustainable natural resource governance (Garau et al., 2021). Incorporating these varied perspectives can enhance the legitimacy and effectiveness of transboundary river basin governance strategies. Hence, the integration of diverse experiences and knowledge systems, each informed by localized interactions with the river, aligns with the principles of IRBM, particularly the emphasis on inclusiveness.

At the same time, the implementation of a whole-of-basin governance integrating stakeholder perspective presents several challenges (Shahjahan & Harvey, 2012). One of the primary difficulties lies in incorporating all relevant elements and layers of the basin into IRBM, particularly at the community level, where engaging stakeholders remains complex. For example, perceptions of communities residing in upstream mountain basins are markedly shaped by distinct physiographic conditions, which are likely to hold perspectives on river management that may differ from those of communities in downstream plains. These divergent viewpoints can be attributed to variations in environmental challenges, livelihood practices, and the underlying cultural beliefs and values that shape local interactions with the river system. Such diversity, sometimes contrasting perceptions, underscores the complexity of transboundary river basin management. Therefore, a key challenge remains: to what extent can the diverse, and at times conflicting, perceptions of upstream and downstream communities be meaningfully integrated into an integrated transboundary river basin management framework? Prior to addressing this question, it is essential to undertake systematic documentation of the diverse perspectives held by multiple communities across the river basin. Such documentation forms the foundation for any inclusive and effective transboundary water governance framework. This study critically examines stakeholder perceptions in transboundary river basin management, addressing key concerns and challenges that hinder the development of an integrated basin-wide management plan. Furthermore, a geographical perspective including spatial diversity, advantages, and limitations is employed to assess the feasibility of IRBM as an effective management strategy. Therefore, this study aims to address how stakeholders' perceptions in transboundary river basins are shaped by geographical perspectives, applying a geographical lens to document the diverse viewpoints of stakeholders across the Tista basin as a case study. By adopting a nuanced approach to IRBM within the Tista basin, the study explores the possibility of stakeholder-driven governance.

THE CONCEPT OF IRBM

A river basin is a naturally delineated geographical unit. The IRBM is a part or subset of Integrated Water Resource Management (IWRM), while multiple transboundary water-sharing disputes adopted IWRM across borders at the river basin scale (Hooper, 2003). The IRBM avoids the siloed approach of traditional multipurpose resource management and integrates wider but relevant issues involving diverse social values, ecosystem services, and multidimensional aspects of water as more than a natural substance. Therefore, it theoretically integrates diverse physical, ecological, and socio-economic aspects essential in managing a geographical region for ecological values and human use (Anderson et al., 2019). It reveals that the aim of the IRBM framework is to manage the entire basin sustainably integrating land and water beyond administrative units. In order to implement IRBM in transboundary river basins, a special focus has been given on the interconnectedness between upstream and downstream riparian nations. Hence, the rationale of the IRBM is to free the flow of river water irrespective of administrative boundaries for implementing management activities in the upstream part of the basin that are likely to have an impact on the downstream part of the basin. Consequently, transboundary river basins are recognized by many researchers as a singular geographical unit having similar hydrological boundaries rather than as fragmented parts delineated by administrative boundaries (Giordano & Wolf, 2003; Hooper, 2003). This approach of IRBM was implemented by many transboundary nations sharing water from the common river basins to resolve water-sharing tension by adopting basin-wide water management through cooperation (Wolf et al., 2003). In the long run, such cooperation and coordination over shared river basins between nations will help build mutual trust, effective resource management strategies, and sustainability. A similar example of successful implementation of IRBM at a national river basin scale is the Tennessee Valley Authority (TVA) managing the Tennessee River in the United States.

Despite widespread support from policymakers, managers, and academics for basin-wide, integrated, and holistic approaches to managing water, land, and human-environment interactions within river basins and watersheds, the practical implementation of the IRBM concept is very challenging. This "implementation gap" can be attributed to the inherent complexity of managing the interconnected and dynamic socio-environmental systems within the geographical boundaries of river basins (Watson, 2004). These complexities are further amplified in transboundary river basins, where factors such as historical legacies, mistrust, power asymmetries, and competing water-centric development goals often exacerbate water-sharing tensions. Also, the implementation of IRBM poses challenges to the concept of "absolute sovereignty," as it necessitates upstream and downstream nations to consider each other's land and water management decisions. This dynamic has become a focal point of political debates at both national and international levels, with nations seeking to assert their sovereignty over resources. Conversely, IRBM struggles to ensure "territorial integrity," a critical element for fostering cooperation between nations in managing transboundary river basins effectively. Furthermore, the IRBM principle within the IWRM framework predominantly emphasizes managing river water, adopting a siloed approach that complicates decision-making in transboundary river basins by focusing primarily on water-centric problems and management. However, transboundary basins are far more complex, with issues extending beyond water management to encompass broader socio-political, environmental, and economic dimensions and most importantly involvement of stakeholders.

THE ROLE OF STAKEHOLDERS IN IMPLEMENTING IRBM

People and place (geography) play a major role in decision-making and implementation of resources management strategies in which people's perceptions play a crucial role, while interactions of people and geography create opportunities for diverse strategies based on human knowledge (culture) and available resources. Hence, the Dublin Conference on Water and the Environment (1992) concluded with a proposal for a participatory approach to water resources development and management that includes users, planners, and policymakers. Similarly, the IRBM is a holistic and coordinated approach to planning and managing water resources across a river basin. It encourages stakeholders to consider the diverse social and environmental interconnections in a basin across borders. Thus, there have been many efforts to integrate stakeholders in river basins at the national and international levels for implementing IRBM (Millington, 2006). Therefore, transboundary riparian nations are encouraged to foster cooperation to achieve integrated management that involves all relevant stakeholders. In this context, Hooper (2005) categorizes the IRBM approach relating stakeholders' participation into five key elements: basin-wide management, participation in decision-making, water demand management, compliance and commitments, and human and financial capacities. Among these elements, participation in decision-making is particularly significant, as it emphasizes the empowerment of local communities and the enhancement of public and stakeholder engagement in river basin management. In practice, IRBM facilitates the collaboration of diverse stakeholders, each with a vested interest in the river basin, to collectively manage activities and mitigate the impacts of resource use. Stakeholder engagement in IRBM is often emphasized, encompassing government entities, community organizations, businesses, industries, and other relevant organizations, as well as individuals with specific interests or concerns. However, effective implementation of IRBM principles also demands involving the broader public/communities, who, despite having a less clearly defined role, are nonetheless crucial stakeholders in the process. Such a participatory approach fosters the development of more holistic strategies that acknowledge interconnections and accommodate diverse objectives. A key challenge remains the need for greater awareness among stakeholders and understanding of effective management methods to improve river basin governance. As such, approaches vary and depend on the local cultural, political, administrative, and institutional context.

The issue persists: mere residency within a river basin does not inherently ensure stakeholder engagement in decision-making processes, nor does it guarantee that their perceptions will be pertinent or valuable in the formulation of basin management strategies and water resource management. Instead, stakeholders with a more direct involvement in water usage, such as irrigation farmers, islanders, fishermen, boatmen, hydroelectric power authorities, water-dependent industries, and urban and rural water service providers, play a more crucial role in decision-making (Hooper & Lloyd, 2011). These groups have a greater vested interest in the basin's water resources compared to general residents, positioning them as the primary decision-makers in water management strategies. At the same time, as these stakeholders are primarily engaged in utilizing water resources and the water-based economy in a river basin, there is a potential for bias regarding water accessibility and utilization. Consequently, capturing an impartial representation of stakeholder perceptions presents a complex and challenging task in the implementation of IRBM decision-making. Furthermore, in transboundary river basins, a significant bias arises due to the competing dynamics between upstream water utilization and downstream water availability and accessibility. This process necessitates a comprehensive and diverse range of perspectives on water resource use and management. Therefore, this study adopts a geographical lens to document the varied perspectives of stakeholders in river water uses and transboundary basin management.

STUDY AREA AND METHODOLOGY

Tista River Basin

The Tista/Teesta River, a transboundary Himalayan River, originates at an elevation of approximately 7,000 m from the eastern Himalayan glaciers in India and descends to merge with the Brahmaputra (Jamuna) River in Bangladesh at an elevation of around 30 m (Figure 5.1). Spanning approximately 180 km through the

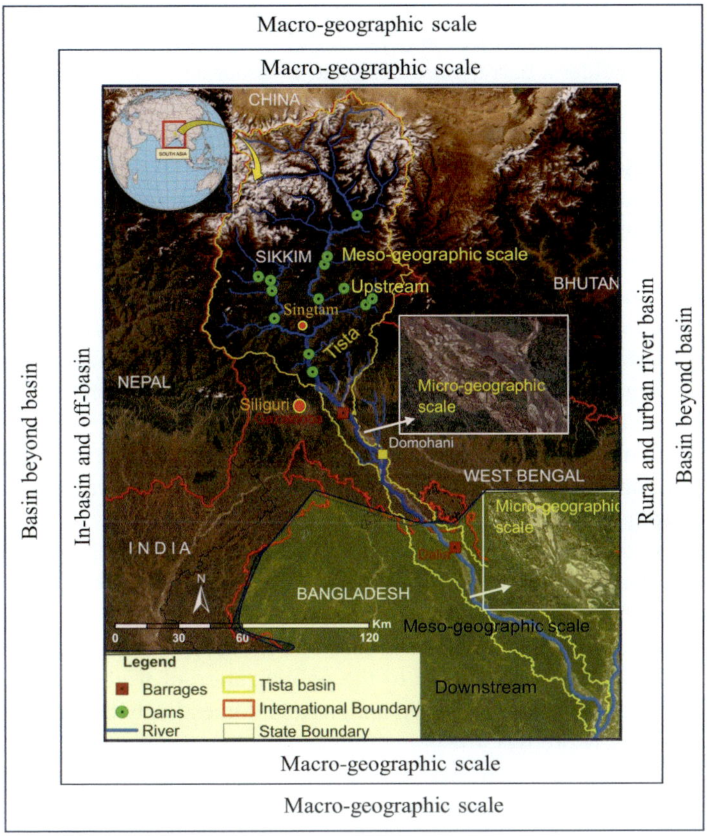

FIGURE 5.1 Study area, transboundary Tista/Teesta River basin, and conceptual framework for the geographic lens/scale including micro-geographic scale, meso-geographic scale, and macro-geographic scale with location and examples. Following Survey of India, this study uses Tista; however, Tista and Teesta is the same river. (*Source:* Prepared by author.)

mountainous terrain of the Sikkim-Darjeeling Himalayas and the northern part of West Bengal, the river continues for another 100 km across the Indian plains before entering Bangladesh, where it flows an additional 132 km (Ghosh, 2014). Thus, the basin covers more than 12,000 km^2 area of which 10,000 km^2 is in upstream India. The hydrological regime of the river closely follows the seasonal rainfall cycle, with nearly 80% of its annual discharge occurring during the monsoon and only 20% during the dry season. The Tista basin supports intensive agriculture, particularly dry-season water-intensive Boro rice cultivation, resulting in significant irrigation demand across both countries. In response, India and Bangladesh have constructed major barrages like Gazaldoba in India, located approximately 25 km from the mountain front, and Dalia in Bangladesh, 100 km downstream of Gazaldoba, to divert Tista water through irrigation canals, collectively targeting 1.5 million hectares of area for irrigation, benefitting over 30 million people (Figure 5.1). Riparian communities along riverbanks, floodplains, and islands are heavily dependent on the river's flow for their livelihoods (Ghosh, 2015). In addition to irrigation, the link canals of the Gazaldoba barrage are designed to generate approximately 67 MW of hydroelectricity. Furthermore, upstream of the barrage, at least 13 major run-of-the-river hydropower projects are operational, collectively producing around 3,500 MW of hydroelectricity (Ghosh & Munoz-Arriola, 2023). These upstream interventions have contributed to growing mistrust and persistent tensions over water-sharing between India and Bangladesh.

In the Tista basin, a significant water-sharing challenge exists, primarily driven by politically motivated proposals for quantitative water-sharing treaties and the ongoing debate over inter-sectoral water distribution, particularly between irrigation and domestic water supply. It fails to bridge the water demand gaps between agricultural irrigation and domestic use. The disagreements over environmental flow (e-flow) emerge as a critical issue, where the ecological needs of water are ignored by both upstream and downstream. Concerns related to the construction of multiple hydroelectric dams by upstream have created tensions regarding water impoundment and maintaining continuous e-flow.

Interstate water-sharing tensions, such as those between West Bengal and Sikkim (however, not prominent but political leaders in West Bengal often raised voice (responsibility of upstream hydropower projects in Sikkim), further complicate governance structures and exacerbate tensions. The failure to effectively address these challenges can largely be attributed to the politicization of water-sharing tensions, both at the international level between nations and domestically between central and state governments. Moreover, the prevailing approach to water management remains excessively quantitative, often overlooking the complexity of the resource itself and the broader challenges involved in its allocation and use (Ghosh, 2021). Additionally, this approach has neglected the inclusion of key stakeholders within the basin in decision-making processes.

METHODOLOGY

Geographic Lens

The geography study encompasses the analysis of spatial processes, including spatial distribution and patterns, and the examination of human-environment interactions. To explore these, it employs a diverse range of methodological approaches, integrating both qualitative and quantitative research techniques. Its human-environment interaction study informs decision-making in river basin management, integrating insights from environmental psychology, ecology, and resource economics to shape modern natural resource management paradigms (Hooper, 2005). A geographic lens in understanding stakeholder perceptions encompasses spatial scales and opportunities related to resource distribution and accessibility. These spatial scales in a river basin can be categorized as micro (e.g. bars, islands, riverbanks), meso (e.g. upstream, downstream), and macro (e.g. the entire basin and areas covered by canal irrigation beyond the basin) (Figure 5.1). The geographical opportunities include locational advantages such as proximity to water sources, surplus water availability, water storage capacity, and the distribution of water through infrastructure and control mechanisms. Conversely, locational disadvantages involve challenges such as flood vulnerability, riverbank erosion, reduced flow during the dry season, limited accessibility, and upstream control over downstream water resources. Hollick (1990) conceptualized this as the "decision space," referring to the domain within which a decision-maker operates, shaped by their perceptions of potential outcomes.

The dynamics of human-environment interactions and stakeholder perceptions across spatial scales within a river basin can be effectively analysed through a geographic lens. For instance, Padło et al. (2021) conducted a comprehensive study on the Danube River in Europe, illustrating how diverse stakeholder perspectives are shaped due to different locations of the respondents in the basin. The findings highlight significant variations in perceptions of the Danube among respondents who share similar demographic characteristics, such as age and education level, but differ in terms of their place of residence within the European region. The study reveals that individuals' perceptions of the Danube's course are highly influenced by their geographic location. Notably, respondents from Eastern Europe tend to view the Danube as an economic and cultural axis of the continent, as well as a symbolic bridge that unites European nations and embodies European integration. This suggests that the river holds a strong pan-European significance in the eastern part of the continent. Conversely, respondents from Western Europe often depict the Danube in a manner that aligns more closely with the course of the Rhine, implying that, from their perspective, the Rhine is the pre-eminent river in Europe. Analysing the maps generated during the study further supports the conclusion that perceptions of the Danube's course frequently diverge from its actual geographic path. These findings underscore the role of regional differences in shaping environmental perceptions and highlight the varying symbolic meanings attributed to the Danube across different parts of Europe.

Focused Group Discussions

This study employed a qualitative research design, conducting five Focus Group Discussions (FGDs) in 2014 using semi-structured, open-ended questions. The primary participants were members of riparian communities situated along the upstream hill regions and downstream plains of the Tista River in India. The upstream FGDs were conducted in Singtam town, Sikkim, with residents living adjacent to the river as the key respondents (Figure 5.1). In the downstream plains, participants were primarily drawn from communities (mostly farmers, fishermen, and boatmen) residing in river islands (chars), riverbanks, and floodplains. Additionally, the perceptions of downstream riparian communities in Bangladesh were incorporated through secondary sources, particularly from Haque et al. (2014). The FGDs were designed to capture local perceptions related to river basin management and transboundary water-sharing. Key thematic areas included direct and indirect uses of river water, perceived minimum flow requirements necessary to sustain livelihoods, and community-recommended strategies for equitable water management. Specific questions addressed whether upstream communities support water-sharing without compromising their needs and how the river could be managed to avoid adverse effects on both upstream and downstream populations. It resembles the novel approach in e-flow science that presents an opportunity to more comprehensively consider the co-constitution of river flows, ecosystems, and society and to more explicitly integrate these interrelationships into river management frameworks (Anderson et al., 2019). The semi-structured format allowed flexibility in discussion while ensuring thematic consistency across groups. This design aimed to capture the spatial variation in community perceptions, which are shaped by both geographic context and socio-political-economic considerations. Portions of the FGD findings have been reported in both published and unpublished sources (Ghosh, 2021; Haque et al., 2014).

RESULTS

Perception of River Basin and Water-Sharing

The perception of the riparian population varies as they are going to get direct and indirect benefits from the river. The perception may also vary with the geographical location of the population, i.e. where they are residing relative to the position of the river. Also, a key concern is how community survival and livelihoods depend on the river's flow? The relationship between communities and the river, particularly in terms of benefit extraction and co-existence, varies significantly across spatial scales defined by geographic location and access to natural resources essential for survival, livelihood, and settlement. This spatial heterogeneity within the river basin can be conceptualized across three geographic scales (Figure 5.1). First, the *macro-geographic lens* encompasses the entire river basin, including areas that benefit from river water diverted canal irrigation systems extending beyond the immediate basin boundary. Second, the *meso-geographic lens* refers to intra-basin divisions, such as the upstream and downstream regions, where ecological conditions and resource access vary significantly. Third, the *micro-geographic lens* focuses on specific fluvial landforms relative to the river channel, such as riverbanks, floodplains, islands (chars), and fluvial bars, where communities interact directly with the physical dynamics of the river. While the micro-geographic analysis is grounded in primary field data derived from the FGDs, insights at the meso- and macro-geographic scales are drawn from existing literature and the conceptual framework proposed by the author.

Macro-geographic Lens

River Basin as Single Unit

River basins constitute the primary geographical units that segment the majority of the global land surface, with transboundary river basins encompassing approximately half of the terrestrial landscape (Wolf et al., 2003). River basins are defined as topographically delineated regions that collect precipitation and channel run-off and sediment towards a common outlet, supporting critical ecosystems, both terrestrial and aquatic biodiversity, and human societies. The development and regulation of river systems for human use have a long historical precedent, dating back thousands of years. Evidence suggests that as early as the sixth century BCE, civilizations in Mesopotamia implemented irrigation and flood control measures along the Tigris and Euphrates rivers. Similarly, the damming of rivers in present-day Jordan occurred approximately 5,000 years ago. However, the conceptualization of the river basin as a distinct unit for development is a more recent phenomenon. In the 1890s, Sir William Wilcocks was among the first to advocate for considering the Nile River basin as a unified entity for development (Fox, 2009). Despite these early initiatives, the scale and impact of contemporary river basin development far exceed historical records. The early twentieth century marked a significant transformation, with large-scale flood control projects, irrigation schemes, hydropower generation, reservoirs, and navigation enhancements reshaping river basins from India to the United States. While these projects have contributed to economic growth and modernization, they have also resulted in profound ecological disruptions and adverse consequences for local communities. These developments necessitated advancements in engineering and a utilitarian perception, resulting in the grossly ignored river basin as a single entity.

According to the Register of International River Basins (1978), a river basin is defined as the geographical area that hydrologically contributes to a first-order stream (headwater streams without any tributary joining similar or large rivers). This contribution encompasses both surface water and groundwater flow. The first-order stream is subsequently characterized by the network of streams and their discharge points, which may be an ocean, a terminal lake, or an inland sea. Therefore, a tributary of some river within one drainage basin does not in itself form a unit for the purpose of the distribution of water resources of the said basin. The entire basin as such has to be taken as a unit for the said purpose. In transboundary river governance, the river basin organizations/institutions (e.g. Joint River Commission (JRC) on transboundary rivers between India and Bangladesh) are established to facilitate consensus-building for managing river basins as unified ecological and hydrological entities. However, these formal institutions often remain predominantly state-centric, limiting their effectiveness in fostering cooperative water management among riparian nations. As a result, decision/policymakers tend to perceive river basins through narrow political boundaries and this perspective is often mirrored among transboundary riparian communities. For example, it led to the partitioning of river basins, as seen in the Indus Waters Treaty in 1960, and the quantitative division of water resources, as exemplified by the Ganga Water Treaty in 1996. In both cases, IRBM approach was grossly ignored. Surprisingly, contemporary discourse advocates for more localized and context-specific governance frameworks, perceiving the river basin as a "problem-shed" due to the rigid "watershed" approach (Woodhouse & Muller, 2017). This trend is evident in the case of the Tista River basin, where the JRC between India and Bangladesh has grossly failed to achieve consensus-based management of the river as a singular unit, rather narrowly focusing on quantitative (50-50 proportional) water-sharing. As water resources continue to be integral to both human development and environmental sustainability in the twenty-first century, effective transboundary water governance based on the river basin as a single unit remains a critical challenge.

In-Basin and Off-Basin

A river basin can be conceptually divided into two distinct categories based on water supply and demand: the in-basin (in-stream) population and the off-basin (off-stream) population. The in-basin (same as the geographical delineations of river basins) population comprises individuals and communities residing within the geographical boundaries of the basin who directly benefit from the water resources available within their proximity. Due to their geographical proximity to the source, in-basin populations are generally better aware of the origin and significance of river flows. In contrast, the off-basin population includes individuals and sectors that rely on water diverted from the basin to areas beyond its geographical boundaries. Water supply to off-basin regions typically occurs through extensive canal irrigation systems (including tunnels, pipes), serving agricultural, domestic, industrial, and hydroelectric purposes. Due to geographical distance from water sources, off-basin beneficiaries may have limited awareness of their water supply origins, while most communities rely on institutions (e.g., municipalities) for water supply rather than directly on river basins.

The Tista River basin exemplifies this dynamic, covering an area of approximately 12,000 km^2. Of this, the upstream region within India comprises around 10,000 km^2, with 8,500 km^2 consisting of mountainous terrain and 1,500 km^2 comprising alluvial plains. The downstream section in Bangladesh covers approximately 2,000 km^2 in the plains. Despite these geographical delineations, both India and Bangladesh have extended water supply infrastructure beyond the basin, increasing the collective command area to approximately 15,000 km^2 in the plains. This expansion has facilitated significant off-basin water use, extending the benefits of the Tista River beyond its immediate geographical area of the basin. In Bangladesh alone, more than 21 million people are estimated to depend on Tista water, with a significant proportion reliant on canal irrigation networks in off-basin command areas. Nevertheless, it is evident that for a majority of off-basin populations, the river is viewed predominantly from an economic perspective, emphasizing water availability and accessibility for hydropower generation, agricultural irrigation, industrial applications, and drinking water supply. It suggests institutional reliance of communities, susceptible to socio-political narratives of water supply in transboundary rivers, exacerbating water-sharing tension. At the same time, the predominantly economy-driven perception contrasts with that of in-basin stakeholders and warrants further investigation to comprehensively understand the socio-economic and environmental implications of river water allocation.

Basin Beyond Basin

The ecosystem services provided by Himalayan rivers extend well beyond the geographical boundaries of their respective basins, provisioning in the form of irrigation water and hydroelectric energy distributed through canal networks and national power grids, respectively. India's Inter-Basin Water Transfer projects are designed to reallocate water resources across basins (Sinha et al., 2020), fostering basin-level integration by connecting over 14 Himalayan and 16 peninsular river systems. These large-scale interventions support multiple objectives, including hydropower generation, water supply to water deficit regions, inland navigation, and the provision of drinking water to approximately 90 million people. However, such expansive redistribution of river flow introduces significant complexity in understanding basin-specific ecosystem services and community perceptions of riverine ecologies. Many beneficiary communities remain unaware that the services they receive originate from distant basins. This disconnection may hinder awareness, affect public perception of river systems, and pose challenges for inclusive and participatory basin management, particularly river basins supplying ecosystem services to geographically distant populations.

A relevant precedent in water resource allocation is the Krishna River interstate water dispute in peninsular India, where the 1973 tribunal ruled that all existing uses of water diversions beyond the basin would be protected (Richards & Singh, 2002). This decision underscores the complexities of water-sharing arrangements across political boundaries, which are also evident in the case of transboundary river basins such as the Tista River system that traverse two nations providing a range of ecosystem services beyond political boundaries, and basin geography.

For instance, one of the most critical aspects of transboundary water management is hydropower generation as a crucial provisioning ecosystem service. The targeted hydropower capacity of approximately 3,500 MW in Tista serves pan India through an extensive power grid network. Also, the India-Bangladesh energy-sharing mechanism further exemplifies benefit sharing approaches in transboundary Himalayan River basins, illustrating the significance of regional energy cooperation, a crucial part of successful implementation of IRBM (Ghosh & Scott, 2025). At the local scale, the Tista–Mahananda Link Canal (TMLC) of the Tista barrage project in India is designed to generate approximately 67 MW of hydroelectric energy by diverting water into the Mahananda River of the Ganga basin (Ghosh, 2021). In such instances, stakeholders/users remain unaware that the energy they utilize originates from these river basins, rather than being influenced by their geographical proximity to the basin. This underscores the importance of examining the significance of individuals' relative distance to the basin in understanding their awareness and engagement with the resource.

Rural and Urban River Basin

The geographical delineation between rural and urban areas is primarily shaped by economic factors, while river water is increasingly governed by the spatial dynamics of economic demand. River flow diversion, water allocation, and fragmentation of river systems are largely driven by the imperatives of a demand-driven economy. Urban centres, characterized by the centre of economic activities, often act as magnets for water resources frequently drawing water from across and even beyond river basin boundaries, thereby creating imbalances that may lead to the marginalization of rural areas in terms of water access.

The TMLC, which operates beyond its designated project command area for rural irrigation, supplies approximately 26,000 m^3 of water to Siliguri city within a 20 km proximity of the basin. Given the rising water demand driven by Siliguri's growing population, this supply may increase to 55,000 m^3 in the future, with a long-term target of providing 5 m^3 per second (cumec) to meet urban water demand. Only recently, when Siliguri faced an acute water crisis due to siltation in the Tista canal and barrage maintenance work in mid-2024 and 2025 that many urban residents became aware of the river's critical role in sustaining urban water supply. In contrast, rural communities, which are heavily reliant on river water for agricultural production, have long been aware of the importance of consistent water availability. This contrast suggests that awareness and perceptions of river basins in urban areas tend to be crisis-driven, whereas in rural regions, such awareness is embedded in the everyday necessity of river-dependent livelihoods, called geographic opportunity.

Bandyopadhyay and Ghosh (2016) characterize this situation as an "inter-sectoral" water dispute, wherein the increasing demand for water from the urban population and service sector conflicts with the original purpose of water allocation, which was intended primarily for agriculture in rural areas. In the Tista basin, agriculture remains the predominant livelihood, with over 78% of the population residing in rural areas. Surface water irrigation is critical for sustaining agricultural activities, particularly during the dry season. The primary objective behind the construction of the Tista barrage in India was to facilitate irrigation through an extensive canal network. However, less than 70% of the intended command area benefits from canal irrigation. The situation is further intensified by the fact that over 80% of farmers within the command area rely predominantly on groundwater for irrigation, rather than surface water (canal irrigation), reflecting a shift towards groundwater dependency for agricultural sustenance, shaping perceptions of the river and the river basin.

Meso-Geographic Lens

Basin and co-basin: In the context of transboundary water management, the terms "basin" and "co-basin" refer to the shared utilization of river water between two or more countries on a large geographical scale. The upstream country is typically referred to as the "basin," while the downstream country or countries are designated as the "co-basin." In the Tista basin, India, as the upstream nation, is considered the basin, whereas Bangladesh, situated downstream, is recognized as the co-basin. This distinction is often framed within the broader classification of upstream and downstream as discussed below.

Upstream (Hills) Basin

Sikkim, located in the upstream region of the basin, is characterized by a mountainous topography that covers its entire area, spanning approximately 6,930 km^2; while the state of West Bengal, situated downstream of Sikkim, encompasses a hilly basin area of 1,121 km^2. The primary water sources in this region include rivers, springs, and aquifers, which provide a year-round water supply. The population distribution within the basin varies with elevation, where high mountain regions exhibit lower population densities, while the lower mountain areas are more densely populated (Chakraborty and Ghosal, 2024). For instance, in the lower mountain region of Sikkim near Singtam town, FGDs indicate that the surface water level of the Tista River decreases by approximately 2 to 3 m during the dry season and 80% of the household plastic water pipes supplying spring water from the hill slope are dried up. This reduction in river flow and water supply is potentially attributed to the presence of upstream dams, as suggested by these communities, such as the Stage V (~3 km upstream) with a hydroelectric capacity of 510 MW. In the high mountain regions of Sikkim, local communities face a complex array of interrelated challenges. These include the imperative to safeguard river ecosystems amid increasing dam construction, as well as the associated risks of displacement, cultural erosion, deforestation, and exposure to natural hazards such as landslides, flash floods, glacial lake outburst floods (GLOFs), and earthquakes. These upstream areas of the basin are particularly sensitive to ecological and socio-economic changes. Community perceptions in these regions are shaped by distinct topographical and cultural contexts, which pose significant constraints to the uniform application of IRBM principles. Nevertheless, over the past few decades, civil society actors across the basin have mobilized around the adverse impacts of dam impoundments, fostering greater upstream–downstream integration and consensus-building among affected communities.

Downstream (Plains) and Co-Basin

The Tista River basin encompasses a total area of approximately 4,108 km^2 in alluvial plains, with about half of this area equally shared by India and Bangladesh. However, in terms of population dependency, the basin serves a significantly larger demographic than its geographical cover. Nearly 28 million people rely on the water resources of the Tista alluvial plains, with approximately 27% of this population residing in West Bengal and around 71% in Bangladesh, either directly or indirectly dependent on the basin for their water needs.

Despite the critical role of the river in sustaining downstream populations, upstream communities in India, particularly in Sikkim and parts of West Bengal, often exhibit limited awareness of the water availability challenges faced by those residing downstream. It is an obvious fact that mountain/upstream communities have year-round access to river water (however, with a different set of challenges), and their livelihoods are not as directly controlled by surface water variations as those in downstream regions. Conversely, the reduction in river flow constitutes a critical concern for downstream regions in both West Bengal and Bangladesh, with significant implications for economic stability and political discourse. However, Bangladesh, as a co-basin state, often perceives upstream dam operations as directly responsible for depriving approximately 120 km of downstream flow within Bangladesh, despite the fact that approximately 75 km of river stretch in alluvial plains lies within Indian territory, downstream of all these 14 upstream dams. Such perceptions reflect a simplified understanding of basin hydrology and overlook climate-induced risks intensifying water scarcity challenges faced by the upstream. These misperceptions of reduced water share undermine the implementation of IRBM principles, which require a holistic and cooperative approach to transboundary river governance.

Upper and Lower Riparian Basin

The geographical delineation of a river basin can be conceptualized geopolitically, based on a nation's relative position and influence over water resources, which is shaped by political, economic, and demographic factors. Consequently, "riparian" or "co-riparian" states can be categorized based on their relative spatial position along the river's course within a geopolitical context which is slightly different from the "basin" and "co-basin" concept. Furthermore, the riparian states can be distinguished into two categories: *upper riparian* and *lower riparian states*. In a transboundary context, the *upper riparian state* encompasses the ecological and hydrological features, including vegetation, topography, landforms, and ecosystems, as well as human populations associated with the river or adjacent water bodies, before the river crosses into the administrative territory of a downstream nation. The *lower riparian state* shares similar ecological and hydrological associations; however, it is distinct in terms of its geographic location downstream from the point where the river enters a neighbouring co-riparian state. Hence, political-administrative boundaries reinforce these distinctions.

Conflicts over water-sharing within transboundary river basins are primarily driven by riparian structures, which influence the spatial and temporal availability of water. The geopolitical advantages or dominance of an upper riparian state over water resources, coupled with the degree of dependence and demand for water by downstream nations, significantly impact human security and livelihoods (Tiwary, 2006). Tiwary (2006) further discusses the sovereignty disputes concerning water resources between upstream and downstream states. He asserts that the principle of "Territorial Sovereignty" or the "Harmon Doctrine" underpins the upper riparian state's claim to absolute control over water resources within its territorial limits. In contrast, co-riparian states advocate for the principle of "Territorial Integrity," emphasizing historical water flows as the basis for their entitlement. The concept of "Restricted Sovereignty" serves as an intermediary framework, suggesting that no single state possesses absolute authority over transboundary water resources, underpinning the IRBM principle. This principle highlights the fundamental basis of water-sharing conflicts and cooperative mechanisms in South Asia.

Water governance within transboundary river basins remains a highly contentious issue. Upper riparian states frequently assert sovereign control over water resources within their territories, whereas lower riparian nations often accuse upstream states of inequitable water diversion, which shapes the upper and lower riparian perceptions. It causes water-sharing tensions, which are further exacerbated by external stressors such as climate change, affecting river water availability. Additionally, restricted access to hydrological data perpetuates mistrust between riparian states. For instance, Bangladesh, as a lower riparian country, has persistently accused upper riparian state India of restricting the Tista water and flow data, thereby exacerbating tensions and complicating the effective implementation of IRBM in the region. The reduced river flow in transboundary rivers can evoke conflict and cooperation which is largely determined by perceptions of their relative benefits (mainly geopolitical and economic) by riparian nations involved (Sadoff et al., 2020).

Micro-Geographic Lens

Perception of River on Different River Reaches

The FGDs enquired the riparian people about the minimum water requirement for their survival and livelihood as well as for the sustaining of the river in terms of e-flow; people along the river and within the river in the alluvial plains for shelter (living) and livelihood respond to it from multiple perspectives which is categorised into different micro-geographical location called river reach.

Shelter and Livelihood in the Char/Island

In the downstream alluvial plains, across the basin, extensive island formations have developed within the river channel belt dissected by multiple fluvial channels (Ghosh, 2015). These fluvial islands, or *chars*, undergo frequent morphological changes due to river channel shifting, although a few have exhibited relative stability and gradual expansion over extended periods (Ghosh & Chakraborty, 2024; Ghosh et al., 2024).

Despite their inherent instability, these islands attract riparian populations due to their agricultural potential. Many settlers have established dwellings on elevated sections of these islands, seeking both livelihood and shelter. The term shelter is deliberately used, as these lands are not formally recognized, documented, or distributed by government authorities; their dynamic nature is shaped and reshaped by channel shifts, which places them outside official land governance systems. The FGDs reveal that over 20,000 people reside seasonally or permanently on these riverine islands only within the Indian stretch of the Tista River, primarily engaging in agriculture. A significant number of these inhabitants are entirely dependent on the islands for both subsistence and habitation. They express a strong belief that upstream flow regulation and the construction of protective infrastructure such as embankments around inhabited islands are essential for sustaining their livelihoods and shelters.

Notably, in the past two decades, island areas proximal to the barrage have eroded by approximately 70%, whereas those located in the distal downstream region have reduced by only 15%, subsequently attracting increased farming activities (Ghosh & Chakraborty, 2022). The demographic profile of these settlers includes marginal farmers from neighbouring regions and in some cases migrants from distant areas such as the Sundarbans, situated nearly 750 km to the south, underscoring the socio-economic significance of these riverine islands.

No Shelter but Livelihood Depends on the Island

A section of the respondent remains heavily dependent on riverine islands for their livelihoods primarily through agriculture, boating, and fishing. Although their dwellings on these islands are typically temporary, they maintain more permanent residences along the embankment-protected riverbanks and floodplains. This group generally supports a limited degree of river regulation through upstream dams and downstream embankments, viewing such interventions as necessary to mitigate catastrophic flooding. At the same time, they recognize the ecological benefits of periodic flooding, which replenishes soil nutrients of these islands, ensures navigable water levels for boating, and sustains aquatic habitats essential for fisheries.

No Shelter and no Livelihood Depends on the Char/Island

Communities that do not directly depend on river water either for shelter or for their livelihood often reside beyond the embankments or protective structures, but remain within the broader geographical boundaries of the river basin. These communities benefit indirectly from the river through ecosystem services and regional ecological functions. Despite their indirect relationship with the river, they often advocate for its unregulated and natural flow within the channel, emphasizing the ecological and cultural significance of a free-flowing river.

Stakeholders in Downstream Bangladesh

Haque et al. (2014) identified a diverse range of ecosystem services derived from the Tista River, encompassing both "*in situ* and *ex situ*" uses. In situ uses, primarily by island (*char*) dwellers, include activities such as bathing, washing, transportation, fishing, spiritual practices, and duck rearing. Ex situ services comprise irrigation, nutrition sourcing from fish and edible plants, fodder collection, and extraction of plants, sand, and wood for domestic use and local trade, as well as spiritual uses involving the transportation of river water for rituals elsewhere. The study further highlights the community's awareness of hydrological conditions such as river flow, water level, and spatial characteristics that support their livelihoods. However, a marked decline in river flow has intensified downstream perceptions of inequitable upstream water withdrawals, fostering a sense of deprivation and mistrust among the affected communities.

From the above perspectives, it becomes evident that human perceptions of rivers are highly diverse even at micro-geographic scales within the same political boundary, largely shaped by livelihood/geographic opportunities and socio-economic contexts. This heterogeneity underscores the inherent complexity in integrating individual perceptions into river management decisions, as doing so may introduce additional layers of conflict and ambiguity. For instance, the documented perceptions of riparian communities in geographical proximity exhibit diverse opinions over river e-flow and river regulation by dams and embankments. Consequently, when transboundary water resources are involved, the divergence in public perception across nations becomes even more pronounced. In such cases, individuals tend to interpret resource sharing through narrow geographic, economic, and political lenses, further complicating cooperative water governance.

DISCUSSIONS

IRBM and Transboundary Water-Sharing Complexities

The management of transboundary river basins is complex, requiring an integrated approach like IRBM to balance the interests of multiple stakeholders while ensuring sustainability and equitable resource distribution. By fostering a shared vision, IRBM facilitates conflict resolution and enhances sustainable development practices. According to Hooper and Lloyd (2011), IRBM is not merely a management technique but an inclusive and long-term process that incorporates multiple perspectives. It follows integrated principles but should not be confused with a comprehensive approach. Instead, it focuses on a few major elements within the system, which can be identified through FGDs among stakeholders. Unlike a comprehensive approach, IRBM is more practical, selective, and stakeholder-driven. The selective nature of IRBM ensures stakeholder participation, particularly decision-makers such as policymakers and planners, who shape management strategies based on their perceptions. However, IRBM has been criticized for relying excessively on rational decision-making, often neglecting the subjective ways individuals perceive risks, which may lead to heuristic-driven assumptions. Furthermore, despite its promise, IRBM's focus on management and technocratic solutions fails to address the underlying inequitable power relations that contribute to unsustainable policies (Fox, 2009).

In South Asia, transboundary river basins such as the Indus, Ganga, and Brahmaputra are shared by multiple countries, making water-sharing negotiations highly complex. Achieving a sustainable future for these basins and extending beyond them necessitates effective management of water resources (Pradhan et al., 2021). This, in turn, requires some form of cooperative or collaborative governance among riparian states within a mutually agreed-upon and rational framework (Wolf, 2005). However, even the bilateral agreements/negotiations on water-sharing cooperation are not straightforward, as upstream and downstream claims often clash, exacerbated by geopolitical tensions and historical disputes. Effective joint management of transboundary river basins requires mutual understanding and equitable sharing of water resources among riparian populations. The hydropolitical landscape, shaped by geographical and political conditions, also affects bilateral cooperation (Giordano, 2002). These geographical complexities extend beyond national boundaries and influence transboundary water cooperation and joint river basin management efforts. A key example is the Tista River basin between India and Bangladesh, where unresolved water-sharing tensions highlight the difficulties in fostering cooperation. Effective management begins with a joint effort to identify and quantify water resources at the basin level, considering both gains (e.g. rainfall, glaciers) and losses (e.g. evaporation, seepage loss) while treating the basin as a single unit. A cascading water allocation approach prioritizes the river's ecological integrity, followed by the basin's catchment needs, where off-basin water distribution is addressed last, factoring in dependency ratio, economy, proximity, and population density.

Geographic Lens and Stakeholder Perceptions

This study explores the potential for joint river basin management following IRBM principles while incorporating stakeholder perspectives. It identifies direct users and indirect beneficiaries of river water, employing a geographic lens to document stakeholder perceptions in the transboundary basin. It is noticed that geography plays a crucial role in determining the relative advantages and disadvantages to access resources/opportunities, shaping stakeholder perspectives on river water management based on their place of residence and livelihood opportunities. For example, Padło et al. (2021) emphasize the role of distance in shaping perceptions of space, and some stakeholders are unaware of the sources of their water benefits. Similarly, geographical distance from the river provides a diverse notion of ecosystem services benefits, resulting in stakeholder involvement in decision-making (Cheng et al., 2021). Such geographical perceptions can lead to contesting views, underscoring the need for policymakers to review and integrate these perspectives into IRBM decision-making. Failure to acknowledge these perspectives can result in unresolved water-sharing conflicts (den Haan et al., 2018).

These geographical perceptions are shaped by diverging mental constructs, each influenced by a range of emotional responses and associated with various biophysical, social, political, and economic dimensions (Garau et al., 2021). They are also linked to subjective values such as aesthetic beauty, tranquility, and cultural significance. Such constructs reflect the individual attitudes and values people assign to specific landscapes or environmental challenges, making them crucial components in participatory planning and decision-making processes. Garau et al. (2021) observed that the ways in which individuals draw or describe a concept are inherently personal and varied. Their findings emphasize that each person interprets concepts such as a river basin differently, incorporating both spatial form (evident in sketches made by stakeholders) and thematic content (revealed in narrative descriptions). This diversity in perception presents a fundamental challenge for policymakers, particularly in the context of transboundary river basins, where geopolitical complexities are already pronounced. Engaging stakeholders at the community level adds further layers of subjectivity, making consensus-building and integrated planning more difficult.

Consequently, the implementation of IRBM often faces obstacles due to the inclusion of diverse community perspectives. While such inclusivity is essential for equitable governance, it can complicate the alignment of objectives and introduce additional layers of complexity to already intricate systems. As observed in this study, community perceptions are shaped significantly by geographic location and geographic opportunities, even in a diverse geographical scale (micro to macro). While social norms surrounding IRBM are often defined by shared attitudes, values, and beliefs stemming from common geographical experiences (Fielding et al., 2005), this definition may be insufficient as social norms are influenced by multiple factors, including social networks, familial beliefs, media discourse, and political ideologies, rather than solely by geographical awareness/proximity within the river basin. Likewise, in Siliguri city, recent media reports on water scarcity have heightened public awareness of the Tista River canal system as a critical water source during the dry season. In contrast, within Bangladesh, various social media platforms frequently attribute reduced downstream flows of the Tista to upstream water control measures, often fuelling anti-upstream sentiment, reinforcing transboundary water-sharing tensions. These complexities are particularly evident among stakeholders in downstream and beyond-basin locations, where perceptions are shaped more by prevailing socio-political influences than by basin-scale realities (for example, Tista River flow is also reduced naturally as reported by Ghosh & Chakraborty, 2022; Ghosh & Munoz-Arriola, 2023). Such divergence often results in conflicting viewpoints among communities that rely on the same resources but reside in distinct geographic locations in reference to the basin. This phenomenon aligns with Tobler's First Law of Geography: "everything is related to everything else, but near things are more related than distant things" (Tobler, 2004). Here, the perceptions of stakeholders in micro-geographic contexts, such as islands and riverbanks, exemplify the fundamental principles of geographical theory. Stakeholders located beyond river basins are often unaware of the ecosystem services provided by these rivers, despite benefiting from them. Such disconnections highlight how geographical scale plays a critical role in shaping perceptions of river systems and their value. However, even within a shared micro-geographical context, communities may exhibit divergent responses and understandings of river-based resources. This complexity is further pronounced in the context of transboundary river basins, where differences between upper and lower riparian states often give rise to mistrust and suspicion. Such tensions are frequently rooted in concerns over perceived control of water resources for geostrategic advantage. Nevertheless, empirical evidence increasingly suggests that both upstream and downstream states are vulnerable to common threats such as water scarcity, exacerbated by climate change (Ghosh, 2022). In this context, the implementation of IRBM likely offers a potential pathway forward. By treating transboundary river basins as single hydrological (geographical) units, IRBM promotes cooperative, data-driven governance that can mitigate conflict and foster equitable, sustainable resource management.

Consequently, capturing the diverse perceptions within these micro-geographic units in managing river basins presents a challenge to the basin-wide implementation of IRBM. It may complicate the strategy and shift focus away from the broader objective of joint transboundary river basin management due to the divergent and sometimes conflicting perspectives of stakeholders. However, a nuanced understanding of these perceptions can provide a critical foundation for developing more effective and inclusive IRBM strategies. Furthermore, defining communities solely based on their river basin location can be misleading, as many residents may be unaware of the status of water resources (limited knowledge of basin-wide condition) within a basin. This lack of awareness makes them susceptible to political influence, fostering mistrust in transboundary river basin management efforts. Looking ahead, the effective implementation of IRBM is crucial for mitigating conflicts over shared water resources and fostering resilience in regional ecosystems. Ultimately, such collaborative frameworks not only benefit the environment but also enhance livelihood opportunities, ensuring that future generations can thrive within a balanced and sustainable developmental paradigm.

CONCLUSION

In conclusion, the effective management of transboundary river basins through IRBM is crucial for promoting sustainable water resources and fostering cooperation among nations. Undoubtedly, the successful implementation of IRBM depends significantly on the engagement of diverse stakeholders, not merely state entities, but also local communities, industries, and individual policymakers. By integrating diverse stakeholder interests such as agriculture, industry, and local communities, IRBM creates a platform for consensus-building and resource sharing.

Also, this study underscores the critical role of geography in shaping stakeholder perceptions of river basin management, particularly within transboundary contexts. It demonstrates that individual and community perspectives are deeply influenced by geographical perspectives like spatial diversity, locational advantages and disadvantages, livelihood dependencies, and socio-political contexts, which can significantly complicate the implementation of IRBM. While inclusive stakeholder engagement is essential, the diversity of perceptions across micro-, meso-, and macro-geographies presents challenges to consensus-building and policy coherence for equitable water governance and joint basin management. Moreover, the disconnect between beneficiaries and the source of riverine ecosystem services highlights the need for greater awareness and communication. Hence, the study concludes that the wholesale integration of all stakeholder perspectives may risk implementing IRBM by overcomplicating transboundary negotiations. Instead, it advocates a more nuanced application of IRBM in the Tista basin, prioritizing resource-sharing, the maintenance of environmental flows, and recognition of the basin's geographical diversity as a means of advancing equitable and sustainable water governance. The findings affirm that IRBM must move beyond technical hydrological integration to incorporate subjective, place-based values and lived experiences. Strengthening collaborative frameworks, supported by data-driven governance and stakeholder capacity building, is vital to achieving resilient and sustainable management of transboundary water resources in the face of climate change and growing water insecurity.

REFERENCES

Agarwal, A., de los Angeles, M. S., Bhatia, R., Chéret, I., Davila-Poblete, S., Falkenmark, M., … & Wright, A. (2000). *Integrated Water Resources Management* (pp. 1–67). Stockholm: Global Water Partnership.

Anderson, E. P., Jackson, S., & Tharme, R. E. et al (2019). Understanding rivers and their social relations: A critical step to advance environmental water management. *WIREs Water, 6*, e1381. https://doi.org/10.1002/wat2.1381

Bandyopadhyay, J., & Ghosh, N. (2016). Hydro-political dynamics and environmental security in the Ganges-Brahmaputra-Meghna basin. *Social Sciences, 24*, 1–25.

Bhowmik, N. G. (1998) River basin management: An integrated approach. *Water International*, 23(2), 84–90. doi:10.1080/02508069808686746

Carr, G. (2015). Stakeholder and public participation in river basin management—An introduction. *Wiley Interdisciplinary Reviews: Water, 2*(4), 393–405.

Chakraborty, P., & Ghosal, S. (2024). An eco-social exploration of tourism area evolution in mountains through stakeholders' perspective. *Environmental Development*, 49(January), 100963. https://doi.org/10.1016/j.envdev.2024.100963

Cheng, P., Tang, H., Zhu, S., Jiang, P., Wang, J., Kong, X., & Liu, K. (2021). Distance to river basin affects residents' willingness to pay for ecosystem services: Evidence from the Xijiang river basin in China. *Ecological Indicators, 126*, 107691.

den Haan, R. J., Fliervoet, J. M., van der Voort, M. C., Cortes Arevalo, V. J., & Hulscher, S. J. M. H. (2018). Understanding actor perspectives regarding challenges for integrated river basin management. *International Journal of River Basin Management, 17*(2), 229–242. https://doi.org/10.1080/15715124.2018.1503186

de Souza, D. T., Hommes, L., Wals, A., Hoogesteger, J., Boelens, R., Duarte-Abadía, B., … & Joy, K. J. (2025). River co-learning arenas: Principles and practices for transdisciplinary knowledge co-creation and multi-scalar (inter) action. *Local Environment, 30*(1), 58–80.

Fielding, K. S., Terry, D. J., Masser, B. M., Bordia, P., & Hogg, M. A. (2005). Explaining landholders' decisions about riparian zone management: The role of behavioural, normative, and control beliefs. *Journal of Environmental Management, 77*(1), 12–21.

Fox, C. (2009). River basin development. *International Encyclopedia of Human Geography*, 414–422. doi:10.1016/b978-008044910-4.00879-8

Garau, E., Torralba, M., & Pueyo-Ros, J. (2021). What is a river basin? Assessing and understanding the sociocultural mental constructs of landscapes from different stakeholders across a river basin. *Landscape and Urban Planning, 214*, 104192.

Ghosh, K. (2014). Planform pattern of the lower Teesta River after the Gazaldoba Barrage. *Indian Journal of Geography & Environment, 13*, 127–137.

Ghosh, K. (2015). Sandy macroforms dynamics and one-dimensional facies characterization of the river Teesta. *Indian Journal of Geography & Environment, 14*, 1–15.

Ghosh, K. (2021). A framework to assess the benefits and challenges of ecosystem services of a reservoir-based wetland in the Himalayan foothills. *Environmental Development, 40*, 100669.

Ghosh, K. (2022). Influence of the anthropogenic and climatic factors on the changing streamflow, sediment and planform pattern of the Tista river in the foothill alluvial plain of eastern Himalaya. Doctoral research thesis. https://shodhganga.inflibnet.ac.in/handle/10603/383726

Ghosh, K., & Chakraborty, T. (2022). Impact of human intervention structures on the rivers: An investigation of the spatiotemporal variation of grain size in the Tista River, Eastern Himalayas. *Earth Surface Processes and Landforms, 47*(9), 2245–2265.

Ghosh, K., & Chakraborty, T. (2024). Study of lithofacies, radar facies and satellite images of the river bars: Implication for regulation structures on the Tista River of Eastern Himalaya. In *Surface Environments and Human Interactions: Reflections from Asia* (pp. 105–137). Singapore: Springer Nature Singapore.

Ghosh, K., Chakraborty, T., & Patel, P. P. (2024). A new braiding index to assess river regulation effects in multi-thread channels: Insights from a highly regulated Himalayan river. *River Research and Applications*, 1–20. https://doi.org/10.1002/rra.4261

Ghosh, K., & Munoz-Arriola, F. M. (2023). Hysteresis and streamflow-sediment relations across the pre-to-post dam construction continuum in a highly regulated transboundary himalayan River basin. *Journal of Hydrology*, *624*, 129885.

Ghosh, K., & Scott, C. A. (2025). Sustainable hydropower challenges for transboundary Himalayan River Tista: Strengthening future water-sharing between India and Bangladesh. *Energy for Sustainable Development*, *89*, 101862.

Giordano, M. A. (2002). *International River Basin Management: Global Principles and Basin Practice*. Oregon State University.

Giordano, M. A., & Wolf, A. T. (2003). Sharing waters: Post-Rio international water management. *Natural Resources Forum*, *27*, 163–171.

Haque, E. A. K., Zaman, M. A., Ghosh, K., & Khan, M. T. (2014). *Revisiting the Concept of Environmental Flows: A perception Study on Riparian Communities on Teesta River Banks in Bangladesh and India. Unpublished report funded by Bangladesh Lawyers Agency (BELA).*

Hein, T., Hauer, C., Schmid, M., Stöglehner, G., Stumpp, C., Ertl, T., … & Wang, C. (2021). The coupled socio-ecohydrological evolution of river systems: Towards an integrative perspective of river systems in the 21st century. *Science of The Total Environment*, *801*, 149619.

Hollick, M. (1990). Land conservation policies and farmers' decision- making. *Australian Journal of Soil and Water Conservation*, *3*(1), 6–13.

Hooper, B. (2005). *Integrated River Basin Governance*. IWA publishing.

Hooper, B. P. (2003). Integrated water resources management and river basin governance. *Journal of Contemporary Water Research and Education*, *126*(1), 3.

Hooper, B. P., & Lloyd, G. J. (2011). *Report on IWRM in Transboundary Basins*. UNEP-DHI Centre on Water and Environment.

Hüesker, F., & Moss, T. (2015). The politics of multi-scalar action in river basin management: Implementing the EU Water Framework Directive (WFD). *Land Use Policy*, *42*, 38–47.

Karambelkar, S., & Gerlak, A. K. (2020). Collaborative governance and stakeholder participation in the Colorado River Basin: An examination of patterns of inclusion and exclusion. *Natural Resources Journal*, *60*, 1.

Millington, P. (2006) Integrated river basin management: From concepts to good practice (English). Integrated river basin management briefing note, no. 15, Water Partnership Program (WPP). Washington, DC: World Bank Group. http://documents.worldbank.org/curated/en/155791468314991547/Integrated-river-basin-management-from-concepts-to-good-practice

Padło, T., Struś, P., & Gil, A. (2021). Danube as a symbol of Europe. Perception of the river from varied geographical perspectives. *PLOS One*, *16*(12), e0260848.

Pradhan, N. S., Das, P. J., Gupta, N., & Shrestha, A. B. (2021). Sustainable management options for healthy rivers in South Asia: The case of Brahmaputra. *Sustainability*, *13*(3), 1087.

Reed, M. S., Vella, S., Challies, E., De Vente, J., Frewer, L., Hohenwallner-Ries, D., … & Van Delden, H. (2018). A theory of participation: What makes stakeholder and public engagement in environmental management work? *Restoration Ecology*, *26*, S7–S17.

Richards, A., & Singh, N. (2002). Inter-state water disputes in India: Institutions and policies. *International Journal of Water Resources Development, 18*(4), 611–625.

Sadoff, C., Grey, D., & Borgomeo, E. (2020). Water security. *Oxford Research Encyclopedia of Environmental Science.* Available online: https://oxfordre.com/environmentalscience/view/10.1093/acrefore/9780199389414.001.0001/acrefore-9780199389414-e-609 (accessed on 16 December 2025).

Shahjahan, M., & Harvey, N. (2012). Integrated basin management for the Ganges: Challenges and opportunities. *International Journal of River Basin Management, 10*(1), 49–64.

Sinha, P., Rollason, E., Bracken, L. J., Wainwright, J., & Reaney, S. M. (2020). A new framework for integrated, holistic, and transparent evaluation of inter-basin water transfer schemes. *Science of The Total Environment, 721*, 137646.

Tiwary, R. (2006). Conflicts over international waters. *Economic and Political Weekly*, 41, 1684–1692.

Tobler, W. (2004). On the first law of geography: A reply. *Annals of the Association of American Geographers, 94*(2), 304–310. https://doi.org/10.1111/j.1467-8306.2004.09402009.x

Warner, J. F., Wester, P., & Hoogesteger, J. (2014). Struggling with scales: Revisiting the boundaries of river basin management. *Wiley Interdisciplinary Reviews: Water, 1*(5), 469–481.

Watson, N. (2004) Integrated river basin management: A case for collaboration. *International Journal of River Basin Management*, 2(4), 243–257. doi:10.1080/15715124.2004.9635235

Wolf, A. T. (2005). Transboundary water conflicts and cooperation. In D.S. Kenney (Ed.), In *Search of Sustainable Water Management: International Lessons for the American West and Beyond* (pp. 131–54), Massachusetts: Edward Elgar Publishing.

Wolf, A. T., Stahl, K., & Macomber, M. F. (2003). Conflict and cooperation within international river basins: The importance of institutional capacity. *Water Resources Update, 125*(1), 31–40.

Woodhouse, P., & Muller, M. (2017). Water governance—An historical perspective on current debates. *World Development, 92*, 225–241. https://doi.org/10.1016/j.worlddev.2016.11.014

6 Hydro-Diplomacy

A Critical Analysis of the Regional Cooperation in South Asia

Ammu S Anil

INTRODUCTION

According to *Bruce Lee*, 'Empty your mind. Be formless, shapeless like water. Now you put water into a cup; it becomes the cup. You put water into a bottle; it becomes the bottle. You put water in a teapot; it becomes the teapot. Now water can flow, or it can crash. Be water, my friend'. This chapter begins with the real stakeholders, i.e. the riverine communities living along the Transboundary Rivers of South Asia (TROSA) creating positive ripples towards building prosperous and resilient futures. The Ganges–Brahmaputra–Meghna (GBM) and Salween transboundary rivers originate in the Himalayas, the tallest mountains, flowing through extensive floodplains, lowlands, wetlands, deltas, estuaries, and bays. They connect nations across political boundaries and are at the heart of the region's rich tapestry of culture, economy, and traditions. For centuries, they have shared waters and river basins and have been a vital resource for the region, their ecosystem services. TROSA is supporting the region's riverine communities, upholding their water rights and building resilience with innovative solutions.

Geeta says about her experience:

> The Mahakali River in Nepal is her lifeline. She depends on the river to irrigate her small field, her only means of survival. But unregulated sand extraction, riverbank erosion, and water pollution make farming challenging. She dreams of starting a farm one day, but like many women in her village, she finds it difficult to break out of her gender role and voice her opinions. Today, as the secretary of the women's empowerment centre, she has seen the successful outcomes of self-organisation, a support network, better irrigation systems, and a purpose in life. Going beyond survival, Geeta is now an aspiring farmer and an emerging water leader in the Mahakali basin. She is also an active member of the community-to-community network of Nepal and India that meets to discuss river issues.

Sushila is from an indigenous community. She adds that:

> For generations, her community has lived along the banks of the Saralbanga river in the India–Bhutan border. They are entirely dependent on the waters brought by the "Jamfwi or Kuhls" (a network of channels) to irrigate agricultural lands and for

DOI: 10.1201/9781003599692-6

household use. Land-use changes and other climatic factors, such as erratic rainfall in upstream Bhutan, often result in insufficient water availability or flash floods during the monsoon for downstream communities. Facing a challenge to survive.

Sushila joined others in community-level initiatives supported by TROSA, which made it possible to revive the traditional 'Jamfwi' practice to irrigate the farms. It created a positive difference in the lives of the indigenous communities, especially women like Sushila, who need a long-lasting solution to the problem. Their purposeful engagement led to better water conversations on both sides of the border and created a favourable environment for dialogues between the members of civil society.

Another is Ambia Begum from Bangladesh, affected by riverbank erosion and seasonal bans on fishing and decreased water flow from upstream in the winter and floods in the monsoon. Ambia's livelihood was all but lost. Determined to overcome the odds, she stepped out of her traditional chores. She began participating in 'Nodi Boithoks' (river meetings) in her village and learnt about her rights to water. She now leads not just in raising awareness about water rights but also works with the local authorities on implementing compensation plans for fisherfolk communities affected by the bans. Her community also successfully worked with local authorities to revive the 'Bandal', an indigenous method of protecting riverbanks with bamboo fences to control erosion. These stories are from the ground and part of the documentation given by Oxfam Cambodia on TROSA (Oxfam in Cambodia, 2020).

A lot of people depend on rivers that cross borders, like the Nile or the Mekong. About 46 per cent of the global population lives in such basins, with more than 90 per cent living in transboundary states. So, people and states must share water that is often scarce due to economic prosperity and population growth, and the demand for water rises further. Now, climate change will impact water availability because there will be less water in some regions and at different times and that will create tensions by building more dams for clean energy, and these dams often become points of contention among neighbouring states. Thus, water becomes a potential source of conflict. But at the same time, water demands cooperation among riparian states, and cooperative planning in a basin offers huge potential benefits in terms of sustainable growth by maximising hydropower potential, for example, and by avoiding the costs of tensions and conflict.

While the national government as an important actor cannot be downplayed, water diplomacy will also involve a process that is potentially intricate or complicated, as many actors (including subnational governments, regional and international entities, and civil society organisations, i.e. non-government organisations (NGOs)) may pursue opportunities to be engaged in water diplomacy. This chapter answers the question, why have SAARC and BIMSTEC failed in ensuring regional water cooperation in South Asia? Why did the South Asian hydro diplomacy efforts remain largely state-centric? How does India's hydro-hegemony in the Indus and GBM basins influence the negotiation outcomes in the region? As you can see, there are many studies on the Track 1 level of diplomacy dealing with traditional security challenges but taking the critical perspectives to analyse the hydro diplomacy of South Asia is lacking. There is a lack of data regarding community-level engagement, and further research on the transboundary context, beyond the geopolitical perspective, is needed. Securitisation of borders will not work in geography or ecology where rivers themselves cut across the borders. Climate change adds more injury to the South Asian vulnerabilities, and thus it becomes so significant to address the water-induced challenges in South Asia. Hence, the topic is highly relevant; reading this chapter you get further insights (Figure 6.1).

The objectives of this study are to examine how, in South Asia, resource nationalism shapes hydro-political tendencies. The novel contribution includes the alternative thinking for the existing complex South Asian political arrangements and deconstructing the Track 1 level negotiation to a multitrack level with more inclusivity and representation of various communities. Viewing a river from a comprehensive perspective, rather than through a narrow focus on securitisation, sovereignty, national interests, and geopolitics, which represent the high politics domains in international relations. A non-traditional security paradigm should be applied while dealing with the transboundary rivers. Rivers are not only geographical or ecological entities but also possessing political significance through the diplomacy conducted by hydrocrats and diplomats. There is an urgent need to revitalise the BBIN grouping to combat the weaknesses of SAARC, BIMSTEC, and other frameworks.

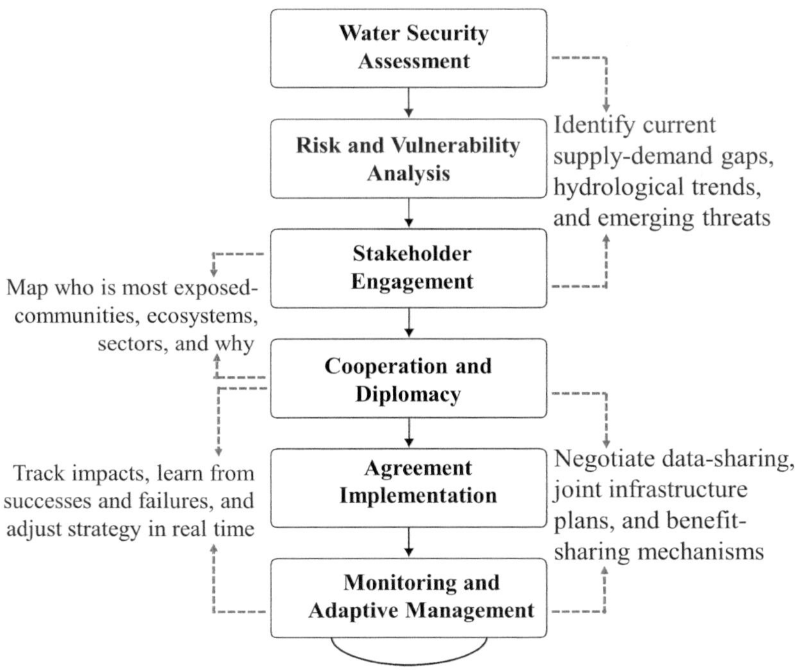

FIGURE 6.1 Hydro diplomacy in transboundary river basins of South Asia. (*Source:* Prepared by author.)

DESCRIPTION ABOUT STUDY REGION

Water does not stick to a particular geography, and it flows through geographies. It means states have only two options: either to sacrifice the national control over water or share, cooperate, or co-exist together. The conversation about hydro-diplomacy, particularly in South Asia, generally lacks this knowledge. Water supplies are seriously threatened by South Asia, which is home to 40 per cent of the world's population and has areas of persistent poverty, late industrialisation, exponential population expansion, and environmental deterioration due to ever rising climate change. India, Pakistan, Bhutan, Bangladesh, Nepal, and Afghanistan are six of the eight South Asian countries that rely on the Ganges, Brahmaputra, and Indus, three significant transboundary river systems (Figure 6.2). China plays the crucial role as the uppermost riparian country where the Indus and Brahmaputra originate from the Third Pole, i.e. Tibetan Plateau. India's geographical location makes it a 'Middle Riparian Power'. China, being an upper riparian state, has the undue advantage of controlling these water resources. The huge population of China causes ongoing water shortages in many parts of China. So, China keeps an eye on the water tanks of the South Asian region. It raises the concerns of lower riparian states even more. All these lead to increasing water scarcity and water insecurities in the region. The transboundary nature of the rivers is causing water disputes as well. It is because of the colonial history of South Asia, where the Partition happened based on religion, not by the geographical features like the flow of transboundary river systems. So, they

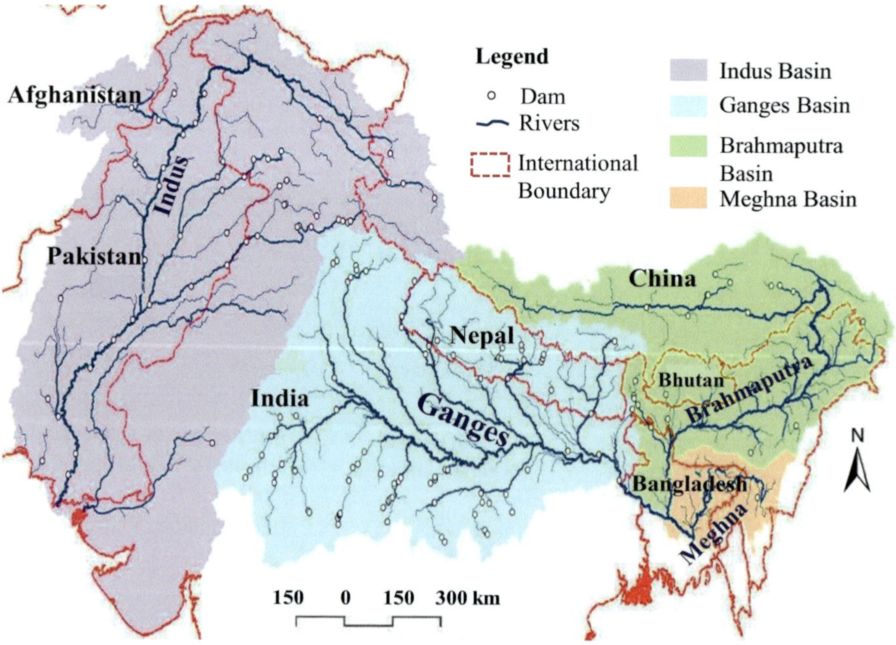

FIGURE 6.2 GBM, Indus Basin, and the proposed dams on the transboundary rivers. (*Source:* Taken from open-source platforms and modified by author.)

face sociopolitical tensions and lack trust and cooperation in dispute resolution. The water demands of over 1.3 billion people in this area are met by these rivers. In addition to providing food for millions of people, they are the region's main source of income.

Most of the economies in South Asia are based on agriculture. These rivers are mostly utilised for agriculture, and dams, canals, and barrages are built out of them to generate electricity. It serves as fuel for machinery. Urbanisation, agricultural growth, industry, and household demands are all contributing to an increase in the need for water. An estimate is that by 2050, India's total water consumption will be 1.4 trillion cubic metres. In South Asia, water insecurity is also a result of climate change. The glaciers in this area will melt in the next 30 years. Observations will occur not throughout the entire year, but rather in specific seasons; these represent the three primary reversals. The issue is further exacerbated by these nations' ineffective storage capabilities and poor management of their water supplies. China is the highest riparian nation, the source of the Indus and Brahmaputra/Jamuna rivers coming from the Tibetan Plateau. The study area is given in Figure 6.2 for further understanding of the relevant issues discussed in this chapter.

RESEARCH QUESTIONS AND MOTIVATION

The primary aim of the chapter is to assess the effectiveness of hydro-diplomacy efforts, critique existing regional institutions, or propose new models of cooperation. This chapter examines the role of hydro diplomacy in South Asia, focusing on how competition for water resources shapes regional politics and the impact of China's growing influence. In this chapter, I analyse the significance of hydro diplomacy and the scope for South Asian regional cooperation by taking various case studies in the region. The key research questions include: Why have SAARC and BIMSTEC failed in ensuring regional water cooperation in South Asia? Why did the South Asian hydro diplomacy efforts remain largely state-centric? How does India's hydro-hegemony in the Indus and GBM basins influence the negotiation outcomes in the region?

In this chapter, I focus on India–Bangladesh water relations along with India–Pakistan, India–Nepal, and Bhutan, as well as China. In the context of Bangladesh, a stipulation of the treaty mandated a revision every five years as necessary; however, this obligation was never fulfilled by the parties. A significant source of tension in Indo-Bangladesh ties is the contention around the allocation of the Teesta River water. Approximately 30 million individuals reside in the Teesta River watershed and depend on the river for their livelihoods. Approximately 21 million individuals in Bangladesh depend on the Teesta River for their livelihoods and essential needs. The problem originated in 1951, but discussions about the dispute commenced in July 1983. The Teesta water dispute in Bengal is not just a domestic issue, but it has transboundary implications in India–Bangladesh relations. As you can see, there are many studies on the Track-1 level of diplomacy dealing with traditional security challenges, but taking the critical perspectives to analyse the hydro diplomacy of South Asia is lacking. There is a paucity of data about community-level engagement, and more research on the transboundary setting, excluding the geopolitical perspective, are lacking. Securitisation of borders will not work in geography or ecology where rivers themselves cut across the borders. Climate change adds more injury to the South Asian vulnerabilities, and thus it becomes so significant to address the water-induced challenges in South Asia. The chapter proposes a multifaceted approach to water management, considering social, political, environmental, cultural, and psychological factors. By exploring new perspectives on river diplomacy, the chapter aims to contribute to finding solutions for sustainable water management in South Asia.

Another theoretical framework lies with hydro-hegemony, which means 'hegemony at river basin level, achieved through water resource control strategies such as resource capture, integration and containment'. Hydro-hegemony occurs in basins characterised by significant power imbalance, where management of water flows is centralised by the hydro-hegemon, hence suppressing competition over water resources. Power asymmetries in a river basin are dictated by the supremacy of one or many governments over the rest. The level of hydro-hegemony is based on 'structural power', 'ideological power', and 'bargaining power'. Menga (2016) came up with the concept of 'circle of hydro-hegemony', and it can be applied in the context of South Asia, where India-centric narratives have more voice than small states and other vulnerable states.

Water disputes in South Asia can be elucidated through Buddhist texts. It is located between the Sakyan and Koliyan tribes on the Rohini River. B. R. Ambedkar stated that Gautam Buddha was compelled to leave his home due to a quarrel over water (Joy and Paranjape, 2007). The largest freshwater resources in the world are found in the vast Himalayas of South Asia, also referred to as the Hindu Kush Himalayas. As many as 20 rivers have their source in the Himalayas, according to the World Bank. The Indus, Ganges, Brahmaputra, and Meghna are the region's lifelines, covering an area of 2,773,700 km^2. Even though the area has a lot of water resources, there has been occasional severe water scarcity. South Asia is one of the world's water-stressed regions, with Pakistan experiencing the most significant water scarcity, according to the Asian Development Bank. According to the Asian Development Bank, South Asia is among the most water-stressed regions globally, with Pakistan experiencing the most significant water scarcity in the area.

India is considered a 'Middle Riparian Power' due to its geographic position. China has an unfair edge in managing these water resources since it is an upper riparian state. There are persistent water shortages in many places in China due to the country's large population. China, therefore, monitors the water tanks across the region of South Asia. It adds to the Lower Riparian States' problems. These all contribute to the region's growing water insecurity and shortage. Water conflicts are also a result of the transboundary nature of the rivers. The reason is South Asia's colonial past, when religion was the primary factor in the Partition rather than physical characteristics like the course of transboundary river systems. To resolve disagreements, it is essential to address sociopolitical tensions, a deficit of trust, and the need for teamwork. Warner et al. contend that governments possess the authority to govern their area, and transboundary collaboration necessitates some concessions from the state. This may incite discord among the riparian states as they strive to achieve a fair accord over their autonomy. Power asymmetry results in an imbalance of authority among the riparian states, which is essential in determining who retains the majority of the sovereignty in the agreement. Nevertheless, contemporary literature and efforts concerning hydro diplomacy appear reluctant to embrace risks (Vij et al., 2020). Power imbalances are seen negatively and associated with problems only when they become evident. However, it must be recognised that power is always used, whether openly or indirectly, in each transboundary choice or non-determination.

Hazarika et al. (2023) cited Zeitoun and Allan (2008), who critiqued the water hegemons like India and China which had the means and authority to cultivate knowledge-building skills in the absence of the international enforcement of the rule of law. By creating the foundational data, formulating investment strategies, generating design concepts, and allocating gains equitably while consulting with other riparian states as needed, they may take the lead among the riparian states. Therefore, it is necessary to recognise power imbalance to comprehend how power influences how the players interact to (re)frame disputes and achieve collaboration. The detrimental effects of power imbalance between the riparian nations of a transboundary waterbody do not have to be the norm. For there to be reciprocal or shared advantages, all the riparian governments would need to sign onto this agreement. Thus, it opens the door for hegemonic riparian nations to be viewed as leaders responsible for justice and order rather than as oppressors or bullies. Biswas asserted that the asymmetric power dynamic between Bhutan and India is perceived as advantageous and symbiotic due to the significant economic benefits derived from their interactions (2011). The conversation about hydro diplomacy must be bolstered by emphasising the beneficial effects of power in reducing conflict, encouraging nations to act, and demonstrating its applicability at the negotiating table.

I also take the case study of Pakistan, which is a non-negotiable neighbour in South Asia, and according to the Asian Development Bank, Pakistan stands at the top 61 position in terms of water scarcity in the region. About 1.3 billion people here depend on these rivers for their water needs. They feed millions of people, the single largest economic resource of the region. South Asian countries are primarily agrarian economies. Mostly used for agricultural purposes, dams and canals are constructed out of these rivers for hydropower. It acts as fuel for industrial units. Water needs are increasing due to industrialisation, agricultural development, urbanisation, and domestic demands. According to an estimate, 'the total water demand in India will reach 1.4 trillion cubic metres by 2050' (Bansal, 2013). Climate change is also causing water insecurity in South Asia. In the coming three decades, all the glaciers of this region will melt. All three major reverses will be seen in a few seasons, not 12 months of the year. Mismanagement of water resources and inefficient storage capacities in these countries are also adding to the problem.

RESEARCH DONE SO FAR

The sharing of water resources across multiple sovereign nations can make matters even more complex, particularly based on the variability in capacity and willingness (for cooperation) in these governments. As a result, the agreements for cooperation that overlook the disparities among the riparian states may result in a detriment for one side, therefore benefiting another, i.e. a zero-sum game (Patricia Wouters et al., 2018). Water Diplomacy understands the limitations of current water-sharing negotiation processes and advocates for a non-zero-sum approach and outcome in which both parties each receive a 'net mutual benefit' (Islam and Susskind, 2012). The concept of water diplomacy is to build relationships to (re)establish trust in fragmented groups susceptible to conflict, and not to address material issues associated with conflict. Some of the potential benefits include averting more disputes and possibly better sustainability of change (Barua, 2018; Wehrenfennig, 2008). This type of specific diplomacy is mostly used across transboundary settings to avert conflict and improve cooperation more broadly (Huntjens and de Man, 2017). Sehring et al., (2022) quotes: "Water diplomacy refers to the deliberative political processes and practices of preventing, mitigating, and resolving disputes over transboundary water resources and developing joint water governance arrangements by applying foreign policy means, embedded in bi- and/or multilateral relations beyond the water sector and taking place at different tracks and levels."

Track-1 is confidential, and although those involved are not allowed to go against their states' interests, there is little opportunity for honest conversations and dialogue, which results in speculation and diminishes issues that are relevant to the weaker party. Track 2 diplomacy, on the other hand, attempts to engage with a larger number of actors, including significant academics, leaders of faiths, leaders from NGOs, and other civil society actors, in addition to government officials. The focus is more on rebuilding relationships between the parties involved than on negotiating specific details. Track 1.5 allows some level of communication between leaders when official contact between the two parties is undesirable or impractical because of some historical context or contradicting circumstances (Barua, 2018, pp. 6–7).

Barua (2018) added that multitrack diplomacy processes may be initiated by a government or third party, such as professionals, corporations, religious groups, media, individuals, education and training organisations, activists, and donors. In the context of transboundary river basins, the transition from formal to informal multitrack diplomacy is beginning to gain recognition, as informal multitrack diplomacy opens discussions and debates that can help frame more reasoned and better-informed positions. Moreover, deliberative processes can enhance equity and effectiveness in water governance and help narrow any power gap amongst stakeholders, leading to previously taken-for-granted discussions becoming open and informed (Dore, 2014). Conventional water diplomacy (intergovernmental) has also been deficient in South Asia. Bisht, M (2019) quotes: 'traditional South Asian water diplomacy, focused on state-level conflicts, misses the crucial role of "water communities" in borderlands, suggesting a "water plus" approach and "culture as a fluid interlocutor" to foster dialogue beyond politics, emphasizing local, shared understandings of water for more inclusive cooperation.'

In South Asia, the riparian states have developed numerous institutional mechanisms (treaties/agreements) that relate to developing transboundary water relations. However, there are no treaties or agreements that engage all riparian states for any shared river. There is no institutional framework for data management or dispute resolution (Bandyopadhyay, 2002; Barua, 2018). Even for the rare occasions where data-sharing agreements exist, transboundary river information is carefully collected and retained by national authorities. For instance, following the 2017 Doklam standoff, China did not release hydrological data about the Brahmaputra water flows to India for a year, despite the existence of a Memorandum of Understanding which mandated China to provide this data to India annually from May 15 to October 15. So, the need for a strong, robust regional framework on hydro diplomacy in South Asia is essential like the Mekong River Commission of Southeast Asia.

METHODOLOGY

The methodology taken is primarily descriptive, utilising qualitative data that relied on treaty analysis, case studies, expert interviews, or secondary literature review to gain insights into hydro-diplomacy from a conceptual perspective rather than an empirical one. The chapter presents a comprehensive strategy for water management, considering various social, political, environmental, cultural, and psychological elements. The exploration of new critical perspectives on river diplomacy in South Asia through the BBIN Framework seeks to contribute to the development of solutions for sustainable water management in the region. The 'Water–Energy–Food '(WEF) nexus and its cross-sectoral dependencies need to be analysed in the light of hydro-hegemony theories by deconstructing discourses, concepts, and the case studies. This chapter is an outcome of my ongoing research journey on hydro diplomacy, and I have chosen discourse analysis to deconstruct narratives around it by analysing the primary and secondary documents. I have taken the BBIN countries, consisting of Bangladesh, Bhutan, India, and Nepal, to validate the research questions of the failure of SAARC and BIMSTEC in South Asian hydro diplomacy. I also included Pakistan, as it is a critical factor in South Asian disintegration due to the multiple terrorist attacks and internal disturbances that make this significant country a troublemaker for regional cooperation to happen.

ANALYSIS AND RESULTS

India shares transboundary rivers with several neighbouring countries, each relationship shaped by geography, hydropower potential, bilateral treaties, and varying levels of conflict and cooperation. The results of this study highlight the complex interplay of hydropower development, water infrastructure, legal frameworks, and geopolitical considerations across India's shared river basins with Bangladesh, China, Nepal, Pakistan, and Bhutan.

Hydropower Potential across Shared Basins

This study indicates that hydropower remains a dominant driver of transboundary water relations in South Asia. Across all relationships, the baseline hydropower potential stands at 150,000 MW (Table 6.1). However, country-specific variations are stark. With China, the potential extends up to 694 GW, largely concentrated on the Yarlung Tsangpo (Brahmaputra) system, making this the most significant and politically sensitive hydropower frontier. With Nepal, hydropower potential is estimated

TABLE 6.1
BBIN Water Frameworks versus China, Pakistan

India–Bangladesh	India–China	India–Nepal	India–Pakistan	India–Bhutan
150,000–775 MW (Hydropower Potential)	150000 MW–694 GW	150,000–42,000 MW	150000 MW–60 GW	150,000–30,000 MW
Water infrastructure and immigration disputes	Conflictual	Disputes over trade, economic, and river treaties	Conflictual	Formal relationship guided by investment, aid, and security arrangements
Ganges Water Treaty (1996–2026)	Hydrological data sharing MoUs for disaster mitigation	Kosi (1954), Gandak Agreement (1959), Mahakali Treaty (1996)	Indus Water Treaty (1960) and in abeyance by 2025	No specific bilateral water treaty
Flooding, excess water intake, unequal water distribution, and exclusion of Nepal from the consultation process	Brahmaputra Dialogue	Tanakpur barrage, Sarada barrage, and Kosi barrage in Nepal and India to divert water for irrigation, flood control, and power	Biased Treaty with contested Indus water allocation rights and the abeyance of IWT by India is a violation of International Water Law	National Water Policy influences Bhutan's approach to water
Teesta is the major irritant due to the objections of West Bengal and Sikkim	China has no legal compliance	Pending Pancheswar Project, update the National Water Resources strategy, and the Electricity Act	Pahalgam terror attacks led India to use IWT as a terror weapon against Pakistan	Water management in Bhutan happens at the basin level like in India, where a basin is treated as the basic hydrological unit

Source: Prepared by author.

at 42,000 MW, reflecting the country's mountainous terrain and perennial river systems. With Pakistan, the potential is 150,000 MW to 60 GW, primarily in the Indus Basin, though this remains contested due to ongoing conflict. With Bhutan, the potential is around 150,000–30,000 MW, already guiding bilateral cooperation and investment. With Bangladesh, potential generation aligns with smaller-scale projects ranging from 150,000 (India) to 775 MW (Bangladesh), reflecting its downstream position and limited topographical advantages (Hussain et al., 2019). Hydropower development serves as both an opportunity and a point of tension, depending on whether states pursue joint investments or unilateral projects.

Water Infrastructure and Disputes

Water infrastructure projects often act as flashpoints for bilateral disputes. The findings of this study highlight how these vary across countries. India–China relations are described as fundamentally conflictual. With no binding legal agreements, China's upstream activities, including large-scale dam construction on the Brahmaputra, remain a major concern for India. India–Nepal relations reflect disputes over trade, economic ties, and river treaties. Infrastructures such as the Tanakpur barrage, Sarada barrage, and Kosi barrage were designed for irrigation, flood control, and power generation, but have often been criticised for excluding Nepal from equitable benefits and decision-making. India–Bhutan relations, in contrast, are more cooperative. Water management is treated at the basin level, aligning with India's own National Water Policy. The relationship is guided more by aid, investment, and security arrangements than disputes. India–Bangladesh relations face tensions over water infrastructure and immigration disputes, with the Teesta River remaining a central irritant due to opposition from West Bengal and Sikkim. Barrages and diversions exacerbate flooding and sedimentation issues in downstream Bangladesh. India–Pakistan relations remain conflictual, particularly over the Indus Basin. Terror attacks, such as the 2016 Pahalgam incident, further politicised water, with India threatening to use the Indus Waters Treaty as leverage (Table 6.1).

Bilateral Water Treaties and Agreements

Legal frameworks reveal the degree of institutionalisation across these relationships: with China, only MoUs for hydrological data sharing during flood seasons exist. These are limited to disaster mitigation and lack the binding force of a treaty, emphasising China's reluctance to adopt international water law. With Nepal, a series of treaties exist: Kosi Agreement (1954), Gandak Agreement (1959), and Mahakali Treaty (1996). However, these have often been criticised for bias in India's favour, especially given Nepal's limited bargaining power. The pending Pancheshwar Multipurpose Project remains a source of contention. With Bhutan, there is no formal water treaty, yet water cooperation is embedded within broader security and investment frameworks. Hydropower projects are developed jointly, with India financing, constructing, and importing electricity from Bhutan. With Bangladesh, the most notable framework is the Ganges Water Treaty (1996–2026), which has provided a baseline for water allocation but remains contested over implementation, particularly during lean seasons. With Pakistan, the Indus Waters Treaty (1960) stands as the most significant water-sharing agreement in South Asia, allocating the eastern rivers to India and the western rivers to Pakistan. However, the treaty's future is uncertain, with India declaring parts of it in abeyance by 2025, raising concerns about violations of international water law.

Major Issues and Challenges

The results indicate recurring patterns of disputes, though the nature of challenges varies. India–China issues revolve around China's unilateral development of Brahmaputra projects without legal compliance or transparency. India has attempted to initiate a Brahmaputra Dialogue, but progress remains limited. India–Nepal challenges stem from infrastructure biases, such as barrages that disproportionately benefit India. Nepalese scholars and policymakers highlight unequal treaties and the need to update strategies under the National Water Resources Strategy and the Electricity Act. India–Bhutan, while more cooperative, challenges lie in balancing hydropower development with ecological sustainability. Bhutan's basin-level management aligns with India's National Water Policy, but long-term impacts on river health remain a concern. India–Bangladesh concerns include unequal water distribution, excessive water intake upstream, and exclusion of Nepal from broader basin consultations. Flooding and the unsettled Teesta issue dominate the bilateral agenda. India–Pakistan treaty's contested allocations and India's threats to suspend cooperation in response to security incidents exemplify the securitisation of water. Pakistan views this as a violation of international law and a serious challenge to its water security.

Comparative Insights

A comparative analysis highlights several key trends, for example:

Conflict vs. Cooperation: Relations with Pakistan and China are primarily conflict-driven, while those with Bhutan and (to a lesser extent) Bangladesh are more cooperative, though not without tensions. Nepal sits in an intermediate position, reflecting both disputes and interdependence.

Legal Institutionalisation: Pakistan and Bangladesh have long-standing treaties, while Nepal has multiple agreements, though all face implementation challenges. In contrast, India–China lacks a legal treaty framework, relying instead on limited MoUs. India–Bhutan relies on informal but structured cooperation mechanisms.

Hydropower Politics: While Bhutan and Nepal view hydropower as a cornerstone of development (albeit often contested), China and India's hydropower rivalry poses strategic and security risks. Bangladesh, as a downstream country, benefits less from hydropower potential and faces higher vulnerabilities.

Colonial Legacies and Contemporary Politics: Many disputes can be traced back to colonial-era arrangements, such as the Indus Basin's partition in 1947 or the Kosi and Gandak agreements. Contemporary challenges, including climate change and demographic pressures, exacerbate these legacies.

The results reveal that India's transboundary water relations are marked by a delicate balance of conflict, cooperation, and asymmetry. Hydropower potential offers opportunities for development but also fuels contestation, particularly with China, Nepal, and Pakistan. Legal frameworks exist in varying degrees, from robust treaties like the Indus Waters Treaty to minimal MoUs with China. Yet, across all cases, the enduring challenge lies in integrating ecological sustainability with geopolitical imperatives. Though not without controversy, these experiences from South Asia and contextually relevant developing nations offer a chance to expand on current understanding to fortify and intensify the conversation on hydro diplomacy in the region. Processes of hydro diplomacy in South Asia must emphasise the interdependencies between riparian countries and push them to share more than just the resource; instead, they should be encouraged to share larger benefits from trade and commerce, cooperative projects, infrastructure, refugee issues, tourism, and security. With the use of hydro-diplomacy, parties can rationally agree to forgo short-term gains in favour of longer-term gains that may or may not be directly related to water. Haran V P (2021) explores challenges and opportunities for water/energy collaboration in BBIN countries, highlighting how hydropower can boost regional trade and stability, but requires political will, policy reforms like opening Indian grids, and managing environmental/social risks for success.

DISCUSSION

In this section, I answer the research questions by structuring them into three main arguments.

The Role of Regional Platforms Like SAARC, BIMSTEC, or the Joint Rivers Commissions

South Asia is the region of immense potential, rich culture, and complex geopolitics. But when it comes to cooperation among nations, two alliances, SAARC and BIMSTEC, have been at the centre of the conversation. One seems to be fading away and the other is gaining momentum. The focus of India is on BIMSTEC, which looks great on paper, but its execution is far from perfect. Critiques argue that it often feels like a symbolic get-together for leaders of member nations rather than a platform for real action. BIMSTEC has the potential, but it got a monumental insult by delaying many projects it promised, and still a long way to go in implementing those in the region. Here comes the wild card story of China and it expressed its interest in joining BIMSTEC by cooperating with its member countries. China sees the Bay of Bengal region as crucial for its Belt and Road Initiative. Omair Ahmad (2023) argued that the power vacuum created by SAARC's failure in South Asia will lead to a greater role by China through the Shanghai Cooperation Organization (SCO) in managing hydro diplomacy in South Asia. The last summit of SAARC happened in Kathmandu in November 2014, and 10 years have already passed. China has already started infrastructure projects in Nepal, Sri Lanka, Bangladesh to transform Teesta River, Maldives, Pakistan, and Afghanistan.

JRC worked better in the context of India–Bangladesh relations, and renewal talks are going to happen, but due to political instability as well as domestic politics, they got stalled. West Bengal CM Mamata Bannerjee in a letter to PM Modi stated:

> I came to understand that the Government of India is in the process of renewing the Indo Bangladesh Farakka Treaty (1996) which is to expire in 2026. It is a Treaty, which delineates the principles of sharing of water between Bangladesh and India, and as you are aware it has huge implications for the people of West Bengal for maintaining their livelihood and that the water which is diverted at the Farakka Barrage helps in

> maintaining the navigability of the Kolkata port. I would like to bring to your notice that river morphology has changed in the eastern part of India and Bangladesh over many years which has deprived West Bengal and negatively impacted the water availability in the state. There has been eastward migration of the Ganges (and Padma in Bangladesh) over the last 200 years disrupting their link with several rivers in West Bengal. For example, Jalangi and Mathabhanga rivers got disconnected from Padma and the freshwater inflow into the Sundarbans has reduced. In fact, the genesis of construction of Farakka Barrage Project is the disconnection of the Bhagirathi from the Ganges. (Atri Mitra, 2025).

This reflects the role of parallel diplomacy in Indian foreign policy while dealing with the transboundary rivers.

Effectiveness of These Institutions in Fostering Hydro-Diplomacy

Multinational donor organisations like the World Bank and the UN have attempted to apply various benefit-sharing frameworks aimed at including local communities and others in the management of a relatively transboundary waterway in South Asia (Hanasz, 2017). However, these attempts have largely failed or had limited impact – in general because they focus on short-term financing arrangements that are typically inconsistent with the long-term collaborative sustainability of transboundary river basins, plus the general disinterest by leaders of the countries involved in continuing or even advancing collaborative processes (Hanasz, 2017; Hazarika et al., 2023, p. 17; Salman and Uprety, 2018), argued that there is a prevailing mindset among South Asian states regarding transboundary relations as a zero-sum game; however, it is also important to recognise that the discussion of hydro diplomacy is significantly underdeveloped, and the parameters remain ill-defined for the countries of South Asia.

The Failures, Shortcomings, and the Reforms Needed

In the meantime, Nepal and India share three transboundary rivers: the Gandaki, the Mahakali, and the Kosi. India built the Sharada barrage, which was the first agreement between the two countries, signed in 1920 along the Mahakali River as part of a land swap between the two countries. Nepal has a huge potential for producing hydropower, but it hasn't been able to fully utilise its water resources because of a lack of resources, including money and technology. Nepal benefits from its upper riparian status with India, while New Delhi possesses the economic and scientific prowess to support Kathmandu's hydropower projects. Although the two countries have signed several accords, academics in Nepal believe that the prior treaties were unfair to Kathmandu and were one-sided. These bilateral agreements have consistently disregarded international water-sharing legislation and draughted clauses that are advantageous to India. The governments of India and Nepal struck a historic agreement to begin construction on the Kosi Project in Nepal following India's independence in 1954 (Jha, 2013). Nonetheless, Kathmandu views the 1954 Kosi Treaty as being biased. Conversely, New Delhi contends that the lack of will on the part of the Nepalese people causes implementation difficulties.

In addition, the treaty stipulated that it would be updated every five years as needed; however, the parties never once attempted to carry out this obligation. The disagreement over Teesta River water distribution is another significant source of friction in ties between Bangladesh and India. However, India did not sign the Teesta agreement in 2011 or 2019, as the conditions of the discussion were not agreed upon by all parties, and the topic of 'rivers' is included in the state list of topics under the Indian constitution. Bangladesh demands 50 per cent access to this river in their dry spell, whereas India claims 55 per cent of it. Bangladesh experiences floods and droughts because of the hydroelectric developments in West Bengal and Sikkim. Bangladesh has further grievances with India about the latter's river water linking project. To make use of its water resources and address the national water issue, India set out to connect all its rivers from north to south.

The power dynamics define South Asia's water relations, particularly upstream dominance and the strategic use of water resources. As a region, there is recognition of climate change; but, due to interconnected geography and ecology, the response to address it collectively among all countries is inadequate. Rivers could be one entry point in terms of regional vision for South Asia in terms of climate change and in terms of water resilience. The entire water cooperation in South Asia seems to be in silos, i.e. fragmented and fractured. We need to move ahead from much of the romanticism with its multi-stroke dispute resolution mechanism of IWT, and as a treaty, it is essential to have an instrument like this between the riparian states to talk, negotiate, mediate, etc. The treaty lacks climate-sensitive roadmap to tackle future problems, and there is a legal space in the treaty to accommodate future cooperation of India, Pakistan, and Afghanistan in securing their water futures.

The BBIN region is significantly dependant on the GBM basin, particularly with India-Nepal water ties, which are integral to its interconnected ecology. Around 6000 rivers trickled down from Nepal to India and that makes the mighty Ganges, which supports the most densely populated region in the world, and therefore the water future for India–Nepal has to be dealt with very carefully. We have 'barrage-centric agreements' in terms of the Sarada, Kosi, Gandak, and Mahakali rivers rather than a water-centric approach. The entire focus is on how to manage barrages, how to operate barrages; the intention is not questioned regarding the need for these instruments according to the time and needs of those times in the 1950s. It is imperative for countries to convene and assess the implementation, extent, and benefits derived from these actions. and what are the possible ways that these could be modified or changed or amended to take care of the future needs because the water scenario has completely changed in these rivers. So, if cases are filed in the Supreme Court of Nepal on pollution in the Ganges and in the Supreme Court of India on the pollution of the same river and in the Bangladesh Supreme Court on the same issue, that means there is something seriously wrong in the entire basin. Therefore, it is imperative for countries to contemplate the hydro-diplomacy between India and Nepal, which is profoundly biased, myopic, and narrowly focused. Water can be a soft power as well as a leverage in the South Asian context, and it is high time to bring all to the same table.

According to Adhikari (2014), 'India intended to divert water from water-surplus areas to water-deficit ones'. Nevertheless, this plan has drawn strong opposition from

Bangladesh and Nepal, among other nearby nations. 'To address its internal water-related problems, the plan calls for the construction of an interconnecting canal system supported by sufficient storage spaces to transfer water to meet the needs of drought-affected areas with each other'. Bangladesh, which is the lower riparian state, is particularly concerned that it may prevent enough water from flowing into their river. In the context of Bangladesh and India, there are 54 transboundary rivers, and the Ganges and Brahmaputra are the main sources of water conflicts. They have a proper institutional set-up like the Joint River Water Commission which was established in 1972. India and Bangladesh signed their first water agreement to establish the Indo-Bangladesh JRC, which aimed to maximise benefits, create and carry out flood management measures, and provide storm warnings. Aktar (2021) quoted: 'The two countries' water relations have been a work in progress since 1972, with both parties seeking to realise equitable water sharing'. The Ganga Water Treaty was signed by Bangladesh and India as a 30-year agreement to settle this problem in 1996. Bangladesh will not receive enough water from this deal when it expires in 2026. The recent political crisis in Bangladesh put the treaty in a standstill mode, leading to tension between India and Bangladesh. Additionally, the environment of these rivers will be adversely affected by this project. Conflicts have also arisen between Dhaka and New Delhi around the Tipaimukh embankment project, which is intended to be constructed near the Barak River in the Indian state of Manipur.

TABLE 6.2
(Bhatia, 2020) SAARC versus BIMSTEC versus JRC

SAARC	BIMSTEC	JRC
South Asian Association for Regional Cooperation (1985)	Bay of Bengal Initiative for Multi-Sectoral Technical and Economic Cooperation (1997)	Joint River Commission (1972)
Afghanistan, Bangladesh, Bhutan, India, Maldives, Nepal, Pakistan, and Sri Lanka (8 members)	Bangladesh, Bhutan, Nepal, India, Sri Lanka, Myanmar, and Thailand (7 members)	Bangladesh–India
Purely South Asian character	Connects South Asia with Southeast Asia	Flood Control and Irrigation
India–Pakistan relations major irritant due to cross-border terrorism, lack of political will, and mutual mistrust	Easier for India to navigate without Pakistan in South Asia and a bridge to Southeast Asia	Mostly bilateral
China has a limited role with only an observer status	China wants to expand its role through the Belt and Road Initiative in the Bay of Bengal region	2026 marks the renewal of the Ganges Water Treaty and the end of its 30-year term

Source: Prepared by author.

Misra et al. (2007) pointed out that due to a lack of understanding and agreement on this project, the two countries have repeatedly postponed construction. There are success stories of the Ganges Water Treaty of 1996 and the Indus Waters Treaty of 1960, which showed resilience during the major wars. At the same time, it failed in recent years due to domestic politics, geopolitics, national security concerns, and so on. Here rises the primacy of politics over water and makes South Asia a paradox in water diplomacy and cooperation. Since India has a significant power shortage, 'the close and friendly ties between Bhutan and India have provided the necessary political will and the market for Bhutan's power' (Tshering and Tamang, 2004).

CONCLUSION

In conclusion, the broad physical terrain, varied cultural origins, and colonial history of South Asia have made governing transboundary river water resources extremely difficult. The major aim of SAARC was to encourage member nations' collaboration and socio-economic development. Cooperation faces both challenges and possibilities because of the SAARC meeting's lack of debate on bilateral conflicts. The intricate nature of water-sharing agreements is brought to light by the historical evolution of hydro diplomacy in South Asia, namely in the Ganga-Brahmaputra-Meghna River Basins and Indus Basins. Certain components of these accords need to be upgraded because there have been discussions about their efficacy and because they must comply with international regulations regulating the sharing of transboundary river water. Similar conflicts have been caused by matters like plans for water diversion and hydropower projects, highlighting the necessity of cooperative and equitable use among riparian governments.

According to several analysts, India dominates the South Asian hydro-hegemony. However, some analysts commend India for its contribution to resolving the water problems in the area. Respect for international law and fair use standards is essential for managing transboundary river water resources to promote integrated regional cooperation. To guarantee prosperity, peace, and sustainable development in the South Asian region, it is essential that a comprehensive hydro diplomatic framework covering the whole basin be developed and that any domestic political developments that could obstruct progress be addressed. To fully realise the potential of their shared water resources, riparian states must work together in cooperative efforts that are supported by communication and the development of trust. Therefore, Salman and Uprety (2018) argue that negotiating new transboundary water treaties amongst South Asia's riparian countries has become a daunting task, and enforcing existing ones remains a real challenge.

The three pillars to build the house of water are knowledge for innovation, financing water, and governance, which must guarantee equitable sharing and total transparency. Water governance must be effectively distributed across the state level, the basin level, and local entities such as municipalities or panchayats. Asia Society (2024) added that there is trust as well as mistrust for community leaders to work together, and there are several initiatives. The civil society engagement is also quite limited in this region, and the Track 1 diplomacy is also quite stuck. It is peculiar to note that South Asia is one of the least connected regions, and the level of institutionalisation of cooperation, be it treaties, MoUs, or RBOs, is the lowest compared to the Middle East or Africa. The treaties are mostly bilateral, even in multilateral basins, mainly focused on infrastructural development, such as India investing in another country, such as Nepal or Bhutan, in infrastructure, which provides very narrow economic benefits to India or, ideally, both countries. The policy recommendations include establishing a South Asian Water Cooperation Council, creating a transboundary early warning system, and fostering Track 2 diplomacy initiatives among water experts.

Multilateral initiatives are extremely rare due to the lack of understanding of the benefits of such cooperation at the basin level and joint basin management in comparison to the perceived possible benefits of unilateral action. So, it seems that those who take decisions are unfortunately often not the communities with their needs but the top-level leadership in a country, which flows from a security and foreign policy perspective rather than a water development community perspective. For them, the benefits will be unilateral, securitised, very politicised, and often short-term nationalist actions, which seem to outweigh the benefits that more long-term water cooperation can bring. This highlights a missed opportunity in water resource management, emphasizing the benefits of enhanced management, efficient infrastructure development, and data sharing for community safety. The neglect of these opportunities harms communities and economies and leads to new security challenges like migration and instability. This can be a vicious cycle in South Asia, and securitisation prevents cooperation on water resources, which in turn produces no benefits, and the vicious cycle continues. It reinforces unilateral action. This needs to be broken to fight against other water problems emerging from climate change. Asia Society (2024) emphasised the role of the 'Women in Water Diplomacy Network' in South Asian water governance which ensures gender equity. Ecological and social value of the water also should be part of the public discourse, not just the economic water value. From dams, we need to move to aquatic water reservoirs to protect ecosystems first. Therefore, water is politics and water needs democracy'.

This chapter explicitly addresses the key issues of sustainability in the context of South Asia's transboundary rivers of BBIN, including China and Pakistan, to get a holistic picture of how hydro-diplomacy operates over here. Water cooperation in a multitrack, multi-scalar, multi-stakeholder level sustains peace and harmony in South Asia amidst the chaos, uncertainty, and conflictual times.

REFERENCES

Adhikari, K. N. (2014). Nepal-India water relations: Time for change in approach. *Pakistan Horizon*, *67*(3/4), 113–132.

Aktar, F. (2021). Water diplomacy and the water sharing problem between Bangladesh and India: A quest for solutions. *International Journal of Research and Scientific Innovation (IJRSI)*, *8*, 223.

Asia Society (2024). 'HD, Policy and the Future of Himalayan Waters Amidst Climate Change', 31 October 2024. https://youtu.be/QUOKqEbfwD0?si=fKY8MeJliq2ESdQc, accessed on 30 November 2024.

Bandyopadhyay, J. (2002). A critical look at the report of the world commission on dams in the context of the debate on large dams on the Himalayan rivers. *International Journal of Water Resources Development*, *18*(1), 127–145. https://doi.org/10.1080/07900620220121701

Bansal, A. (2013). *South Asian Security – 21st Century Discourses*. Routledge.

Barua, A. (2018). Water diplomacy as an approach to regional cooperation in South Asia: A case from the Brahmaputra basin. *Journal of Hydrology*, *567*, 60–70. https://doi.org/10.1016/j.jhydrol.2018.09.056

Bisht, M. (2019). From the edges of borders: Reflections on water diplomacy in South Asia. *Water Policy*, *21*(6), 1123–1138.

Biswas, A. K. (2011). Cooperation or conflict in transboundary water management: Case study of South Asia. *Hydrological Sciences Journal*, *56*(4), 662–670.

Dore, J. (2014). An agenda for deliberative water governance arenas in the Mekong. *Water Policy*, *16*(S2), 194–214.

Hanasz, P. (2017). *Transboundary Water Governance and International Actors in South Asia: The Ganges-Brahmaputra-Meghna Basin*. Routledge.

Haran, V. P. (2021). Water and hydropower cooperation in BBIN countries: Policies and way forward. *International Journal of Water Resources Development*. *37*(3), 424–438. https://doi.org/10.1080/07900627.2018.1503076

Hazarika, N., Barua, A., Vij, S., Deka, A., & Salame, L. (2023). Collective deliberation or just the state (in)action: How do we change the hydrodiplomacy landscape in South Asia? *Water Policy*, *25*(1), 15–22.

Huntjens, P., and de Man, R. (2017). Water diplomacy: Making water cooperation work. The Hague Institute Policy Brief, 10, 1–12. Accessed from https://www.planetarysecurity-initiative.org/sites/default/files/2017-04/PB_Water_Diplomacy_WG_4.pdf.

Hussain, A., Sarangi, G. K., Pandit, A., Ishaq, S., Mamnun, N., Ahmad, B., & Jamil, M. K. (2019). Hydropower development in the Hindu Kush Himalayan region: Issues, policies and opportunities. *Renewable and Sustainable Energy Reviews*, *107*, 446–461. https://doi.org/10.1016/j.rser.2019.03.010

Islam, S., & Susskind, L. E. (2012). *Water Diplomacy: A Negotiated Approach to Managing Complex Water Networks*. New York: Routledge.

Jha, H. B. (2013). Nepal–India cooperation in river water management. *Strategic Analysis*, *37*(2), 217–230.

Joy, K. J., & Paranjape, S. (2007). Understanding water conflicts in South Asia. *Contemporary Perspectives*, *1*(2), 29–57.

Menga, F. (2016). Reconceptualising hegemony: The circle of hydro-hegemony. *Water Policy*, *18*(2), 401–418. https://doi.org/10.2166/wp.2015.063, accessed on 31st October 2024.

Misra, A. K., Saxena, A., Yaduvanshi, M., Mishra, A., Bhadauriya, Y., & Thakur, A. (2007). Proposed river-linking project of India: a boon or bane to nature. *Environmental Geology*, *51*(8), 1361–1376.

Salman, S. M., & Uprety, K. (2018). Shared watercourses and water security in South Asia: Challenges of negotiating and enforcing treaties. *Brill Research Perspectives in International Water Law*, *3*(3), 1–100.

Sehring, J., Schmeier, S., Horst, R., Offutt, A., & Sharipova, B. (2022). 'Diving into water diplomacy exploring the emergence of a concept'. *Diplomatica*. *4*(2), 200–221.

Tshering, S., & Tamang, B. (2004, October). Hydropower – Key to the Sustainable Socio-economic Development of Bhutan. In *United Nations Symposium on Hydropower and Sustainable Development* (pp. 27–29).

Vij, S., Warner, J. F., Biesbroek, R., & Groot, A. (2020). Non-decisions are also decisions: Power interplay between Bangladesh and India over the Brahmaputra River. *Water International*, *45*(4), 254–274.

Wehrenfennig, D. (2008). Multi-track diplomacy and human security. *Human Security Journal*, *7*(Summer), 80–89.

Wouters, P., Chen, H., & Nickum, J. E. (2018). *Transboundary Water Cooperation*. Routledge.

Zeitoun, M., & Allan, J. A. (2008). Applying hegemony and power theory to transboundary water analysis. *Water Policy*, *10*(S2), 3–12.

Web Sources

Ahmad, O. (2023), 'SCO Acquires Greater Say in South Asian Water Diplomacy as SAARC Withers'. *The Wire*, 28 April. https://thewire.in/diplomacy/sco-south-asia-water-diplomacy-saarc-wither, accessed on 16 May 2025.

Bhatia, R. (2020) 'BIMSTEC or SAARC', *Gateway House*, 27 February, https://www.gatewayhouse.in/bimstec-or-saarc/, accessed on 15 May 2025.

Mitra, A. (2025), 'Meeting not for any conclusion': Joint River Commissions of India, Bangladesh meet in Kolkata to discuss Ganga Water Treaty. *The Indian Express*, 7 March. https://indianexpress.com/article/cities/kolkata/indo-bangla-meet-on-ganga-water-treaty-begins-in-kolkata-ahead-of-renewal-talks-9873038/, accessed on 6 May 2025.

Oxfam in Cambodia (2020), Transboundary Rivers of South Asia (TROSA) – Positive Ripples, August 12, 2020. https://youtu.be/6T1ijZd6jW4?si=u_YYwomo8S1gsyCV, accessed on 20 May 2025.

Map Sources

Dams: https://www.globaldamwatch.org/

River lines: https://www.hydrosheds.org/

Basin: https://www.hydrosheds.org/

India Administrative boundary: https://onlinemaps.surveyofindia.gov.in/Digital_Product_Show.aspx

Other countries' administrative boundary: https://data.jrc.ec.europa.eu/dataset/jrc-10112-10004

7 Transboundary Rivers of South Asia

Interplay between Dispute and Diplomatic Deadlock under Climate Change Conditions

Sufia Khanom

INTRODUCTION

The term 'Hydro-diplomacy or Water diplomacy' attracts the utmost attention at present in South Asia. This emerging concept and practical approaches started to connect with climate security, environmental security, conflict, and peace building in the 1990s (Farnum, 2018). The diplomatic initiatives involve the process to facilitate communication between two sovereign states to prevent water conflicts and enhance mutual trust. Although theoretically diplomatic process should include all the relevant levels/stakeholders related to water resources management, in practice only the state-centric tools are utilised (Keskinen et al., 2021; Van Genderen & Rood, 2011). Hydro-diplomacy can be defined from process oriented and goal-oriented perspectives as follows: *Water diplomacy refers to the deliberative political processes and practices of preventing, mitigating, and resolving disputes over transboundary water resources and developing joint water governance arrangements by applying foreign policy means, embedded in bi- and/or multilateral relations beyond the water sector and taking place at different tracks and levels* (Sehring et al., 2022, p. 212).

The tension over transboundary water sharing and the urgency of clean hydro energy are the centre of attention for the policymakers, academicians, and activists of this region. Therefore, stakeholders are more aware of the use and challenges of transboundary resources in the changing climatic conditions that are usually embedded in the intrinsic political decisions. About 40 percent of the world population live in South Asian countries, suffering from chronic poverty, slow industrialisation, huge population growth, and the impacts of environmental changes in their everyday life. The six South Asian countries among eight, Afghanistan, Bangladesh, Bhutan, India, Nepal, and Pakistan, share three major transboundary river systems such as the Ganges, the Brahmaputra, and the Indus. These three transboundary rivers originate from Tibet and connect China as an upper riparian. Therefore, the two major

DOI: 10.1201/9781003599692-7

powers, India and China, along with the other small states, are the major stakeholders in the state-level diplomatic process, and in South Asia, it recognises the power asymmetry in the realm of hydro-diplomacy.

The transboundary water management in South Asia is narrated by nationalistic, securitised, and narrow technocratic frameworks. The decreasing trends of freshwater availability and the increasing diverse demands for various countries making the competition over existing transboundary water sharing arrangements (Nepal & Shrestha, 2015). Among the South Asian countries, Pakistan and India are facing challenges for water scarcity; inadequate water management system is the main hurdle for Bhutan and Nepal; Bangladesh is struggling with managing too many and too little water during monsoon and dry season, respectively, as the surface water availability completely depends on the upstream countries (Khalid et al., 2014; Shan et al., 2020). In this context, state governments' plan to construct hydropower projects for their own water and energy security without any collaborative approach in terms of designing, financing, or any sort of benefit sharing with the riparian countries further enhances the existing mistrust among the countries (Salman & Uprety, 2018; Vij et al., 2020a). Additionally, not all countries are signatories to the international water laws, and due to the flexibility of the guiding principles, which might work as the backbone of the successful agreements or treaties, are significant characteristics of the hydro-diplomacy of South Asia. Then again, the media plays a vital role in shaping the narratives, building trust between policymakers and scientists and strengthening people-to-people contact. Finally, the present trend of hydro-diplomacy of the South Asian countries is significantly influenced by the impacts of climate change.

This region consists of wide geographic and climatic variations, ranging from mountain ranges to deserts and vast distributions of plain agricultural land. According to a report published on Sustainable Development, the top-ten climate vulnerable countries in the world are mostly situated in South Asia. The vulnerability accelerates due to poverty, social inequality, and malnutrition. Frequent disasters and stress over freshwater availability aggravate the risks of food and livelihood insecurity in the agriculture-based economy like South Asia. So far, countries in South Asia are working on reducing the impacts of climate change and disaster risk reduction and mitigation in isolation rather than the other regions like the African Union or Small Island Countries. Thus, climate-induced risks and vulnerabilities pose further uncertainty for a win-win transboundary water resources management of this region. In this backdrop, the main objective of the chapter is to identify the major contested or disputed issues in terms of transboundary water sharing and explore the factors for diplomatic deadlocks of successful transboundary water management in South Asia. The chapter consists of four sections, including the introduction and conclusion sections.

DATA AND METHODOLOGY

This study adopts a qualitative research approach to examine the dynamics of transboundary rivers in South Asia, with particular attention to the interplay between disputes and diplomatic deadlocks under climate change conditions. The choice of a qualitative framework is grounded in the need to understand the complexities of hydro-politics, which are often shaped by historical legacies, institutional arrangements, and shifting sociopolitical contexts that cannot be adequately captured through quantitative data alone. The research is primarily based on secondary sources. Data were collected from a wide range of academic journals, books, policy documents, official treaties, institutional reports, and credible online repositories, including government and non-governmental organisation websites. Particular emphasis was placed on literature that addresses international water law, hydro-diplomacy, climate change impacts on river basins, and the political economy of transboundary water governance in South Asia.

Following data collection, materials were systematically reviewed and organised according to the key objectives of the chapter: (i) to identify major disputes over transboundary rivers in South Asia, (ii) to examine the reasons behind recurring diplomatic deadlocks, and (iii) to assess how climate change exacerbates these challenges. Thematic categorisation was used to cluster information under broad analytical headings, such as legal mechanisms, institutional arrangements, hydrological variability, and regional cooperation frameworks. This facilitated comparative insights across different river basins, including the Ganges, Brahmaputra, Indus, and Teesta systems. The analysis employs a critical interpretative method, drawing connections between historical treaties, contemporary conflicts, and projected climate scenarios. By synthesising perspectives from diverse sources, the study aims to highlight the structural barriers to cooperation and propose directions for future research and policy.

HYDRO-DIPLOMACY IN SOUTH ASIA: DRAWING NARRATIVES FROM THE WATER COOPERATION

Hydro-diplomacy is defined as discussed earlier, an integrative and innovative approach where management strategies, environmental policies, and technical solutions are used as tools for the diplomats to resolve the conflicting issues over transboundary water resources management. According to Harold Nicholson (1939), diplomacy is 'the art of conducting dialogue between and among states'. Diplomacy is action rather than theory. Therefore, the water diplomacy framework has been defined as value-focused approach and techniques for managing water issues. These tools and techniques should be assisted by cost-benefit or benefit-sharing analysis and context analysis for taking decisions. Even hydro-diplomacy helps to resolve the conflicts within states in terms of water shared use, allocation, and availability (Islam & Repella, 2015; Honkonen & Lipponen, 2018). So, hydro-diplomacy is termed as 'relational practice' in terms of shared benefits through sustainable water management and to establish balance among social, economic, and ecological aspects related to water. This broader meaning of hydro-diplomacy on shared resources shapes greater geopolitical narratives in South Asia (Figure 7.1). This geopolitical narrative is very important for South Asia as most of the shared water governance is bilateral in nature without any consideration of a regional or basin-wide approach. According to Bisht (2023a), 'the penchant for bilateralism over multilateralism has made the transboundary rivers vulnerable to emerging geopolitical discourses, often a non-starter for effective water

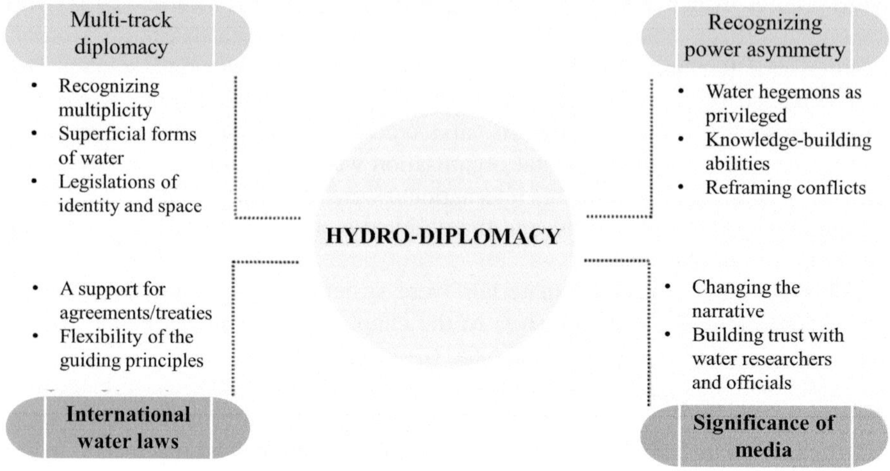

FIGURE 7.1 Hydro-diplomacy on shared river resources shaping broader geopolitical narratives in South Asia. (*Source:* Hazarika et al., 2023)

diplomacy'. To overcome the challenges of geopolitical shadows, it is necessary to analyse the various discourses for shaping the narratives around hydro-diplomacy for shared resources.

The old discourses were heavily influenced by the Cold War politics, mistrust among the countries, lack of regional economic integrity, and domestic political instability. These were burdened with various narratives: (i) competition over regional supremacy between India and Pakistan; (ii) criticism over a land-centric regional approach by Sri Lanka and the Maldives; (iii) unequal benefit-sharing agreements by Bhutan and Nepal (Bisht, 2023b). Thus, hydro-diplomacy in South Asia was considered as a piecemeal in foreign policy in regards to transboundary water sharing.

After the Cold War, economic liberalisation discourse gained momentum, which was reflected through the Ninth SAARC Summit in Male in 1997. New approaches (for example, South Asia Growth Quadrangle (SAGQ), the South Asian Subregional Economic Cooperation (SASEC), the Mekong Ganga Cooperation, the Sub-regional Cooperation between Bangladesh, Bhutan, India, and Nepal (BBIN)) focus on new regionalism approaches extended up to South East Asia, emphasising the regional connectivity and economic liberalisation. Water–Energy–Food–Ecology and Integrated Water Resource Management (IWRM) are two broader aspects of contemporary hydro-diplomacy focusing on hydroelectricity and connectivity. Global goal on sustainable energy transition and Indo-Pacific discourse initiated by the United States identified the South Asian countries as ocean-centric connectivity and economic hubs. Therefore, connectivity and hydropower energy are prominent hydro-diplomacy discourse in South Asia at present. The following subsections discuss the existing cooperation and agreements among the South Asian countries.

Water Cooperation between India and Bangladesh

Bangladesh and India share a total of 54 transboundary rivers. Still, the water sharing discussion revolves around the Ganges (renegotiation or continuation), Teesta, and Barak rivers. The discussion is going on with Manu, Muhuri, Feni, Dharla, Dudhkumar, and Khowai rivers for cooperation. The Ganges water treaty of 1996 has been considered as the only successful agreement. The Farakka Barrage over the Ganges River created multiple narratives among different states in India and Bangladesh. Bangladesh stressed on to maintain minimum environmental flow for the river and to follow the regional approach by adding other riparian countries like Bhutan and Nepal in this water sharing treaty. The Farakka Barrage was built to ensure the navigability of the Kolkata port. This created huge sediment deposition and backwater flow resulted in flood almost every year in Bihar. Then again, the hydraulic structure like a barrage has always had some contested issues around flood management and sediment blockage in the downstream although some discussions were held in August 2022 by the Technical Committee of the Joint River Commission (JRC) on sustainable Ganges water sharing. Both countries have agreed to conduct joint studies, river flow data sharing, transboundary pollution management, technical cooperation for river bank erosion, and protect the Hilsha habitat near the coastlines of the Bay of Bengal by ensuring freshwater flow in the Sundarbans delta.

Water Cooperation between India and Bhutan

Bhutan and India have successful projects on hydropower generation agreements on the Wangchu, Sankosh, and Trosa with the assistance of the World Bank. They have produced about 20,000 MW of hydroelectricity up to 2007. Most of the produced electricity has been exported to India. That was the progressive era of the India–Bhutan relationship. At the second phase of the water cooperation, the MoU on the Kurichu and Tala projects was signed in 2007 with an ambition of generating 10,000 MW of hydroelectricity by 2020 (Bagale & Adhikari, 2020). In 2022, around hydel projects were stalled due to heavy sediment loads and long-term pay-off uncertainty to Bhutan that opened up more opportunities like solar and wind energy for the country. Natural disaster like glacier outburst is a common phenomenon in the Northern Bhutan. This glacial outburst cause floods with heavy sediment flow. One of the studies estimated that Bhutan has '677 glaciers and 2.674 glacial lakes in which 25 glacial lakes' are potentially vulnerable (Choden, 2009). Bhutan also has a water shortage during the lean season. So, the hydropower production reduces by around one-sixth of its total production capacity. Again, Bhutan is a signatory to the Paris Climate Agreement and is committed to the deployment of solar energy. Although India proposed that Bhutan join the cross-border solar connectivity through the 'One Sun One World One Grid' project for transboundary energy sharing, Bhutan has signed an agreement with Japan to install a solar power plant for producing around 180 kW of electricity.

Water Cooperation between India and Nepal

India and Nepal have multiple water cooperation agreements on the Koshi, Mahakhali, and Gandak transboundary rivers. The framework for cooperation on transboundary water management between India and Nepal is different from the other countries in this region. It is primarily focused on getting optimal benefits from cooperation through energy and food security, flood control, and navigation. Nepal perceived the water cooperation with India as 'iniquitous-non-reciprocal in nature' (Bisht, 2023b, p. 16). For example, cooperation over the Koshi and Gandak rivers was made for power generation, irrigation needs, flood management, and inland water trafficking. However, there are narratives on unfair and delayed compensation packages for Nepal and inequitable water sharing for the irrigation needs. The locations of the dams and the mismanagement of the dams are also a matter of frustration to Nepal. Another disputed water agreement between Nepal and India is the Mahakali Treaty of 1996. The treaty has not been translated into reality and has mass population objections from Nepal considering the indirect socio-economic loss. This treaty is also termed as the most controversial agreement, and the Kalapani border dispute added a new dimension of controversy to this agreement. Besides these, improper groundwater and spring-shed governance as a result of improper land use and water management policies influence ecological disturbance in the Ganges Basin. Additionally, the shareholding rights of community people in terms of benefit sharing are another internal challenge for Nepal.

Water Cooperation between India and Pakistan

In 1960, India and Pakistan negotiated over Indus water sharing under the third-party arrangement of the World Bank. The Indus water treaty is regarded as conflict management through thin mediation rather than conflict resolution or conflict transformation. It is commonly referred to as the 'river distribution treaty' between Pakistan and India rather than a water sharing treaty. Three rivers of both Eastern and Western sides have been distributed between India and Pakistan, respectively, through this agreement. Conflicts specially on border issues between Pakistan and India do not necessarily affect the functional aspects of the treaty. In recent years, political conflict over this agreement has shifted from water sharing to social and environmental challenges. According to the 12th Standing Committee Report of the Indian Government, India proposes to renegotiate the Indus Water Treaty due to the emerging issues such as climate stressed conditions with reference to Article 7 for initiating the discussion (Bisht, 2023b). In terms of water allocation and sharing, Pakistan is dependent on a third-party mediator due to the mutual mistrust. On the contrary, India considers the engagement of a third-party mediator for sharing the Ravi River as a delaying technique of Pakistan for renegotiation. However, both countries agree on the deteriorating water quality and quantity due to the upstream pollution; disconnection among water policies; legal structures; community needs; and unfair distribution of development initiatives.

From the above discussions, it has been shown that countries in this region have agreements and cooperation on water for various purposes. Although there are macro and micro narratives on the water cooperation, most of the cooperation/agreements are beneficial for both parties. Over the years, new discourses have emerged to define regionalism and grasp centre of attention for cooperation. In recent days, climate change impacts, global warming, environmental degradation, ocean-centric cooperation have added new dimensions as well as challenges for cooperation among the South Asian Countries. The above-mentioned challenges, along with the geopolitics, power asymmetry, and geographic location, work as diplomatic deadlocks for any win-win cooperation on water sharing.

DIPLOMATIC DEADLOCKS IN TRANSBOUNDARY WATER SHARING: SOUTH ASIAN CONTEXT

The diplomacy scholarship on water cooperation or hydro-diplomacy has been animated in so many ways over the period of time. The existing literature has been grouped into three broad categories: strategic, normative, and critical (Bisht, 2019). To minimise the conflict and influence the track-1 diplomatic behaviour while negotiating on water sharing, different approaches of strategic diplomacy need to be explored. This approach is based on international Water Laws. Practitioners and scholars are divided on whether water is an exploratory or an independent variable for negotiation. Unfortunately, this scholarship offers policy focused tools and is applicable for diplomats to resolve the conflicts (Butts, 1997; Gleditsch et al., 2006; Homer-Dixon, 1994; Wolf, 2007; Zeitoun et al., 2019). The normative diplomacy is based on soft laws and received attention from the Dublin Conference in 1992. It emphasises public–private partnership and initiated the debate over 'water as an economic good' vs 'fundamental human right' (Cullet, 2013, p. 57) vs 'water service needs to be perceived as public service' (Simoes, 2017, p. 273). This approach also recognised the importance of global water governance for achieving the Millennium Development Goals and Sustainable Development Goals (Klimes et al., 2019, p. 1363). However, the hydro-diplomacy in South Asia is still locked in riparian claims and immune to the international policy context (Bisht, 2019). The third broad category of literature focused on the importance of engaging the community and human development indicators for water sharing. It is argued that 'resource centric idea on water' needs to be redefined by the multiple meanings in terms of place and context (Blatter & Ingram, 2001). Water Diplomacy Framework (Islam & Susskind (2013), which is a combination of various complex issues, and negotiation theory and ecosystem perspective (Bandopadhyay & Ghosh, 2009) are main focus of this stream of knowledge. Therefore, hydro-diplomacy considers water sharing to be a subject of politics assisted by diplomatic methods. It also has a normative contest which values mutual trust, interaction, and openness to discussion among the stakeholders by acknowledging the conflicts of interest and abiding by the key international policies formulated by the United Nations (UN, 1997; UNECE, 1992). In such a backdrop, the hydro-diplomacy is confined to the following deadlocks in terms of sharing the transboundary water resources.

Perception on Water

Perception of water is the most important deadlock for hydro-diplomacy and the diplomatic process. There are three major aspects that need to be considered for overcoming the diplomatic deadlocks. First, water is not a static, rather a mobile resource. It cannot be stored in a larger volume permanently. This will follow the hydrologic cycle. The physical structure can make an obstacle for water flows, but it does not obey the sovereignty of countries. No country can claim or own the transboundary water resources. Second, the amount of available water in the transboundary rivers, lakes, aquifers, or wetlands varies. It depends on the upstream diversion, seasonal variation, and weather patterns. Impacts of climate change made the uncertainty about available water more acute. Third, only physical (quantity) and mental (quality) waters are considered during the water sharing negotiation. But water has emotional and spiritual connections which are often overlooked during diplomatic discussions. Therefore, hydro-diplomacy struggles with the complex and uncertain aspects of water for equitable and effective water sharing.

The Power Interplays

The power is defined by different disciplines with different connotations, meanings, and interpretations. Realists focus on the importance of material resources, whereas neo-institutionalists highlight the significance of both material and non-material or ideational resources utilised by actors during diplomacy (Daoudy, 2009; Turton & Ashton, 2008). Resources are linked to factors such as economic development, potential for hydropower generation, geographic location, military stability, and access to external political and financial aid (Menga, 2016a). States justify their actions and seek advantage over less powerful nations by leveraging material resources (Luttwak, 1990).

Then again, the central state has limited control over the complexities of institutions and their interdependencies over intra-state actors. The influential states also use the ideational powers, i.e. constructing narratives, knowledge, rules, and ideas to influence the diplomatic process (Cascão & Zeitoun, 2010; Nye, 2009). This ideational power plays a vital role in inspiring certain actions through shaping certain norms, choices, and values at the transboundary negotiation level. This power interplay has been referred as 'hegemon' and 'non-hegemon' by various scholars (Evans & Newnham, 1998; Zeitoun & Warner, 2006; Zeitoun et al., 2011). 'Such power interplay can result in zero-sum or non-zero-sum outcomes, but there is also a possibility of partial or complete status quo' (Vij et al., 2020b, p. 257). The status quo refers to a scenario where no definitive agreements are reached between the hegemonic and non-hegemonic parties on matters like water allocation, resource development, flood and erosion control, navigation pathways, dam construction, or collaborative research.

On the contrary, domestic power plays are also less focused in the transboundary research (Menga, 2016b; Petersen-Perlman & Fischhendler, 2018; Warner & Zawahri, 2012). Multiple levels of governance system in a hegemonic state create power interplay which shapes the transboundary negotiations. This limits or challenges the power of hegemonic states and is referred to as 'hegemonic vulnerabilities' (Petersen-Perlman & Fischhendler, 2018). The non-hegemonic or less powerful state can use these hegemonic vulnerabilities by utilising their ideational resources as a strategic tool to create a suitable ground for negotiation.

Bilateralism over Multilateralism

The transboundary water sharing or agreements among South Asian countries are bilateral. India has bilateral cooperation with Bangladesh, Bhutan, Pakistan, and Nepal. India always prefers bilateral agreements to establish supremacy in the region. Therefore, a regional approach or basin-wide ecosystem management is absent in South Asia. Although there are lots of criticisms over India's bilateralism in academic literature, the country is trying to limit the contestation and riparian coalition possibility. This condition increases the scepticism and mutual trust in a region. On the other hand, India pursues the status quo by having ideational resources, practicing strategic autonomy, and capturing a major share of the benefits from transboundary water resources.

Inadequate Research on Hydro-Diplomacy in South Asia

There was very little academic literature on hydro-diplomacy till 2009. The number started to increase gradually. The highest number of academic literature was published on hydro-diplomacy in 2020. Figure 7.2 shows the academic literature on hydro-diplomacy (Figure 7.2). From the figure, it is found that only 2 percent of academic research focuses on South Asia and Southeast Asia. Lack of academic research on water diplomacy is one of the limitations for successful negotiation.

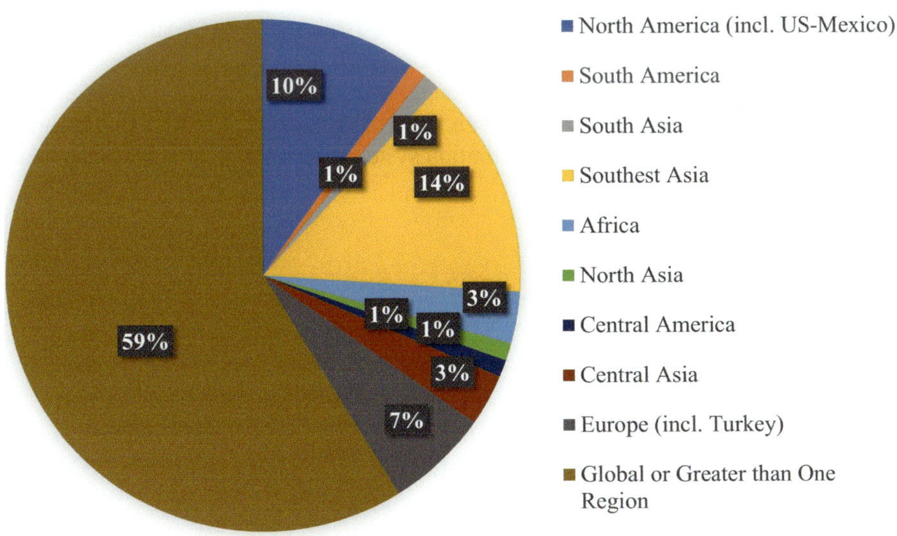

FIGURE 7.2 Distribution of academic literature on hydro-diplomacy at the global scale, highlighting the comparatively limited share of studies focused on South Asia. (*Source:* Sehring et al., 2022.)

Water Insecurity

Rivers offer a vital source of freshwater and ecosystem services, such as food production, energy generation, and navigation. In addition, they play a vital role in human civilisation and have the ability to foster collaboration by distributing benefits among the people living in a specific geographical area. Conflicts may occur as a result of resource strain, including factors such as population expansion, economic activities, unsustainable urbanisation, and climate change. 'More than 260 river basins are shared by two or more countries, encompassing 145 states that collectively comprise 90 percent of the world's population … Therefore, water is a humanity's greatest contemporary challenge' (Uexkull & Buhaug, 2021, p. 3). According to a report, 'in 2020, 6.13 billion people were living in critically water-insecure or water-insecure countries, including 4.31 billion people in the Asia-Pacific region, 1.34 billion in Africa, 415 million in the Americas, and almost million people in Europe' (MacAlister et al., 2023). Climate–security nexus literature predicts the additional threats to the availability and governance/management of transboundary water resources. Improper functioning and unstable centrality in both natural and human systems are influencing to securitise water (Bogardi et al., 2016). Based on the literature, abundance, stability, quality, and accessibility of water are considered the dimensions of water to measure water security (Pahl-Wostl et al., 2016a; Stringer et al., 2021).

Both climate change and water thefts are contributing to water insecurity (Figure 7.3). According to Felbab-Brown, water theft 'is highly controversial because there is no common definition as to what constitutes water theft and smuggling, or, for that matter, whether such phenomena exist at all' (Felbab-Brown, 2017) and 'any taking of water in violation of existing regulations' (Felbab-Brown, 2017). Unauthorised water usage, including illegal connections, siphoning, and unapproved pipelines, can hinder cities or countries from collecting enough revenue for water usage. These widespread violations and significant theft can lead to insufficient funds for maintaining, upgrading, and expanding water distribution systems, as well as for addressing water scarcity challenges.

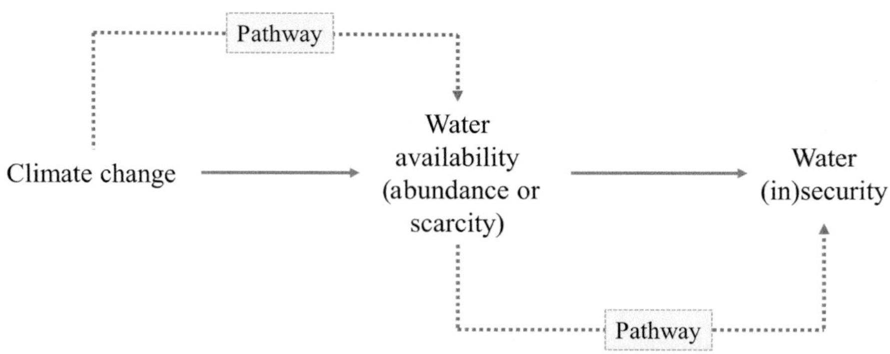

FIGURE 7.3 The nexus among climate change, water availability, and water (in)security. (*Source:* Asakaa et al., 2024.)

These can work both ways. For example, people can be motivated to engage themselves in water theft in a water stressed condition. 'Water theft is a common occurrence for individuals experiencing water insecurity on a daily basis. Thousands of years of overuse, extraction and exploitation of freshwater resources have contributed to diminished availability and accessibility of freshwater in contemporary society worldwide, causing both rich and poor to steal water for agriculture/irrigation and domestic use, for religious and spiritual purposes, and for profit' (Baird & Walters, 2020, p. 379). Therefore, the water scarcity is influenced by two factors. Firstly, water theft directly facilitates water scarcity, which is a key pathway connecting climate change to water theft. Secondly, climate change indirectly encourages water theft by weakening state or institutional capacity, which creates conditions for such activities to thrive.

Divided Climate Diplomacy

Countries negotiate for establishing climate justice at the UNFCCC platform through several blocs. All the South Asian countries are only members of the G77 group. Besides, Bangladesh, India, Pakistan, and Sri Lanka are also members of the Like-Minded Group. India and the Maldives are members of BASIC and AOSIS, respectively. South Asian countries do not have any common platform for negotiation to minimise the climate risk or financial assistance like the African Union (AU). National interests or Bloc's interest gets priority over geographic proximity in climate diplomacy. Therefore, it is difficult for the South Asian countries to minimise the climatic adversity through regional cooperation.

CONCLUSION

Hydro-diplomacy in South Asia exemplifies the intricate challenges of managing transboundary water resources amid diverse geopolitical, economic, and environmental constraints. The region, characterised by its significant reliance on shared rivers like the Ganges, Brahmaputra, and Indus, has struggled to foster cooperation, largely due to power asymmetries, fragmented governance, and bilateral rather than regional approaches. These dynamics have perpetuated mistrust among nations and hindered effective water resource management, particularly in the context of increasing demands and the adverse impacts of climate change. One of the core challenges lies in the entrenched perception of water as a static and sovereign resource. This limited view fails to recognise water's dynamic and shared nature, leading to unilateral actions that exacerbate disputes. Additionally, hegemonic states in the region often utilise their material and ideational power to maintain control over agenda-setting and decision-making processes, creating a status quo that prioritises national interests over equitable resource sharing. Non-hegemonic states, lacking the leverage to challenge these dynamics, are often compelled to accept unfavourable arrangements, further entrenching existing disparities.

Bilateral agreements dominate the landscape of transboundary water governance in South Asia, with countries like India preferring such frameworks to maintain regional supremacy. While bilateralism has resulted in some successful agreements, such as the Ganges Water Treaty between India and Bangladesh, these arrangements often lack inclusivity and fail to address basin-wide or regional concerns. As a result, opportunities for holistic, multilateral solutions remain untapped, and scepticism regarding mutual intentions persists. The lack of regional platforms for negotiation further complicates the scenario. Unlike regions such as Africa, where frameworks like the African Union provide a cohesive front for addressing transboundary water issues, South Asia lacks a unified mechanism to advocate for collective action. National interests and the pursuit of individual benefits often overshadow the potential for regional cooperation, particularly in the face of climate-induced challenges that demand integrated responses.

Climate change exacerbates water scarcity and insecurity, posing significant threats to regional stability. Unpredictable weather patterns, glacial melt, and seasonal variations impact water availability and quality, heightening tensions over resource allocation. Furthermore, issues like water theft and inadequate infrastructure undermine efforts to ensure sustainable water management. These challenges, combined with insufficient academic and policy focus on the interconnectedness of climate change and water governance, limit the region's ability to develop adaptive and resilient strategies.

Addressing these issues requires a paradigm shift in the approach to hydro-diplomacy in South Asia. First, embracing a regional and ecosystem-based framework is crucial. This involves moving beyond bilateral agreements to adopt basin-wide strategies that consider the interconnectedness of social, economic, and environmental factors. Establishing a regional platform for dialogue and negotiation, akin to existing frameworks in other parts of the world, can facilitate trust-building and promote equitable sharing of benefits. Second, integrating scientific research, technological advancements, and data-sharing mechanisms into water governance can enhance transparency and informed decision-making. Joint studies on water flows, pollution control, and climate impacts can foster mutual understanding and create a foundation for collaborative action. Involving non-state actors, including communities and civil society organisations, can further democratise the negotiation process and ensure that diverse perspectives are considered. Third, South Asian countries must align their hydro-diplomacy efforts with international water laws and climate goals. By adhering to global frameworks like the United Nations Convention on Non-Navigational Uses of International Watercourses, nations can strengthen the legal basis for cooperative agreements and ensure accountability. Simultaneously, integrating water governance into broader regional initiatives, such as those focused on sustainable energy and economic connectivity, can create synergies and amplify the benefits of collaboration.

Hydro-diplomacy in South Asia stands at a critical juncture. The region's shared reliance on transboundary water resources, coupled with the escalating impacts of climate change, underscores the urgent need for innovative, inclusive, and integrated approaches to water governance. By transforming water from a source of contention into a catalyst for cooperation, South Asian nations can foster regional stability, enhance resilience to environmental challenges, and pave the way for sustainable development. The road ahead demands collective effort, political will, and a commitment to prioritising the common good over narrow national interests. With the right strategies and frameworks in place, hydro-diplomacy can serve as a powerful tool for building peace and prosperity in one of the most water-stressed regions of the world.

REFERENCE

Asakaa, J. O., Argomedob, D. W., & Jones, N. P. (2024). Climate change risks to water security: Exploring the interplay between climate change, water theft, and water (in)security. Water Policy, 26(4).

Bagale, D., & Adhikari, K. (2020). Mahakali Treaty: Delay in implementation and resulting impacts from Nepal's perspective. Water Policy, 22, 15.

Baird, A., & Walters, R. (2020). Water theft through the ages: Insights for green criminology. Critical Criminology, 28, 371–388.

Bandopadhyay, J., & Ghosh, N. (2009). Holistic engineering and hydro-diplomacy in the Ganges-Brahmaputra-Meghna basin. Economic & Political Weekly, 44(45), 1–10.

Bisht, M. (2019). From the edges of borders: Reflections on water diplomacy in South Asia. Water Policy, 21, 1123–1138.

Bisht, M. (2023a). Water Diplomacy in South Asia: Towards a Relational Approach. Friedrich Nauman Foundation for Freedom, South Asia, p. 06.

Bisht, M. (2023b). 'Let Diplomacy Flow'. *Indian Express*, March 11. https://indianexpress.com/article/opinion/columns/on-the-indus-water-treaty-hedging-and-entrenched-attitudes-from-india-and-pakistan-are-a-reminder-that-technical-agreements-are-only-a-partial-solution-8413832/

Blatter, J., & Ingram, H. (2001). Reflections on Water, New Approaches to Transboundary Conflicts and Cooperation. Massachusetts Institute of Technology.

Bogardi, J., Spring, U. O., & Brauch, H. G., (2016). Water security: Past, present and future of a controversial concept. In Handbook on Water Security. Pahl-Wostl, C., Bhaduri, A. & Gupta, J. (eds.). Edward Elgar, Northampton, MA, USA, pp. 38–58.

Butts, K. H. (1997). The strategic importance of water. Parameters Spring, pp. 65–83.

Cascão, A. E., & Zeitoun, M. (2010). Power, hegemony and critical hydropolitics. Transboundary Water Management: Principles and Practice, 27, 42.

Choden, S. (2009). Sediment Transport Studies in the Punatsangchu Basin. Lund University, Sweden, p. 44. http://lup.lub.lu.se/luur/download?func=downloadFile&recordOId=1415878&file-OId=1415879

Cullet, P. (2013). Right to water in India – Plugging conceptual and practical gaps. International Journal of Human Rights, 17(1), 56–78.

Daoudy, M. (2009). Asymmetric power: Negotiating water in the Euphrates and Tigris. International Negotiation, 14(2), 361–391.

Evans, G., & Newnham, J. (1998). The Penguin Dictionary of International Relations (pp. 235–236). Penguin Books, London.

Farnum, R. L. (2018). Drops of diplomacy: Questioning the scale of hydro-diplomacy through fog-harvesting. Journal of Hydrology, 562, 446–454.

Felbab-Brown, V. (2017). Water Theft and Water Smuggling: Growing Problem or Tempest in a Teapot? Brookings, Washington, DC.

Gleditsch, N. P., Furlong, K., Hegre, H., Lacina, B., & Owen, T. (2006). Conflicts over shared rivers: Resources scarcity or fuzzy boundaries? Political Geography, 25(4), 361–382.

Hazarika, N., Barua, A., Vij, S., Deka, A., & Salame, L. (2023). Collective deliberation or just the state (in)action: How do we change the hydrodiplomacy landscape in Southa Asia? Water policy (1). doi: 10.2166/wp.2023.166

Homer-Dixon, T. F. (1994). Environmental scarcities and violent conflicts: Evidence from cases. International Security, 19(1), 5–40.

Honkonen, T. and Lipponen, A. (2018). Finland's cooperation in managing transboundary waters and the UNECE Principles for Effective Joint Bodies: Value for water diplomacy? Journal of Hydrology 567, 320–331.

Islam, S., & Repella, A. C. (2015). Water diplomacy: A negotiated approach to manage complex water problems. Journal of Contemporary Water Research & Education, 155, 1–10.

Islam, S., & Susskind, L. (2013). Water Diplomacy: A Negotiated Approach to Managing Complex Water Networks. Routledge, New York.

Keskinen, M., Salminen, E., & Haapala, J. (2021). Water diplomacy paths – An approach to recognise water diplomacy actions in shared waters. Journal of Hydrology, 602, 126737.

Khalid, I., Mukhtar, A., & Ahmed, Z. (2014). Water scarcity in South Asia: A potential conflict of future decades. Journal of Political Studies, 21(1), 259.

Klimes, M., Michel, D., Yaari, E., & Restiani, P. (2019). Water diplomacy: The intersect of science, policy and practice. Journal of Hydrology, 575, 1362–1370.

Luttwak, E. N. (1990). From geopolitics to geo-economics: Logic of conflict, grammar of commerce. The National Interest, 20, 17–23.

MacAlister, C., Baggio, G., Perera, D., Qadir, M., Taing, L., & Smakhtin, V. (2023). Global Water Security 2023 Assessment. United Nations University Institute for Water, Environment and Health, Hamilton, ON, Canada.

Menga, F. (2016a). Reconceptualizing hegemony: The circle of hydro-hegemony. Water Policy, 18(2), 401–418.

Menga, F. (2016b). Domestic and international dimensions of transboundary water politics. Water Alternatives, 9(3), 704–723.

Nepal, S., & Shrestha, A. B. (2015). Impact of climate change on the hydrological regime of the Indus, Ganges and Brahmaputra river basins: A review of the literature. International Journal of Water Resources Development, 31(2), 201–218.

Nicholson, H. (1939). *Diplomacy*. Harcourt Brace. Accessed from https://archive.org/details/in.ernet.dli.2015.14552/page/n7/mode/2up. Also available at: https://books.google.co.in/books/about/Diplomacy.html?id=uJcIAAAAMAAJ&redir_esc=y.

Nye, J. S., Jr. (2009). Get smart: Combining hard and soft power. Foreign Affairs, 88(4), 160–163.

Pahl-Wostl, C., Bhaduri, A., & Gupta, J. (2016). Handbook on Water Security. Edward Elgar, Northampton, MA, USA.

Petersen-Perlman, J. D., & Fischhendler, I. (2018). The weakness of the strong: Re-examining power in transboundary water dynamics. International Environmental Agreements: Politics, Law and Economics, 18(2), 275–294.

Salman, S. M., & Uprety, K. (2018). Shared watercourses and water security in South Asia: Challenges of negotiating and enforcing treaties. Brill Research Perspectives in International Water Law, 3(3), 1–100.

Sehring, J., Schmeier, S., Horst, R., Offutt, A., & Sharipova, B. (2022). Diving into water diplomacy – Exploring the emergence of a concept. Diplomatica, 4, 200–221.

Shan, V., Singh, S. K., Haritash, A. K. (2020). Water crisis in the Asian countries: Status and future trends. In: Kumar, M., Munoz-Arriola, F., Furumai, H., Chaminda, T. (eds) *Resilience, Response, and Risk in Water Systems* (pp. 173–194). Springer Transactions in Civil and Environmental Engineering. Springer, Singapore. https://doi.org/10.1007/978-981-15-4668-6_10

Simoes, D. F. (2017). The role of investment arbitration in water services governance. Water Policy, 19(2), 271–285.

Stringer, L. C., Mirzabaev, A., Benjaminsen, T. A., Harris, R. M., Jafari, M., Lissner, T. K., Stevens, N., & Tirado-von Der Pahlen, C. (2021). Climate change impacts on water security in global drylands. One Earth, 4, 851–864.

Turton, A. R., & Ashton, P. J. (2008). Basin closure and issues of scale: The southern African hydropolitical complex. International Journal of Water Resources Development, 24(2), 305–318.

Uexkull, N., & Buhaug, H. (2021). Security implications of climate change: A decade of scientific progress. Journal of Peace Research, 58(1), 3–17.

UN (1997). Convention on the Law of the Non-navigational Uses of International Watercourses. Accessed from https://legal.un.org/ilc/texts/instruments/english/conventions/8_3_1997.pdf.

UNECE (1992). The Convention on the Protection and Use of Transboundary Watercourses and International Lakes. Accessed from https://unece.org/environment-policy/publications/convention-protection-and-use-transboundary-watercourses-and.

Van Genderen, R., & Rood, J. (2011). Water Diplomacy: A Niche for the Netherlands. Netherlands Institute of International Relations 'Clingendael', Netherlands Ministry of Foreign Affairs and the Water Governance Centre. https://www.clingendael.org/sites/default/files/2016-02/20111200_cling_report_waterdiplomacy_rgenderen_jrood.pdf.

Vij, S., Warner, J., & Barua, A. (2020a). Power in water diplomacy. Water International, 45(4), 249–253.

Vij, S., Warner, J. F., Biesbroek, R., & Groot, A. (2020b). Non-decisions are also decisions: Power interplay between Bangladesh and India over the Brahmaputra River. Water International, 45(4), 254–274.

Warner, J., & Zawahri, N. (2012). Hegemony and asymmetry: Multiple-chessboard games on transboundary rivers. International Environmental Agreements: Politics, Law and Economics, 12(3), 215–229.

Wolf, A. (2007). Shared waters: Conflict and cooperation. Annual Review of Environment and Resources, 32(3), 1–29.

Zeitoun, M., Mirumachi, N., & Warner, J. (2011). Transboundary water interaction II: The influence of 'soft' power. International Environmental Agreements: Politics, Law and Economics, 11(2), 159–178.

Zeitoun, M., Mirumachi, N., Warner, J., Kirkegaard, M., & Cascão, A. (2019). Analysis for water conflict transformation. Water International, 1–21. https://doi.org/10.1080/02508060.2019.1607479.

Zeitoun, M., & Warner, J. (2006). Hydro-hegemony – A framework for analysis of transboundary water conflicts. Water Policy, 8(5), 435–460.

8 How India Manages Its Transboundary Waters

Analyzing Ethical Considerations

Japungsar Basumatary and Dipmala Roka

INTRODUCTION

The starting point of this chapter argues that environmental sustainability in transboundary water governance requires integrating ethical considerations, particularly in geopolitically sensitive and water-scarce regions like South Asia. This sustainability lies in India's ethical considerations, as it is a pivotal riparian state located at upstream, midstream, and downstream geographical positions and navigating intricate interplays of geopolitical, sociocultural, and environmental imperatives. Analyzing through the lens of normative ethical theories such as *utilitarianism*, *deontology*, *Rawls'justice*, *virtueethics*, and *environmentalethics* provides an insight into India's moral dimension; decision, action, outcome, and challenges, informing a sustainable water governance that would be framed or re-framed. Evaluation of policy formulation on five integrated ethical principles aligned with existing water ethic discourse and policy framework advances India's ethical gaps and challenges, particularly between technical, political, socio-cultural, and environmental challenges, contributing to sustainable decision-making, action, and outcomes. It also contributes to existing scholarship: its debate and discussion on environmental sustainability in and through water management/governance, particularly in South Asia.

India, with its diverse river system, faces such critical challenges in managing transboundary waters comprising a wide range of upstream-downstream power politics, upstream activities and downstream impacts, a lack of a cooperative framework, water scarcity, human and environmental costs, pollution, climate change, flood, and so on. Transboundary rivers such as Brahmaputra, Barak, Ganges, Indus, Mahakali, and Teesta, altogether tributaries of the Ganga-Brahmaputra-Meghna (GBM) River Basin flowing in and through India crossing its borders, offer opportunities for cooperation and conflict due to these challenges. India shares 20 major rivers with China, Bangladesh, Bhutan, Nepal, Pakistan, and Afghanistan (Figure 8.1). Existing bilateral efforts, such as the Indus Water Treaty 1960 (IWT), Ganges Water Treaty 1996 (GWT), Mahakali Water Treaty 1996 (MWT), and the Joint River Commission with Bangladesh 1972 (JRC), reflect India's attempts to manage shared water resources. These efforts are designed to provide certain management guidelines for equitable water sharing, infrastructural development, and largely conflict resolution (Ho,

DOI: 10.1201/9781003599692-8

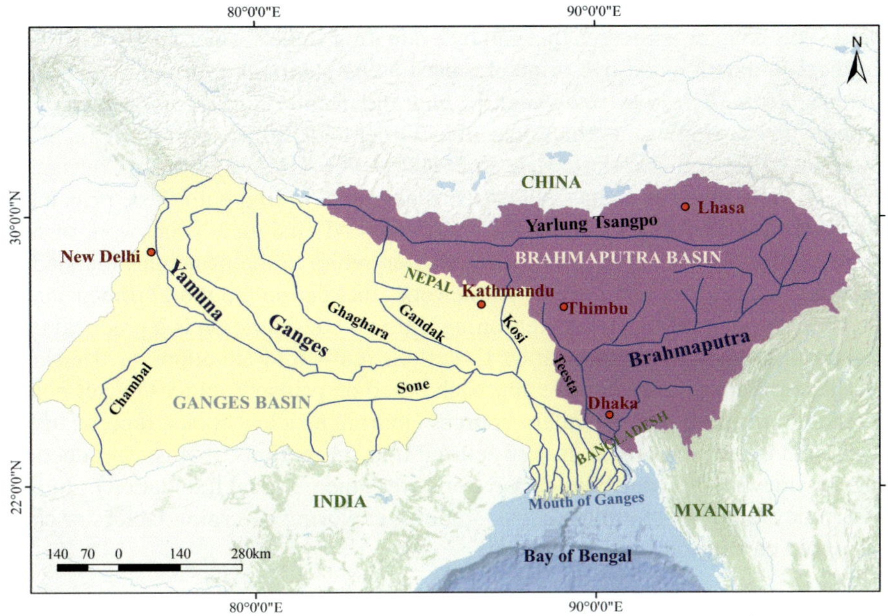

FIGURE 8.1 India shares major rivers with riparian countries and their transboundary nature. (*Source:* Prepared by authors.)

2016). A win-win approach between India and Bhutan can be brought as a good example of how transboundary water is managed, and cooperation can be achieved sustainably and successfully (Biswas, 2012), for ensuring how ethical cooperation can foster sustainable water management and regional stability. Despite irregularity, India–China relations are marked by the progress of flood-data sharing agreements through the Memorandum of Understanding (MoU) since 2002 (Holslag, 2011).

However, persistent conflicts over the Barak, Brahmaputra, and Teesta Rivers have remained key challenges, making India a critical riparian state for the downstream rights of Bangladesh (Hassan, 2019). Despite IWT with Pakistan, India has always been a critical country, often perceived as a hydro-hegemony for equitable water sharing (Hussain et al., 2021; Qureshi, 2017). At the same time, India is also a critical downstream country for Nepal, for instance, with the Karnali project and Saptakoshi Dam (Tandan, 2021). This complexity offers a critical inquiry into India's ethical foundation and considerations. What ethical considerations (universal, sociocultural, and political) justify India's approaches to ethical transboundary water governance? How do these align with existing global water ethics and foundational ethical frameworks? What challenges does India face in ethical governance? To what extent do India's policies foster environmental sustainability and regional cooperation? This chapter is an attempt to answer these critical questions by reviewing existing scholarship. While few scholars have analyzed the legal and institutional aspects of India and China, comparing their riparian approaches (Araral & Wu, 2016; Ho, 2016; Momen & Rahman, 2023; Wouters & Devlaeminck, 2024), the ethical dimensions of India's transboundary water governance remain underexplored. This chapter addresses this gap by evaluating India's water policy to explore its ethical foundations on five principles such as *no significant harm*, *cooperation*, *humanrightsanddignity*, *environmentalvalues* and *accountability,andtransparency*, aligned with existing water ethics discourse (Groenfeldt, 2019; Jennings et al., 2016) and global policy frameworks (e.g.Helsinki Rules 1966, UN Watercourse Convention 1997). Then, it analyses India's ethical considerations in line with these principles, through the theoretical lens employing an integrated analytical framework of normative ethical theories. Here, ethical foundation refers to the principles founded or established through existing policy frameworks that define the moral dimension of right and wrong, guiding ethical reasoning. Ethical consideration is a practical one, weighed in applying this foundation to certain real-world situations or decisions. The chapter employs a qualitative approach based on primary sources (water policy documents and acts) and secondary sources (journal articles, books, reports, newspapers, and online materials). It is structured into six sections, that is, introduction, ethical framework for transboundary water governance, ethical foundation of India's water policy, analysis of ethical considerations, exploring diverging factors or challenges, and conclusion.

ETHICAL FOUNDATION OF TRANSBOUNDARY WATER GOVERNANCE: A FRAMEWORK FOR SUSTAINABILITY

Transboundary waters—rivers, lakes, and aquifers—comprise approximately 60 percent of the world's freshwater, sustaining millions of lives, livelihoods, and ecosystems globally (UN-Water, 2021). Its effective management fosters international trade, climate adaptation, economic prosperity, food security, and regional integration, making it pivotal to environmental sustainability (UN-Water, 2018). Recent scholarship underscores the need for legitimate, democratic, and value-driven water governance to address the complex interplay of ecological, social, and political challenges (Meisch, 2017). Ethical considerations, particularly those protecting water ecosystems (Postel, 2008) and guiding decision-making (Groenfeldt, 2021), have gained prominence in both academic discourses. The discourse on water ethics has evolved significantly driven by institutions like UNESCO and scholars advocating for value-based governance. UNESCO's seminal work, *Water and Ethics: Overview* (Priscoli et al., 2004), introduced six foundational principles—human dignity, participation, solidarity, equality, common good, and stewardship—shaping ethical water management. In 2018, UNESCO's Commission on the Ethics of Scientific Knowledge and Technology (COMEST) expanded this framework to include frugality, sustainability, justice, gender equality, research integrity, knowledge sharing, and considerations for oceans and transboundary waters (UNESCO-COMEST, 2018). Concurrently, the Botin Foundation has emphasized environmental ethics in water management, aligning with broader academic contributions. Scholars like Jennings et al. (2009) outlined six principles—human dignity, equity, solidarity, common good, responsible stewardship, and participation—while Groenfeldt (2019) proposed five dimensions: environmental, social, cultural, economic, and governance ethics (Groenfeldt,

2019; Jennings et al., 2016). These principles collectively underscore the multidimensional nature of ethical water governance.

While the term "ethics" is rarely reflected in water legislation, ethical principles underpin global conventions and declarations. The UN Resolution 64/292 (2010) declared water access a human right, elevating its ethical significance to encompass dignity and equity (Bogard et al., 2021). Historical milestones, from the Mar del Plata Conference (1977) to the UN International Decade for Action (2018–2028), including the Dublin Statement (1992), have embedded principles like cooperation, environmental and economic values, participation, and public awareness (Desai, 2019). Legally binding frameworks, such as the UN Watercourse Convention (1997) and Helsinki Rules (1966), provide ethical foundations through principles of equitable and reasonable utilization, upholding the principle of *No Significant Harm*, and integrated management, addressing flood control, pollution, navigation, and conflict resolution. These frameworks, notably the UN Watercourse Convention 1997, though it is not universally adoptednor ratified by either China or India, offer normative ethical guidance for sustainable transboundary water governance (Wouters & Devlaeminck, 2024).

On the other hand, there is a plethora of normative ethical theories that also provide a robust lens to evaluate transboundary water governance. Starting with *Utilitarianism* rooted in scholarship (J.S. Mill 1861 and Sidqwick 1874), it emphasizes actions that maximize benefits for the greatest number, assessing whether water-sharing agreements deliver widespread benefits like irrigation, drinking water, and livelihoods. *Deontology*, rooted in Kantian ethics, stresses the duty to uphold agreements and principles like no significant harm, regardless of outcomes. *Rawls'Justice* advocates decisions made under a "veil of ignorance," ensuring fairness, particularly for weaker or downstream riparian states, to achieve equitable water distribution. *VirtueEthics* focuses on fostering cooperative behaviours and trust among stakeholders, promoting long-term collaboration. *Environmental Ethics* prioritizes basin-wide ecological preservation, safeguarding river flows, biodiversity, and wetlands for sustainable ecosystems. This chapter proposes a comprehensive framework for transboundary water governance, integrating these theories to analyze India's riparian ethics to govern transboundary waters with its neighbourhood. Five foundational principles guide this analysis, such as (a) *No Significant Harm* aligned with Rawls' Justice—this principle ensures equitable water sharing to minimize downstream impacts, fostering fairness across riparian states. (b) *Cooperation* rooted in Utilitarianism and Virtue Ethics—cooperation through bilateral and multilateral agreements prevents conflicts and promotes regional peace, security, and prosperity. (c) *Human Rights and Dignity* reflecting UN Resolution 64/292—this principle recognizes water as a human right, prioritizing access for individuals and communities, particularly marginalized groups. (d) *Environmental Values* grounded in Environmental Ethics—this principle emphasizes ecological balance, ensuring policies and treaties protect ecosystems and maintain natural flows. (e) *Accountability and Transparency* drawing from Deontology and Justice perspectives—this principle demands responsible decision-making and open data sharing to equitably distribute benefits and burdens. These principles offer an integrated approach to addressing ethical challenges, evaluating policy formulation, stakeholder actions, and governance outcomes in transboundary contexts. It contributes to environmental sustainability by bridging the gaps between governance, human and environmental perspectives and informing ethical decision-making and action of riparian states. For instance, no significant harm ensures upstream activities such as infrastructure projects, and water diversion minimizes ecological and sociocultural damages to downstream riparian states. Cooperation fosters collaborations on shared rivers, enabling equitable water sharing and conflict resolution and enhancing regional stability and sustainability. Human rights considerations prioritize access to water for marginalized communities, ensuring social equity. Environmental values emphasize from human dimension to non-human by ensuring eco-centric stewardship, ecosystem protection, and water purification. Accountability and transparency build trust among all water users, including state and non-state actors, by ensuring open access to data and inclusive decision-making.

INDIA'S WATER POLICY: SEARCHING ETHICAL FOUNDATION FOR TRANSBOUNDARY WATER GOVERNANCE

India's National Water Policy (NWP) (2002, 2012) provides a foundation for transboundary water governance, emphasizing bilateralism as a cornerstone of hydro-diplomacy. The policy conceptualizes rivers as integrated units for development, supporting hydropower, irrigation, and drinking water projects while advocating for cooperative mechanisms to address transboundary challenges. Key bilateral treaties, such as the IWT with Pakistan, GWT with Bangladesh, MWT with Nepal, and MoUs with China (2002–2018), reflect India's efforts to balance national interests with regional obligations (Table 8.1). These frameworks aim to promote equitable

TABLE 8.1
India's Existing Agreements and Treaties Signed with Other Riparian States

Riparian State	Agreements and Treaties
Bangladesh	Joint River Commission (1972)
	Ganges Water Agreement (1977)
	Ganges Water Treaty (1996)
	MoU (2019) on 1.82 cusec water withdrawal from Feni River for Tripura, India
	MoU (2022) on 153 cusec water withdrawal from Kushiyara for consumption required in both countries
Bhutan	Comprehensive Scheme for Establishment of Hydro-meteorological and Flood Forecasting Networks
China	MoU (2002) on Hydrological Information about Brahmaputra
	MoU (2004) on Hydrological Data Share about the Sutlej River
	Expert Level Mechanism (2006)
	MoU (2013) on Strengthening Cooperation on Trans-border Rivers
Nepal	Mahakali Treaty (Pancheshwar Multipurpose Project) (1996)
	Joint Project-Sapta-Koshi Dam Project (2004)
	Joint Project-Kamla-Bagmati Multipurpose Project
Pakistan	Indus Water Treaty (1960)

water sharing, hydrological data exchange, and conflict resolution, aligning partially with ethical principles of water governance.

No Significant Harms

This principle, central to ethical water governance, requires minimizing adverse impacts on downstream states. India's NWP supports extensive infrastructure development, with 548 projects on the Ganges, 146 on the Indus, over five on the Teesta, and 12 on the Barak and its tributaries (Government of India, 2014a, 2014b, 2014c). These projects, including dams and reservoirs, drive hydropower, irrigation, and water supply but often affect downstream ecosystems and communities. For instance, the Teesta project located at Gajoldoba (West Bengal, India) has raised concerns about water scarcity in Bangladesh (Basumatary, 2018), while Indus developments, such as the Baglihar and Kishanganga dams, have sparked disputes with Pakistan over water flow manipulation (Iqbal, 2018). From a Rawlsian perspective, India's policies partially uphold fairness through treaties like the GWT, which ensures minimum flows for Bangladesh's Sundarbans, but fall short when national priorities, such as hydropower, override downstream rights. The policy does not explicitly embed the no-harm principle, relying instead on bilateral agreements to mitigate impacts, which limits its ethical alignment.

Cooperation

Cooperation, rooted in Utilitarianism and Virtue Ethics, is a cornerstone of India's transboundary water governance, aiming to prevent conflicts and foster regional stability, embedding the provision of bilateralism in NWP. India has been engaging in bilateral talks regarding the signing of agreements and treaties with both upstream and downstream riparian states. The important treaties on the Indus (between India and Pakistan), Ganges (between India and Bangladesh), and Mahakali (between India and Nepal) provide important provisions to the riparian states to share and manage water resources. These treaties are designed to ensure equitable water sharing, hydrological data exchange, and largely conflict resolution and good neighbourhood relations for the maintenance of regional peace and security. The IWT equitably allocates eastern rivers (Beas, Ravi, Sutlej) to India and western rivers (Chenab, Indus, Jhelum) to Pakistan, averting conflict despite geopolitical tensions (Siddiqui, 2021). The GWT regulates water sharing at Farakka Barrage, ensuring for 30 years water sharing at the Barrage near their border (Hanasz, 2014). It provides India and Bangladesh with water distribution between January 1 and May 31, considering the rights and responsibilities of India to manage the river basin during the critical period. For instance, India is allocated a flow of 990 cubic meters per second, (equal 990,000 liters) during any critical period (International Policy Digest 2013). India is permitted to withdraw up to 40,000 cusecs of water at the Barrage annually between 1 January and 31 May. If water flow falls below 70,000 cusecs, each country receives equally, at least 35,000 cusecs per year over alternating 10-day periods between 11 March and 10 May (Hanasz, 2014; Sands, 1997). India has also engaged in equitable water sharing with Nepal through the MWT signed in 1996, and the Chukha Hydro-Electric Project with Bhutan exemplifies project-specific cooperation, delivering mutual benefits like hydropower and flood control (Wouters & Devlaeminck, 2024).

For the hydrological data exchange, India has also initiated MoUs with the uppermost riparian state, i.e. China. Both countries signed the first MoU in 2002 for five years over the hydrological information to determine the Brahmaputra River, particularly during the flood season. Accordingly, the hydrological information is being shared with India regularly from 15th May to 15th October every year. Both countries renewed the MoU in 2008, 2013, and 2018 (Ministry of Jal Shakti, n.d.). There is also another separate MoU, i.e. "Strengthening Cooperation on Trans-Border Rivers" signed on 23rd October 2013, which enhances the scope of the provision of hydrological information of three hydrological stations (Nugesha, Yangcun, and Nuxia). Following it, another MoU was also initiated on the Sutlej River or Langpen Zangbo (in China) requiring trans-border cooperation for an early warning system. This MoU was designed for disaster mitigation and flood control in the downstream areas, subsequently renewed in 2010 and 2015 for another five years. It was followed by the signing of "Implementation Plans" on 13th June 2019 at Ahmadabad and on 13th April 2016 in Delhi respectively (Ministry of Jal Shakti, n.d.). Moreover, India and China have established a joint expert-level mechanism for interaction and cooperation on flood-related hydrological data, emergency management, and transboundary river-related issues (Government of India, 2006). These mechanisms reflect a utilitarian approach, maximizing regional benefits, but unresolved disputes, such as the Teesta with Bangladesh, highlight challenges in sustaining cooperative commitments from a deontological perspective.

Human Rights and Dignity

Aligned with UN Resolution 64/292 (2010), which recognizes water access as a human right, this principle emphasizes equitable access, particularly for marginalized communities. India's NWP prioritizes drinking water and sanitation, ensuring allocations for domestic use and targeting marginalized groups like women and farmers (NWP, 2012, Section 4.2). However, its transboundary application is limited, lacking explicit provisions to safeguard downstream communities' rights. For example, water scarcity in Bangladesh's Teesta basin affects livelihoods, undermining human-centric ethics (Hassan, 2019). This gap reflects a disconnect between domestic and transboundary priorities, misaligning with the ethical imperative of universal access and dignity.

Environmental Values

Environmental ethics demands policies that prioritize ecological balance, ensuring sustainable ecosystems alongside human needs. India's NWP advocates Integrated Water Resource Management (IWRM), recognizing water's role in the eco-hydrological cycle and mandating environmentally sound planning at local, regional, and national levels (NWP, 2012, Section 5.1). The GWT supports environmental sustainability by maintaining minimum flows for Bangladesh's Sundarbans delta (Hanasz,

2014). However, treaties like the IWT lack provisions for emerging challenges like climate-induced glacial melt and altered rainfall patterns, which threaten river health (Sachwani & Sharma, 2025). Experts recommend maintaining 15–20 percent minimum environmental flows during lean seasons, a standard not yet integrated into India's transboundary agreements (Sachwani & Sharma, 2025). This partial alignment with environmental values highlights a prioritization of developmental goals over ecological preservation.

Accountability and Transparency

The principle of accountability and transparency directly asserts responsible and reasonable decision-making and action, along with its open access to not only water sharing but also hydrological information and data sharing. India's policy framework supports hydrological data exchange, as seen in flood forecasting data shared with Bangladesh for the Barak, Brahmaputra, Ganges, and Teesta during monsoons (NWP, 2012). The IWT mandates transparency, requiring India to provide Pakistan with daily discharge and project details. Regular meetings on Farakka Barrage operations and data sharing with Nepal and Bhutan reinforce accountable governance (Government of India, 2023). However, geopolitical tensions, particularly with China and Pakistan, limit public access to data, prioritizing national security over full transparency (Vater, 2021). This selective transparency undermines the ethical principle of open and equitable decision-making.

While critically examining India's existing policy framework indicates a partial incorporation of water ethics and a quite challenging for ensuring sustainable management of the existing river basins. Evaluating its existing policy framework reveals that there is a huge gap at the juncture of simple water management and transboundary water. Partially, India has explicitly adopted riparian ethics always aligned with the principle of no significant harm to the riparian state, achieved through a cooperative mechanism. However, there has been criticism on the basis of its ethical stand on environmental values, and human rights to water resources applying to riparian rights. Accountability and transparency fail to ensure the public domain of hydrological data and information. Transboundary water governance has become a state-centric and nationalist perspective of water use. The river basin has become a single unit for developmental purposes, often failing to address the key approach, such as IWRM, making it almost limited to regional cooperation (bilateral rather than multilateral). There has been a series of hydro-diplomacy and bilateral agreements and treaties; however, these efforts still lack its holistic approach. It only acknowledges the water sharing with seemingly state-centric and national security interests, largely conflict resolution, without a holistic approach for sustainable river basin management in light of climate change, population growth, and industrialization. On the other hand, bilateral over multilateral may bring inequitable and inefficiency of water management/governance, addressing only one-on-one basis, often neglecting the potential benefits of all the riparian states, particularly the downstream and weaker riparian states and non-state actors. It offers India more negotiating power on a case-by-case basis rather than engaging in a multilateral or basin-wide framework. India can control the negotiation process as there is often upstream and political power creating potential imbalance. Weaker riparian states will have less bargaining power, making fairness more rhetorical than real. This backdrop requires further analysis of India's ethical considerations through existing analytical theories.

ANALYZING INDIA'S ETHICAL CONSIDERATIONS: A THEORETICAL EXAMINATION

Ethics plays a critical role in transboundary water governance, especially when balancing the rights of upstream and downstream countries. India's unique position as an upstream, midstream, and downstream state in South Asia places it at the heart of complex ethical challenges in transboundary water governance. With a population of 1.45 billion, India's water policies are driven by the demand for socioeconomic development, hydropower generation, irrigation, and safe drinking water, while navigating climate change, sovereignty, geopolitics, and national security. It reflects a delicate balance of ethical principles, national interests, and geopolitical realities. Utilitarianism fuels its domestic development agenda for maximizing benefits for its citizens through infrastructure projects like the Farakka Barrage (on the Ganges), Teesta Barrage, and Tipaimukh Dam (on the Barak), which support agriculture, hydropower, and drinking water. However, these projects often lead to conflicts with downstream states, notably Bangladesh, creating ethical dilemmas as they fail to deliver the greatest good for the greatest number across borders, pitting national interests against regional equity. Rawlsian justice exposes this inequity perceived by downstream states (Bangladesh), and deontology underscores its commitment to international treaties, though constrained by domestic and geopolitical factors. Rawlsian justice highlights the perception of India as a hydro-hegemonic player, particularly by Bangladesh and Pakistan as well, who view India's control over shared river basins as unfair. From a deontological perspective, India strives to uphold international obligations, such as the GWT and IWT, but struggles to balance these duties with its development goals, complicating equitable water-sharing. India fosters cooperation with Nepal and Bhutan, highlighting equitable water-sharing, transparent data exchange, and robust regional cooperation to align national priorities with the rights of riparian states.

India's relationship with Pakistan over the Indus River exemplifies tensions despite there being mandate of a bilateral treaty mediated under the World Bank. India, as an upstream state controlling six major rivers, is seen as a hydro-hegemony. The IWT (1960) allocates eastern rivers (Sutlej, Beas, Ravi) to India and western rivers (Indus, Jhelum, Chenab) to Pakistan, with India permitted limited use of the latter for domestic, agricultural, and hydroelectric purposes, including storage up to 3.6 million acre-feet. Pakistan has repeatedly challenged India's projects, such as the Baglihar and Kishanganga dams, alleging violations through water diversion (Figure 8.2). International rulings, including the Hague Arbitration Court, have upheld India's rights while mandating minimum river flows (Dawn, 2011; Iqbal, 2018). From a deontological perspective, India adheres to the treaty, but utilitarian concerns underscore Pakistan's vulnerability to water scarcity, revealing an ethical divide between India's national priorities and Pakistan's downstream rights. There has been a recent analysis claiming that India is not fully responsible for the water

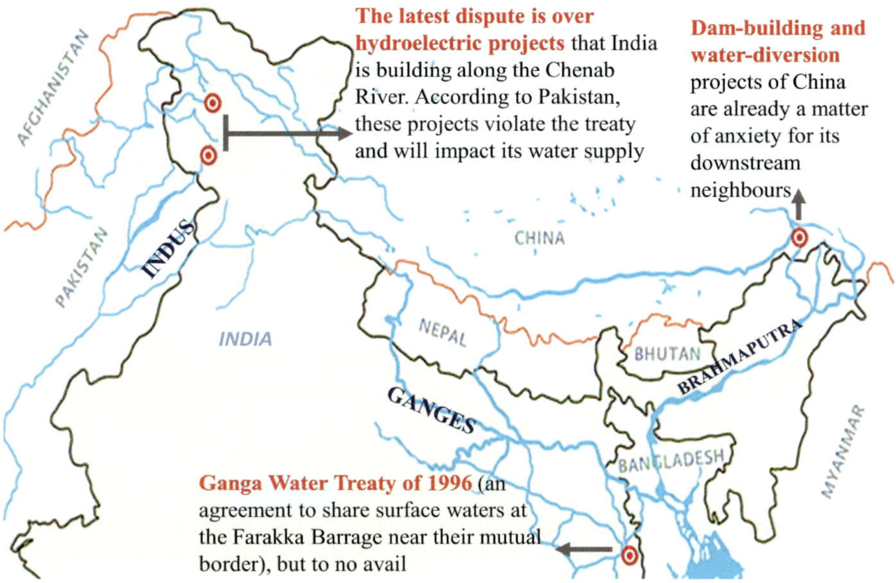

FIGURE 8.2 Various issues in transboundary river management. (*Source:* Prepared by authors.)

issue when it comes to the principle of cooperation and no harm principle. India's riparian ethics towards the Indus water do not violate the rights and sovereignty of Pakistan at all. India agrees to make an effort to resolve with Pakistan (Siddiqui, 2021). The IWT grants India exclusive control over the eastern rivers, while Pakistan has unrestricted use of the western rivers, which India must allow to flow, subject to limited permitted uses. India is also allowed to utilize western river waters for domestic, non-consumptive, agricultural, and hydroelectric purposes and can construct storage up to 3.6 million acre-feet for specified purposes. Despite Pakistan's objections, the Hague Arbitration Court upheld India's right to divert water for power generation, provided a minimum flow is maintained (Iqbal, 2018). Regarding the Baglihar project, India was permitted to construct and store up to 32.58 million cubic meters, excluding flow control measures (Dawn 2011).

Similarly, Rawlsian justice again reveals that India is under criticism for balancing its riparian ethics to equitable water sharing with downstream Bangladesh. India is the upper stream country of Bangladesh with an advantage of 54 shared river courses. Due to its geographical position and India sharing all these river courses, Bangladesh feels vulnerable to water-related hazards, mainly floods, droughts, and scarcity or crisis. The Teesta issue is an urgent need for equitable water management based on downstream rights, urging environmental sustainability and regional peace. The Teesta River is the lifeline for two countries, feeding millions of people in the two Indian states of Sikkim and West Bengal. India's water-sharing with Bangladesh, involving 54 shared rivers like the Teesta, is fraught with certain challenges. Bangladesh faces severe water scarcity during the lean season, perceiving it as exacerbated by India's hydroelectric power projects and water diversion through irrigation projects. An adhoc 1983 agreement allocated 36 percent of Teesta water to Bangladesh and 39 percent to India, but domestic pressures in West Bengal have stalled further progress, notably derailing a 2011 interim deal (Noolkar-Oak, 2017). Rawlsian justice underscores Bangladesh's demand for equitable sharing, while deontology highlights India's intent to negotiate fair solutions, though constrained by domestic politics. Utilitarian reveals India's approach to benefit its citizens through existing infrastructure projects; however, it failed to address the maximum population across borders, particularly in downstream Bangladesh. Recent bilateral discussions on the Teesta and other rivers, such as the Feni, adding to six other rivers Gumti, Jaldhaka, Khowai, Manu, Muhri, and Torsa, reflect mandates of India's Neighbourhood First policy, aiming for agreements that benefit all the parties (Ministry of Jal Shakti, 2022), yet the unresolved Teesta issue underscores the ethical tension between national and regional priorities. West Bengal has more concerns regarding the Teesta for livelihood in the Northern region. The decision-making in the future is therefore paramount concerning the problems of West Bengal, and the policymaking process should acknowledge the perceptions of people and the state. The problems of Bangladesh also must need to be acknowledged in the policymaking process. Since India has been maintaining good neighbourhoods through the Neighbourhood First policy, Bangladesh should be on the first list.

In contrast, India's relations with upstream states Nepal and Bhutan are characterized by stable cooperation, facilitated by India's economic and military dominance. Collaborative projects, such as the Pancheshwar initiative with Nepal and hydro-meteorological networks with Bhutan, demonstrate mutual benefits. The absence of significant geopolitical tensions supports a sustainable framework where water serves as a source of cooperation. India's ethics for maintaining cooperation with Nepal has an integrated manner in project management, particularly at the Pancheshwar and Rupaligad sites. Similarly, India, with Bhutan, funded a scheme, i.e. "Comprehensive Scheme for Establishment of Hydro-meteorological and Flood Forecasting Networks,"on common rivers including Aie, Jaldhaka, Manas, Pagladiya, Puthimari, Raidak, Sunkosh, andTorsa. Both countries established a joint expert team consisting of senior officials appointed by the Governments of both India and Bhutan to continuously review the progress and other necessities of 32 networks (Government of India, 2023). However, Rawlsian justice suggests that India's dominance may raise concerns about equitable benefit-sharing with these weaker states, hinting at an ethical imbalance driven by power asymmetry. It can be rightly argued that India's overall relationship with both Bhutan and Nepal is much better due to the relations between the weak and strong in terms of economic and military power. The national interest of the stronger state (India) over the weaker (Bhutan and Nepal) has led to the development of stable management. Ethical stances that India meets reflect this power imbalance, often shaped by the water resource dimension of national interest. India has an interest in utilizing water resources flowing from these two upper riparian states to meet its domestic needs, unlike other downstream riparian states, including Bangladesh and Pakistan, and the upstream China.

With the uppermost riparian state of China, India's relationship over the Brahmaputra River is different, marked by conflict and mistrust. Long-standing geopolitical tensions, particularly over border disputes, limit cooperation, with hydrological data-sharing and reluctance to a legal framework (UN Watercourse Convention) having a very deep connection to the present and future water governance. As a result, water has always been perceived as a source of conflict rather than a source of cooperation. China, being the most upper riparian state to share the Brahmaputra River, has a bigger geopolitical position. India, as a lower and mid (if including Bangladesh), always perceives that China has a bigger geographical position to capture all major river systems originating in Tibet. China's hydropower projects and dam constructions, including its north–south water diversion, pose significant threats to India's water security. Despite a global policy framework and the UN-water initiatives (such as the General Assembly Resolution as mentioned earlier), China's state-centric approach prioritizes development over human rights, making it more or less a hydraulic mission of construct, command, and control. The International Campaign for Tibet report (2024) finds a significant impact of the Chinese rapid hydropower dam on the Tibetan civilization, including irreparable damage to the environment. It is noted that China disrespects human costs, particularly the rights and dignity of the Tibetan people. The report finds that 1.2 million people will be displaced, affecting their sociocultural aspects (International Campaign for Tibet, 2024). As a midstream state, India faces ethical challenges in advocating for equitable governance while fulfilling its downstream obligations to Bangladesh, with China's actions influencing India's own water policies towards other riparian states.

DIVERGING FACTORS/CHALLENGES TO ETHICAL CONSIDERATIONS

Lack of Common Framework

There is still a lack of common international and global governance on transboundary water management. The UN Watercourse Convention still needs to go long for implementation. Even this framework was adopted in 1997 and came into force only in 2014. Some important riparian states like China even have not signed it. In fact, these frameworks are limited to dialogue and discursive, which do not contribute at all to shaping individual nations and regions. Anarchical society, what we call an absence of global governance, stands at this point of lack and weakness of a common management framework. With the value of water resources at present and future, there is an urgent need to have a common legal and institutional framework like the UN for the specific management of water resources globally. The recent UN Initiatives, such as SDGs, Resolution, UN-Water, etc., have also failed to bridge the gap between ethics and politics, particularly in the context of transboundary water. These initiatives have mentioned the ethical issues related to drinking water, sanitation, agriculture, and largely the environmental concerns but failed to connect with the transboundary waters for addressing power politics. Transboundary water has been considered as simple water management, primarily focusing on these ethical issues without looking deeply from a political perspective. Water has only been considered as a human dimension of ethical norms rather than the political and security dimensions.

Geopolitical Dynamics

History always tells us how water has been used as a cause, trigger, and weapon in many international conflicts between states and non-state actors. The geopolitical tensions have a deep connection to water resources making them as weapons. India has had border and territorial disputes with China and Pakistan since the very beginning. History indicates a long-standing trouble of geopolitics on the boundary or territorial dispute between India and its neighbourhood. These tensions have replicated the current water-related issues. The transboundary water issue always complicates territorial disputes and vice versa. In the case of China, there has always been tension over the Brahmaputra River water sharing, replicating their geopolitical issue, particularly the border tension. There are ongoing concerns that water issues could exacerbate the issue and hinder regional stability (Rahman, 2017). China's behaviour of weaponization of the Brahmaputra River when the Doklam Crisis was at its peak by refusing to share hydrological data with India (Singh, 2020) was a critical example. Again, in the recent clash at the Galwan Valley in eastern Ladakh, China weaponized the Galwan River flows to India (Tembey, 2020).Similarly, the Indus Water is shaped and becomes more complicated due to geopolitical tensions and as it flows through Kashmir, a strategic nerve centre for India and Pakistan(Khalid, 2023). Although the Indus Treaty plays a significant role in equitable water sharing and mitigating conflict, both countries remain locked in complex issues on territorial issues of Kashmir and terrorism. The water issue has been perceived as a security threat and the source of conflict is often interlinked to these issues. For instance, following the terrorist attack in India, Indian PM Modi proclaimed that "blood and water can't flow together," leading to the suspension of the review meeting of the Indus Commission. India was under huge pressure from other officials to fully exploit the Eastern Rivers (Vater, 2021). India was pressured to give a befitting and early response to Pakistan, including the scrapping of the treaty. This decision was short of abrogating the Treaty but hardened its full use of legal rights (Chaudhury, 2016). During the Pulwama terrorist attack in 2019, the then water resources minister Nitin Gadkari threatened to stop water flows to Pakistan. PM Modi also stated that waters should be diverted to Haryana for farmers. These are all reported as an act of aggression (Vater, 2021) which should not be part of decision-making at all. The recent terrorist attack in Pahalgam, Kashmir, has also forced India to move towards abeyance of the Treaty temporarily until Pakistan's initiative to stand against cross-border terrorism. This somehow contributed to the escalation of the bilateral conflict when India moved with military operations against terrorism.

Again, the changing geopolitics in South Asia, with a huge influence of China, plays a crucial role in shaping India's riparian ethics and transboundary water governance with Bangladesh. China is one of the biggest trading partners of Bangladesh with zero duty on 97 percent of imports from Bangladesh (Ghosh, 2020). As parts of the Chinese BRI project, quota-free programmes for the Least Developed Countries and infrastructure investment in Bangladesh are all significant factors that play crucial roles. The recent decision of Bangladesh to discuss Chinese loans to manage the Teesta River is leading to more trouble between the two countries. This decision could have a reverse implication for Bangladesh overwhelmingly. Being the immediate neighbor of India, Bangladesh should be conscious of India instead of relying on such crucial geopolitical options.

Climate Change

Climate change exacerbates water scarcity and variability, complicating India's transboundary water management efforts. Fluctuations in monsoon patterns and droughts not only affect domestic water security but also have implications for neighbouring countries that share water resources. Even, it has a huge impact on the provision of equitable and reasonable management. More scientific and advanced technological adaptive strategies should be developed for their altered precipitation patterns and their impacts on river flows. Particularly, adaptive management strategies are necessary to collaborate effectively on mutual challenges such as floods and sustainable ecosystems (Asdak & Yulizar, 2024). The water flows from these glaciers are the lifeline for millions of people in the region and the availability of the river basins depends on the fluctuation of the climate. South Asia is one of the most vulnerable regions to climate change (Mousa et al., 2021). The Sixth Assessment Report (2023) by the UN Intergovernmental Panel on Climate Change (IPCC) presents a troubling forecast of predicting hotter weather, longer monsoon, and increasing droughts due to the increasing global warming by 1.5 degrees Celsius in the next two decades. This change worsens water scarcity complicating the likelihood of tensions among India, Pakistan, and China over the transboundary rivers (Kugelman, 2024). It is vulnerable to long-term changes in run-off as climate change is affecting glaciers (Chandrashekhar, 2022). In the case of the Brahmaputra River, the recent floods have worsened human livelihood in India and Bangladesh. Floods in 1998, 2007, 2010, and 2020 resulted in serious human displacement in Bangladesh with three million people rendered homeless (Krajick, 2020). The implications of climate change should be considered within the mandate of such treaties and institutional arrangements collaboratively.

DEMOGRAPHY, TECHNOLOGY, AND ECONOMY

The interlinking of these three aspects also plays in determining water management for ensuring sufficient, equitable, efficient, effective, and sustainable management. It is interlinked since there is a need for water security for demographic change such as rapid population growth and demand for water resources, and the weak technological capacity to ensure water management contributes to the current unsustainable water management practice. The economy plays in shaping how weak technological development ensures water insecurity triggered by population growth. In this regard, India, with the second-largest population in the world, is responsible for providing water resources. As the demand is increasing without advanced technological capacity, the management practice of water resources is dependent on conventional methods, and the economically unsound leads to underdeveloped and unsustainable management practices. At the same time, the technological use should be developed in a collaborative approach so that the construction of the dam, hydroelectric projects to empower the economy, diversion of water, etc. within one country would not have implications for other countries. Uncollaborated use of technology and management of water will be the driving factor leading to water conflict between India and the neighbourhood since these countries are also dependent on the same methods. Pakistan and Bangladesh are two lower riparian states that depend on water flows from upper riparian states India and China. These two countries are economically unsound, with a larger population and an agriculture-oriented economy, and unscientific methods of water management lead to more water insecurity in the region. Therefore, there is a need for a technologically advanced approach to water security with climate overcoming potential among the riparian states. Instead of playing downstream and upper-stream power politics, despite individual projects, one investment with all-around potential application needs to be undertaken by all the riparian states.

CONCLUSION: THE WAY FORWARD

India's transboundary water policies, as articulated in its NWP (2002, 2012) and operationalized through bilateral treaties and agreements, demonstrate a complex interplay of ethical commitments, national interests, and regional dynamics in managing transboundary waters. These policies contribute to environmental sustainability and regional cooperation to a moderate extent, achieving successes through frameworks like the IWT (1960), GWT (1996), MWT (1996), and cooperative hydropower projects with Bhutan, which align with principles of equitable water sharing, hydrological data exchange, and conflict resolution. The India–Bhutan model, in particular, exemplifies how ethical cooperation can foster sustainable water management and regional stability. However, significant challenges such as geopolitical tensions, climate change, demographic pressures, and technological limitations undermine India's ability to fully embody ethical principles of no significant harm, human rights, environmental values, and accountability. Other important factors are also influencing the general ideas, thoughts, and discourse regarding transboundary water governance. These are the disciplinary problems associated with the study of transboundary water, often being studied in International Relations, Political Science, and Peace and Conflict and Security Studies. The state actors and seemingly state interests and problems are the subject matters that lead to the development of state-centric policy guidelines and legislation. Despite global popularity, the ethical dimensions of human and ecological balance have been missed from being included. The regulatory mechanisms are also under criticism for their failure to deal with such critical issues.

From a utilitarian perspective, India's policies prioritize national benefits like hydropower and irrigation, often at the expense of downstream ecosystems and communities, as seen in conflicts over the Teesta and Brahmaputra rivers. Rawlsian justice highlights inequities in water allocation, particularly with downstream states like Bangladesh and Pakistan, where India is perceived as hydro-hegemony creating such imbalances. From a deontological perspective, India adheres to treaty obligations but faces dilemmas when domestic priorities, such as West Bengal's concerns over the Teesta, clash with international commitments, for instance, with Bangladesh over the Teesta River. Environmental ethics reveals a gap in prioritizing ecosystem health, with treaties like the Indus lacking provisions for climate-induced challenges like glacial melt. The absence of a multilateral framework, compounded by geopolitical rivalries with China and Pakistan, further limits cooperative and sustainable outcomes. To enhance its contribution to environmental sustainability and regional cooperation, India must integrate adaptive management strategies to address climate change, foster multilateral basin-wide frameworks, and prioritize transparency and inclusivity in data-sharing and decision-making. By aligning more closely with global standards like the UN Watercourse Convention (1997) and embedding ethical principles into policy implementation, India can transform transboundary water governance from a state-centric, nationalist endeavour into a holistic, sustainable, and equitable model that ensures ecological integrity, human dignity, and regional harmony in South Asia.

India can only respond to these challenges effectively by embracing an integrated water management strategy at the regional scale, supported by a multilateral agreement among South Asian nations. Such an accord would not only facilitate fairer distribution of water but also help address soil erosion, unsustainable farming practices, and the overuse of natural resources. In particular, India and its neighbours must establish and regularly update a regional database of freshwater resources, enabling evidence-based policymaking and contingency planning to confront water scarcity. The worrying increase in pollutants contaminating freshwater supplies, together with their adverse impacts on agriculture, human health, and soil fertility, demands urgent attention. Legal frameworks play a crucial role in shaping the norms and mechanisms of water governance at local, national, regional, and global levels, setting out the rights and responsibilities of contracting states in relation to the allocation and use of shared water resources.

ACKNOWLEDGMENT

The authors declare that there is no conflict of interest. The authors want to acknowledge the Indian Council of Social Science Research for the grant. The authors declare that there is no conflict of interest. They gratefully acknowledge the Indian Council of Social Science Research (ICSSR) for providing the grant supporting this work. This chapter is part of ongoing Ph.D. research conducted under the two-year Doctoral Programme funded by the ICSSR. The authors also extend their sincere thanks to the external reviewers for their invaluable comments and suggestions, which greatly contributed to the revision.

REFERENCES

Agarwal, Ruchir, Balasundharam, Vybhavi, Blagrave, Patrick, Cerutti, Eugenio M., Gudmundsson, Ragnar, and Mousa, Racha. 2021. "Climate Change in South Asia: Further Need for Mitigation and Adaptation." IMF.https://www.imf.org/en/Publications/WP/Issues/2021/08/20/Climate-Change-in-South-Asia-Further-Need-for-Mitigation-and-Adaptation-464333.

Araral, Eduardo, and Wu, Xun. "Comparing water resources management in China and India: Policy design, institutional structure and governance." *Water Policy*, 18, no. S1 (2016): 1–13.

Asdak, C., and Yulizar, Y. "Transboundary water resources management and food-security: An Indonesian perspective." *World Water Policy*, 10, no. 2 (2024): 553–565.

Basumatary, Japungsar. "Bangladesh-India water issue: A study on Teesta River." *International Journal of Research and Analytical Reviews*, 5, no. 4 (2018): 20–23.

Biswas, A. "Cooperation or Conflict in Transboundary Water Management: Case Study of South Asia." In *Asian Perspectives on Water Policy*, Cecilia Tortajada and Asit Biswas(eds.). London: Routledge, 2012.

Bogard, Janos J., Bharat, Luna, Foster, Stephen, and Dhaubanjar, Sanita. "Water and Its Management: Dependence, Linkages and Challenges." In *Handbook of Water Resources Management: Discourses, Concepts and Examples*, Janos J. Bogardi et al. (eds.). Dordrecht: Springer, 2021.

Chandrashekhar, Vaishnavi. 2022. "As Himalayan Glaciers Melt, a Water Crisis Looms in South Asia." Yale E360. October 3, 2022. https://e360.yale.edu/features/himalayas-glaciers-climate-change.

Chaudhury, Dipanjan Roy. 2016. "Blood and Water Can't Flow Together: PM Modi at Indus Waters Treaty Meeting." *The Economic Times*, September 27, 2016. https://economictimes.indiatimes.com/news/politics-and-nation/blood-and-water-cant-flow-together-pm-modi-at-indus-waters-treaty-meeting/articleshow/54525240.cms.

Dawn. 2011. "2008: 'Illegal' Filling of Baglihar Dam Led to Water Scarcity in Pakistan." DAWN.COM. July 2, 2011. https://www.dawn.com/2011/07/02/2008-illegal-filling-of-baglihar-dam-led-to-water-scarcity-in-pakistan/.

Desai, Priya. 2019. "Applying Ethics to Water Management, to Ensure Equity in Access to All." India Water Portal. https://https://www.indiawaterportal.org/articles/applying-ethics-water-management-ensure-equity-access-all.

Ghosh, Palash. 2020. "China Grants Tariff Exemptions to 97% of Bangladesh's Imported Products." International Business Times. June 19, 2020. https://www.ibtimes.com/china-grants-tariff-exemptions-97-bangladeshs-imported-products-2997286.

Government of India. n.d. *Ministry of Jal Shakti, Govt. of India*. https://www.jalshakti-dowr.gov.in/offerings/international-collaboration (accessed August 30, 2022).

Government of India. 2006. "India-China Cooperation." Department of Water Resources, River Development and Ganga Rejuvenation. https://www.jalshakti-dowr.gov.in/offerings/international-collaboration (accessed February 24, 2024).

Government of India. 2014a. *Barak and Others Basin*. Ministry of Water Resources, New Delhi: Government of India.

Government of India. 2014b. *Ganga Basin*. Ministry of Water Resources, New Delhi: Government of India.

Government of India. 2014c. *Indus Basin*. Ministry of Water Resources, New Delhi: Government of India.

Government of India. 2023. "India-Bhutan Cooperation." Department of Water Resources, River Development and Ganga Rejuvenation. January 3, 2023. https://www.jalshakti-dowr.gov.in/offerings/international-collaboration.

Groenfeldt, David. *Water Ethics: A Values Approach to Solving the Water Crisis*. London and New York: Routledge, 2019. https://doi.org/10.4324/9781351200196.

Groenfeldt, David. "Water Ethics." In *Handbook of Water Resources Management: Discourses, Concepts and Examples*, 5. Cham: Springer International Publishing, 2021. https://doi.org/10.1007/978-3-030-60147-8.

Hanasz, Paula. "Power flows hydro-hegemony and water conflicts in South Asia." *Security Challenges*, 10, no. 3 (2014): 95–112.

Hassan, A. E. "Indian Hegemony on water flow of the Ganges: Sustainability challenges in the southwest part of Bangladesh." *Sustainable Futures*, 1 (2019): 100002.

Ho, S. "'Big brother, little brothers': Comparing China's and India's transboundary river policies." *Water Policy*, 2016.

Holslag, Jonathan. "Assessing the Sino-Indian water dispute." *Journal of International Affairs*, 64, no. 2 (2011): 19–35.

Hussain, M., Khalid, M., and Imran, S. Hydro-Hegemony Framework: A study of the India Pakistan trans-boundary water competition and cooperation. *Liberal Arts and Social Sciences International Journal (LASSIJ)*, 5, no. 1 (2021):537–553.

International Campaign for Tibet. December 5, 2024. https://savetibet.org/chinas-hydropower-dam-expansion-destroys-tibetan-homes-and-temples-stands-to-displace-1-2-million-people-and-drives-climate-change/ (accessed May 6, 2025).

International Policy Digest. December 20, 2013. https://intpolicydigest.org/2013/12/20/india-bangladesh-river-water-sharing-politics-cooperation/ (accessed July 4, 2021).

IPCC, 2023: Climate Change 2023: Synthesis Report. Contribution of Working Groups I, II and III to the Sixth Assessment Report of the Intergovernmental Panel on Climate Change [Core Writing Team, H. Lee and J. Romero (eds.)]. IPCC, Geneva, Switzerland, 184 pp., https://doi.org/10.59327/IPCC/AR6-9789291691647.

Iqbal, Anwar. 2018. "Explainer: What Is the Kishanganga Water Dispute." *DAWN*, May 20, 2018. https://www.dawn.com/news/1408795.

Jennings, Bruce, Paul Heltne, and Kathryn Kintzele. "Principles of water ethics." *Minding Nature* 2, no. 2 (2009): 25–28

Khalid, Ozer. 2023. "India's Weaponisation of the Indus Water Treaty." *The Express Tribune*. February 7, 2023. https://tribune.com.pk/story/2399677/indias-weaponisation-of-the-indus-water-treaty.

Krajick, Kevin. 2020. "Future Brahmaputra River Flooding as Climate Warms May Be Underestimated, Study Says." News.Climate.Columbia.Edu. November 30, 2020. https://news.climate.columbia.edu/2020/11/30/brahmaputra-river-floods-climate/.

Kugelman, Michael. 2024. "South Asia Is on the Front Lines of the Climate Crisis." *Foreign Policy* (blog). February 22, 2024. https://foreignpolicy.com/2021/08/12/south-asia-climate-ipcc-report-front-lines/.

Meisch, Simon. "What Is Water Ethics and to What End Do We Study It? Lessons for the Water Ethics Charter." In *Global Water Ethics: Towards a Global Ethics Charter*, 49. London and New York: Routledge, 2017.

Ministry of Jal Shakti. "38th Meeting of Ministerial level Joint Rivers Commission of India and Bangladesh." Government of India, Delhi, 26 August 2022.

Momen, N., and Rahman, M. "The impoverishment of political discourse and legal rights in transboundary Teesta water management."*World Water Policy*, 9, no. 1 (2023): 11–28.

National Water Policy. Government of India, Ministry of Water Resources. 2012. Accessed from https://www.nwm.gov.in/sites/default/files/national%20water%20policy%202012_0.pdf.

Noolkar-Oak, Gauri. 2017. "Geopolitics of water conflicts in the Teesta River Basin." Amsterdam: Both ENDS 57 (2017).

Postel, Sandra. 2008. "The Missing Piece: A Water Ethic." *The American Prospect*. https://www.internationalwaterlaw.org/bibliography/articles/Ethics/Postel-AmProspect.pdf.

Priscoli, J. D., Dooge, J. C., and Llamas, R. Water and ethics: Overview. UNESCO. (2004). Accessed from https://www.internationalwaterlaw.org/bibliography/articles/Ethics/Overview.pdf.

Qureshi,Waseem Ahmad. "Indus waters treaty: An impediment to the Indian hydrohegemony." *Denver Journal of International Law& Policy*, 46, no. 1 (2017): 45.

Rahman, M. Z. "Geopolitics of Sino-Indian transboundary water management in the Yarlung Tsangpo and the Brahmaputra." *Mondes en Développement*, 45, no. 1 (2017): 63–77.

Sachwani, K., and Sharma, S. 2025. "A Pivotal Moment in Water Sharing Norms." *Outlook India*, June 2, 2025. https://www.outlookindia.com/national/a-pivotal-moment-in-water-sharing-norms.

Sands, Philippe. "Bangladesh-India: Treaty on sharing of the Ganges waters at Farakka." *International Legal Materials*, 36, no. 3 (1997): 519–528.

Siddiqui, Naveed. 2021. "Pakistan, India Agree to Make Efforts to Resolve Indus Water Issues, Conduct Inspection Tours." *DAWN*, March 25, 2021. https://www.dawn.com/news/1614554.

Singh, Mayank. 2020. "China's Weaponisation of Water." *The Times of India*, April 25, 2020. https://timesofindia.indiatimes.com/blogs/voices/chinas-weaponisation-of-water/.

Tandan, P. Challenging hydro-hegemony of India: Resistance of Nepal in the upperKarnali and Saptakoshi dam project. *Sustainable Water Resources Management*,7, 106 (2021). https://doi.org/10.1007/s40899-021-00580-2

Tembey, Urvi. 2020. "China's Weaponization of River Water." Statecraft. July 25, 2020. https://www.statecraft.co.in/article/china-s-weaponization-of-river-water.

UNESCO. 2018. "Water Ethics: Ocean, Freshwater, Coastal Areas." UNESCO-COMEST. https://unesdoc.unesco.org/ark:/48223/pf0000265449.

UN-Water. 2018. "Transboundary Waters." UN-Water. https://www.unwater.org/sites/default/files/app/uploads/2018/10/WaterFacts_transboundary_water_sep2018.pdf(accessed June 7, 2025).

UN-Water. 2021. "Progress on Transboundary Water Cooperation – 2021 Update." UN-Water. https://www.unwater.org/publications/progress-transboundary-water-cooperation-2021-update(accessed June 7, 2025).

Vater, John. 2021. "The Indus Waters Treaty: Prospects for India-Pakistan Peace." The Institute of South Asian Studies. https://www.isas.nus.edu.sg/wp-content/uploads/2021/06/WP-345.pdf.

Wouters, Patricia K., and Devlaeminck, David J. "Comparing China and India's transboundary water governance: Insights from international law." *International Journal of Water Resources Development*, 41, no. 4 2024: 697–716.

9 Examining Legal Frameworks for Sustainable Transboundary Water Governance amidst Climate Change between India and Bangladesh

Nyamat Sekhon, Siddharth Prabhakar, Naresh Prajapati, Taranjot Singh, and Kumkum Sharma

INTRODUCTION

Nature inherently strives to maintain balance; however, human activities disrupting this equilibrium have led to climatic changes and a range of environmental challenges. Such disruptions frequently call for robust transboundary water management, particularly where geopolitical boundaries intersect with shared water sources (Varady et al., 2023). The disputes between India and Bangladesh illustrate how these challenges can further complicate the governance of river basins. This has long been a major source of contention, as much of the territory lies along river channels and fertile floodplains, which are frequently affected by seasonal flooding. Such events not only alter boundary definitions but also cause the displacement of local populations (Yan et al., 2019). The 4,000 km border between India and Bangladesh presents unique obstacles, intensified by varied terrain and contested islands (Figure 9.1). Effective transboundary water governance is therefore indispensable for ensuring water security, regional stability, and sustainable development (Pasricha, 2022). Climate change exacerbates these challenges, making it imperative to devise new solutions that uphold justice in the sustainable management of water resources (Douville et al., 2023). Although collaboration between India and Bangladesh has often been emphasized as essential, progress has been hindered by competing national interests and historical conflict (Baten & Titumir, 2015). While the issues remain politically sensitive, disputes over water-sharing, hydropower development, and navigation rights continue to dominate bilateral relations, underscoring the need for an inclusive framework that integrates climate change adaptation and mitigation strategies (Thankappan, 2024).

DOI: 10.1201/9781003599692-9

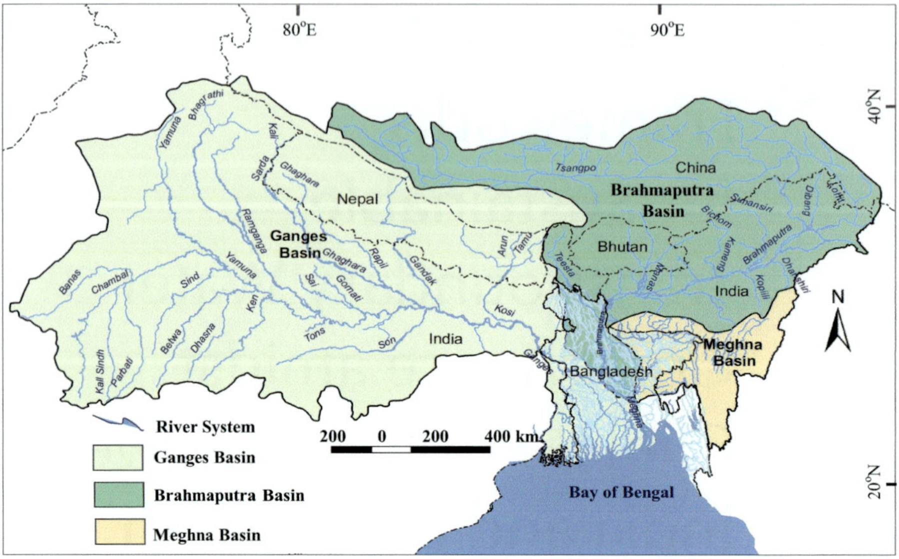

FIGURE 9.1 Transboundary river basins between India and Bangladesh. (*Source:* After BDP 2100, Knowledge Portal.)

Climate change is significantly intensifying hydrological stress across the transboundary river basins shared by India and Bangladesh, particularly in the Ganga and Teesta River systems. These river basins, originating in the Himalayan region and flowing through densely populated areas, are increasingly vulnerable to climatic disturbances such as altered precipitation regimes, rising temperatures, glacial melt, and extreme weather events. The Hindu Kush Himalayan (HKH) region, which serves as the primary water source for the Ganga and Teesta rivers, is warming at a rate almost double the global average (Pradhan et al., 2021). This region, known as the "Third Pole" due to its extensive glacial reserves, supports the hydrology of South Asia's most important river systems, including the Ganges and Brahmaputra, and affects the water security of over 1.3 billion people. One of the most visible impacts of climate change in this region is the rapid retreat of glaciers. Studies have shown that glaciers in the HKH region could lose up to one-third of their volume by the end of this century even if global warming is limited to 1.5°C (Bolch et al., 2019; IPCC, 2023). This retreat is altering river flow patterns by increasing run-off and sedimentation in the short term and decreasing water availability in the long term, particularly during dry seasons. As a result, both flood risks during monsoons and water scarcity during lean periods are intensifying, especially in downstream areas such as Bangladesh. This dual threat is of particular concern for rivers like the Ganga and Teesta, which exhibit strong seasonal variability in flow.

The Teesta River Basin (TRB), shared between Sikkim and northern West Bengal in India and the Rangpur division in northern Bangladesh, has witnessed a dramatic decline in dry-season water flow. According to estimates, the Teesta's discharge has declined by over 75% in the dry months since the early 2000s (Obaidullah &

Howlader, 2025). This decline has severe implications for irrigation-dependent livelihoods in Bangladesh, where large agricultural areas in the northern districts rely on Teesta waters during the rabi cropping season. The increasing unpredictability in river flow, driven by both climatic changes and upstream water diversions, has led to significant crop failures, drinking water shortages, and social distress among local populations (Arfanuzzaman, 2025). In addition to declining base flows, the frequency and intensity of monsoon-induced floods have increased across the region. The Ganga and Brahmaputra rivers are experiencing more frequent high-discharge events due to erratic monsoon rainfall and glacial meltwater surges. For example, the Brahmaputra basin witnessed five "once-in-a-century" flood events in just two decades (Sharma et al., 2025).

In Bangladesh, where over 70% of the land is flood-prone, this results in large-scale displacement, economic loss, and heightened food and water insecurity (Bhatt et al., 2020). The Ganga River, which flows through the Indo-Gangetic Plain and into Bangladesh's southwestern delta, has also recorded increased variability in flow and extreme flood events. These disturbances are further aggravated by infrastructure such as embankments and barrages, which are not designed to accommodate these new hydrological extremes. Shifting monsoon patterns and warming temperatures are having cascading effects on regional ecosystems. The altered flow regimes are disrupting wetland systems, fisheries, and groundwater recharge patterns. In the lower Teesta floodplain, declining surface flows have reduced wet-season replenishment of aquifers, which are heavily used for irrigation and domestic consumption. Simultaneously, flood-induced erosion along riverbanks is contributing to the loss of arable land and habitat fragmentation. Recent climate projections indicate that both the Ganga and Teesta basins will continue to face rising hydro-meteorological risks. According to the World Bank (2021), South Asia could incur climate-related economic losses of up to $160 billion annually by 2030 due to floods, droughts, and cyclones. In Bangladesh alone, the cost of climate-induced disasters could reduce GDP by 6.8% by 2050 if no adaptive measures are undertaken (ADB, 2020). Such vulnerabilities are compounded by the lack of adaptive legal mechanisms in existing transboundary water governance arrangements between India and Bangladesh.

The existing water shortage will get even more severe for millions with the intensification of climate change. With about 16% of the world's population, India has only 4% of its water resources and is already experiencing acute water stress. The situation further deteriorates with an average temperature rise of 0.7°C over the past century (NITI Aayog, 2021). Further, the rising sea levels enhance the salinity of the freshwaters in Bangladesh, affecting agricultural practices and human drinking (IPCC, 2019). The United Nations projects that, if current warming trends continue, by 2050, these countries could experience a reduction in the availability of their water by 40% (UN-water, 2021). The greater need arises for integrated water resource management programmes and investment in sustainable technologies as soon as possible, which will reduce these impacts.

Figure 9.2 illustrates the estimated number of flood-affected people (in millions) across South Asia as of 2010. Figure 9.3 overlays this flood-affected population with simulation results for the Himalaya, Karakoram, and Hindu Kush (HKHK) region. The maps present three key hydrological indicators for the Indus, Ganges,

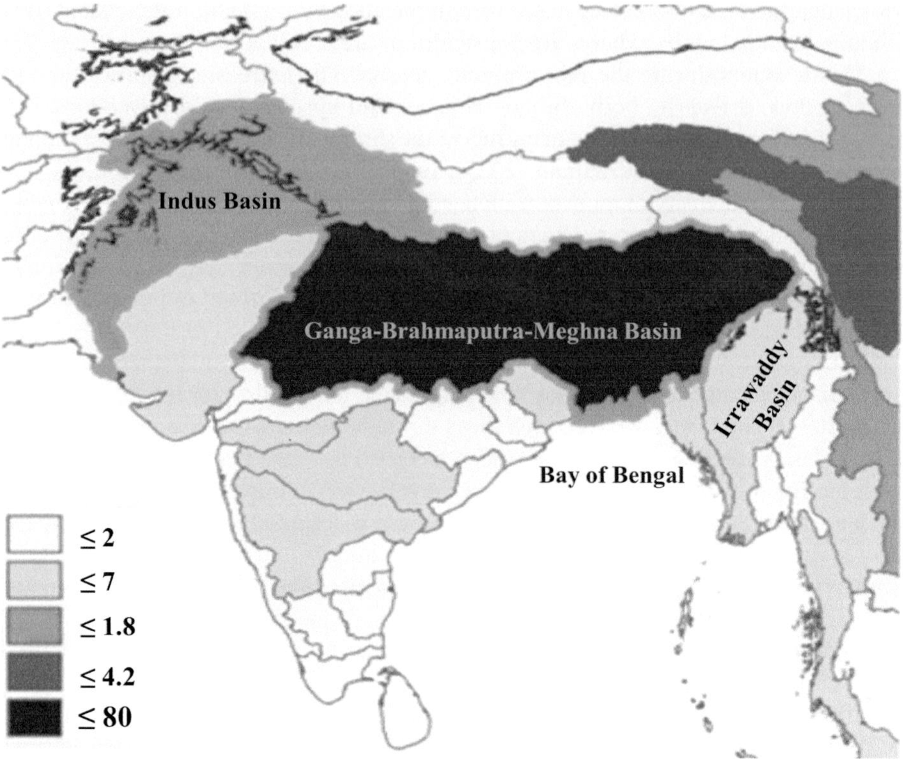

FIGURE 9.2 Estimated number of flood-affected population (millions) by basin 2010. (*Source:* Prepared by authors.)

and Brahmaputra River basins: (a) rainfall that results in immediate run-off, (b) total water production or run-off, and (c) water released from snowpack. In parallel with indicators of glacier melt, the flood-affected population in these three basins, particularly the Ganges and Brahmaputra, totals approximately 78 million people. The Indus basin also emerges as a high-risk area, with an estimated 15 million people affected by floods (World Bank, 2021). Climate change is no longer a future risk but a present reality, disrupting the hydrological stability and equitable access to shared river systems. This environmental vulnerability is not uniformly distributed but disproportionately affects downstream, lower riparian regions like Bangladesh, where communities already face structural limitations in coping with such crises. Addressing these vulnerabilities requires not only scientific monitoring and infrastructural adaptation but also legal frameworks that are responsive to changing climatic and ecological realities.

METHODOLOGY

This study employs doctrinal legal methodology to analyse India–Bangladesh transboundary water governance. Secondary sources, such as legislation, statutes, and treaties, are systematically examined. Furthermore, sources like reports,

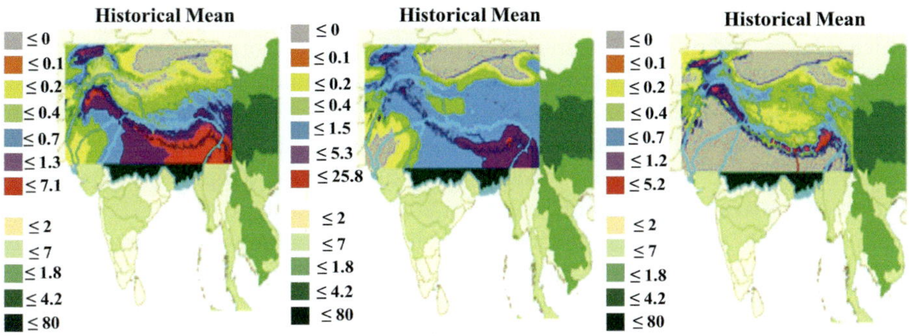

(a) Rain that runs off immediately **(b) Total water production or run-off** **(b) Water released from snowpack**

FIGURE 9.3 Historical mean indicators of glacier melt and estimated flood-affected population in South Asia. (*Source:* Based on Winsemius et al., 2015.)

Note: Winsemius et al. (2015) use hydrological data for 1960–1999 to generate flood inundations for the 10 years of a flood-return period. The estimation follows the baseline socio-economic and hydrological scenarios. The light blue line denotes the Indus, Ganges, and Brahmaputra basins. The dark blue line denotes 2,000 m of altitude in the Himalaya, Karakoram, and Hindu Kush region.

commentaries, and research papers have been analysed. The present study outlines the basic principles of India's neighbourhood policy, coupling it with water sharing principles for transboundary rivers, along with legal frameworks set at different levels of management across the globe and between India and Bangladesh, in order to correlate the same with climate change impacts and the need for modifications in the existing frameworks for adapting to the present circumstances.

LEGAL AND INSTITUTIONAL FRAMEWORKS IN GOVERNING TRANSBOUNDARY RIVERS

Transboundary rivers between India and Bangladesh are governed by a combination of bilateral treaties, domestic water governance systems, and international legal principles. However, the fragmented nature of these frameworks and their limited capacity to address climate-induced hydrological variability have increasingly exposed structural weaknesses in ensuring equitable and sustainable cooperation. The Ganga and Teesta rivers, two of the most significant transboundary systems in the region, demonstrate both the potential and limitations of current legal and institutional arrangements. The underlying legal instruments remain largely static in the face of evolving environmental realities, including those driven by climate change.

INTERNATIONAL WATER LAW PRINCIPLES AND REGIONAL APPLICATION

While neither India nor Bangladesh is a party to the 1997 United Nations Convention on the Law of the Non-Navigational Uses of International Watercourses (UNWC), the Convention's core principles, such as equitable and reasonable utilization, no significant harm, and prior notification, continue to influence normative expectations in South Asian transboundary governance (Uprety & Salman, 2011). These principles were earlier elaborated in the Helsinki Rules of 1966, adopted by the International Law Association, and remain the most widely recognized customary law principles concerning shared watercourses (McCaffrey, 2019). The principle of equitable and reasonable utilization implies that all riparian states must share benefits fairly, based on factors such as population dependency, climatic conditions, and contribution to the water source. In the India–Bangladesh context, this principle often collides with political asymmetries, where the upper riparian state (India) exercises more leverage (Xenarios et al., 2021). The "no significant harm" principle, which mandates that states avoid actions that cause appreciable harm to co-riparian countries, remains under-enforced, especially where upstream infrastructure projects are involved. Despite this international legal context, South Asia lacks a regional multilateral river basin treaty akin to the Mekong Agreement or the EU Water Framework Directive. As a result, legal governance of rivers remains bilaterally negotiated and politically fragile.

There are a number of internationally agreed-upon rules that govern the sharing of river water across borders. These rules are meant to make sure that shared water resources are used fairly, sustainably, and peacefully. The principle of fair and reasonable use is one of the most important norms. It encourages all riparian countries to use water fairly and sensibly, taking into account things like population, geography, ecological needs, and economic growth. The no harm rule, also known as the no impairment rule, is another important rule. It says that states must not do anything that could seriously hurt another State's environment, public health, or economic interests. Another important principle is the duty to cooperate, which stresses how important it is to keep talking, sharing information, and working together to manage shared waters well. Respect for sovereignty and territorial integrity is still very important. It means that each country's territorial rights and control over its own resources are recognized. The human right to water is important because it shows how important it is to make sure that everyone has access to clean and enough water as a matter of human dignity and basic rights. These principles form the moral and legal basis for international, regional, and national water laws. They are often used to make treaties, settle disputes, and come to agreements about water that crosses borders (Norouzi & Ataei, 2022). Based on these principles, all international, regional, and national water laws are framed and used in dispute resolution mechanisms.

WATER SHARING NEGOTIATIONS BETWEEN BANGLADESH AND INDIA

The divergent perspectives on water sharing between Bangladesh and India provide considerable hurdles to forging a fair deal. Bangladesh promotes multilateral discussions with Nepal and Bhutan, emphasizing their position as basin sharers. However, India favours bilateral discussions, expressing fears that Nepal's aims may not be compatible with Bangladesh's demands (Rasul, 2015). Another source of dispute is basin management, with Bangladesh favouring separate resolutions for each river basin and India arguing that all major rivers should be treated as a linked system. India's plan to connect rivers via a Brahmaputra dam exemplifies this crow's nest; however, Bangladesh and others contend that controlling the GBM rivers as a unified system is impracticable given its size, complexity, and international character (Faisal, 2002).

THE GANGES WATER-SHARING

Unresolved political tensions and inequality in the Ganges water-sharing problem led to poverty and environmental degradation. Bangladesh has attempted to address these concerns, but no equitable and viable solution has emerged. India has generally agreed with the Awami League since 1971, when it backed Bangladesh's independence (Faisal, 2002). The Farakka Barrage, which opened in 1975, strained relations. In 1977, a five-year agreement was struck, but India's plan to transfer Brahmaputra water and Bangladesh's proposal to construct dams in Nepal and Bhutan were rejected. After renewing the agreement in 1982 and submitting fresh suggestions in 1983, disagreements remained over diversion plans, with each nation anticipating severe social, environmental, and economic consequences (Rahman et al., 2019). The 1977 Bangladesh–India Ganges Water Sharing Agreement expired in 1988 after one renewal. From 1989 until 1996, there was no functional agreement. Bangladesh had water shortages due to India's control over flow downstream from the Farakka Barrage, especially during the dry season (Table 9.1). Bangladesh ratified the Ganges Treaty, a 30-year agreement, in 1996 after eight years of bilateral negotiations (Thomas, 2017).

The 1996 agreement used flow data from 1949 to 1988 to determine Farakka's water availability. It devised a complicated formula for water sharing during the dry season (January 1–May 31) to guarantee that each nation receives at least 35,000 cusecs of water every ten days, with Bangladesh receiving 27,633 and India

TABLE 9.1
Comparative Analysis of the Ganga Water Sharing Agreements between India and Bangladesh

Characteristics	1977 Agreement	1996 Agreement (Ganga Treaty)
Duration	Five years	30 years
Water availability data	Not specified	1949–1988
Water sharing formula	Not specified	Complex formula for the dry season
Minimum guaranteed water	Not specified	35,000 cusecs (conditional)
Focus	General water sharing	Lean season water sharing
Flood management	Not addressed	Neglected monsoon floods
Dispute resolution	Not specified	Joint Committee, Joint Rivers Commission
Outcome for Bangladesh's water supply	No significant increase	No guarantee of historical share

Source: Government of India, Ministry of External Affairs.

receiving 25,992. Water would be distributed equitably if the Farakka flow was reduced to 50,000–60,000cusecs. Despite practical criticism, the Ganges Treaty is widely regarded as a model transboundary water-sharing arrangement that fosters justice, equity, and mutual benefit. It also encourages both nations to address water-sharing difficulties in other rivers and use the Farakka Barrage water resources more evenly. However, unequal water distribution, generally 45:55 or 30:70, sparks heated debate in Bangladesh. Low water discharges during shallow flows are not guaranteed (Article III, the Ganges Treaty 1996).

A study of the 1977 and 1996 accords revealed that neither considerably increased Bangladesh's lean season water supply. The 1996 agreement failed to guarantee Bangladesh's historical water share. The agreement prioritizes water sharing during the lean season, neglecting the Ganges' importance in monsoon floods. Despite Bangladesh's desire for further water cooperation, India has limited sharing of the lean season and has provided SAARC flood forecasting data (Nishat & Faisal, 2000). As mentioned in Article 7 of the Ganges Treaty 1996, the Joint Committee shall be responsible for implementing the provisions of this Treaty and investigating any difficulties that may arise due to the preceding provisions and the functioning of the Farakka Barrage. Any discrepancy or dispute in this respect that the Joint Committee does not address would be forwarded to the Indo-Bangladesh Joint Rivers Commission (JRC). If the disagreement or issue remains unresolved, it will be reported to the two governments, who will meet immediately at the appropriate level to settle it by mutual conversation.

THE LINKING OF THE INDIAN RIVER PROJECT

Bangladesh and India have signed water-sharing agreements, but finding a compromise has proven difficult due to various factors. Bangladesh is concerned about the Indian River Linking Project (IRLP), one of the contentious topics. Indira Gandhi proposed the project in 1982 to transport water from the Ganges and Brahmaputra basins to southern India via pipelines, reservoirs, and dams to alleviate water scarcity. Until 2002, the idea to connect 30 rivers and build thousands of storage buildings and canals spanning hundreds of kilometres saw little progress. In response to a public interest petition in 2012, the Indian Supreme Court resurrected the project, highlighting national interest and drought and flood assistance (Bandyopadhyay & Perveen, 2008). The IRLP plans 30 river linkages, 14 Himalayan and the remaining Peninsular. Despite their objections, Bangladesh was excluded from feasibility studies for 14 Peninsular and two Himalayan linkages. Bangladesh says the plan breaches the 2010 Joint Communiqué, the 2011 Framework Agreement for Cooperation and Development, the 1996 Indo-Bangladesh Ganges Water Treaty, and the 1992 UN Convention on Biological Diversity (Mirza et al., 2008).

The project displaces communities, harms the environment, and affects Nepal and Bangladesh. Environmentalists criticize the planned linkages' capacity to address remote, severe water issues. Changing river flows may destabilize fishing habitats, endangering lives in India and Bangladesh. Despite these concerns, neither government has conducted a comprehensive social and environmental impact assessment on the impacted people (Rashid, 2012a). With the Treaty set to expire in 2026, both countries face mounting pressure to reformulate their agreement in light of growing climate stress, basin-wide ecological degradation, and increased dependence on surface water for agriculture and municipal supply in Bangladesh.

THE STALLED TEESTA WATER SHARING AGREEMENT

The Teesta River remains the most contentious transboundary river issue between India and Bangladesh (Figure 9.4). Despite extensive negotiations, a comprehensive water-sharing agreement has not been finalized. In 2011, both countries nearly reached a consensus on sharing Teesta's dry-season flow, reportedly proposing a 42.5% share for India and 37.5% for Bangladesh—but the deal was halted due to opposition from the West Bengal state government, which holds de facto veto power in India's federal structure. This underscores the complexity of Indian water federalism, where states retain constitutional authority over water resources under Entry 17 of the State List in Schedule VII of the Indian Constitution.

Bangladesh has persistently claimed that India's upstream diversions through the Gazoldoba Barrage have severely reduced Teesta flows, particularly in the lean season between December and March. Studies confirm that the river's dry-season flow has declined by over 75% in some years, exacerbating agricultural losses, groundwater depletion, and drinking water shortages in northern Bangladesh. In contrast to the Ganga Treaty, there is no existing binding legal framework governing the Teesta River. This absence of a formal agreement creates institutional uncertainty, heightens political tension, and exposes Bangladesh's northern districts to chronic water insecurity. Climate change further complicates negotiations: Teesta's flow is highly sensitive to both monsoonal shifts and glacial melt, which have become increasingly erratic over the last two decades. Additionally, the Teesta issue illustrates the broader problem of scale mismatch in transboundary governance; negotiations are conducted at the national level, but implementation and impact are local. Without participatory

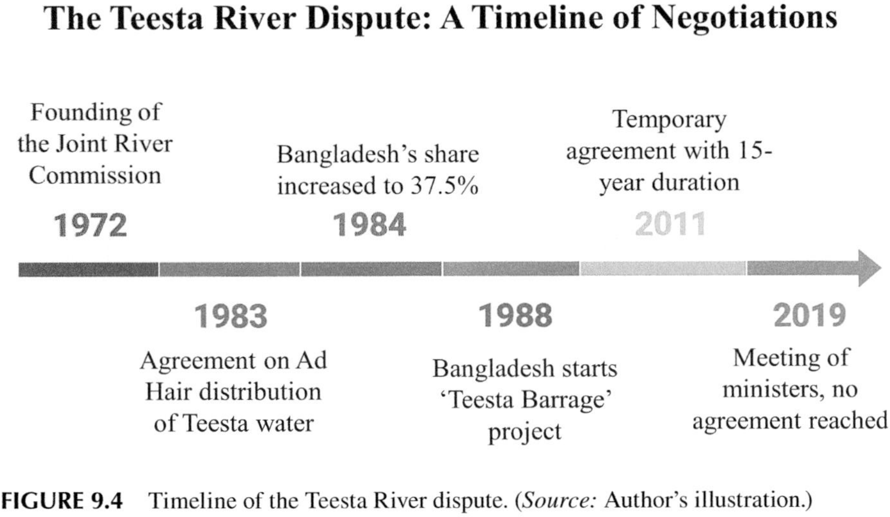

FIGURE 9.4 Timeline of the Teesta River dispute. (*Source:* Author's illustration.)

institutional mechanisms involving affected communities, basin-wide data sharing, or climate-informed modelling, the legal vacuum over Teesta continues to generate conflict rather than cooperation.

THE TIPAIMUKH DAM PROJECT

The Indian Tipaimukh dam, built in 1972 to avoid flooding, has been enlarged to produce energy. The Tipaimukh dam will control the Surma and Kushiyara rivers, affecting Tanguar Haor and millions of Bangladeshis. Bangladesh is concerned about dam impacts on water flow and the basin ecosystem (Nishat & Faisal, 2000). The two countries have bilaterally agreed to increase water flow and sharing, but Bangladesh may not get enough (Figure 9.5). India plans to dam the Barak River to regulate floods and create electricity, upsetting Bangladesh. The Barak (Surma) flows from Manipur, India, into Bangladesh as the Meghna. Despite early agreements and coordinated research by India and Bangladesh on flood scenarios and socio-economic consequences, execution proved difficult. Bangladesh requested Indian project bids since the dam is weak and seismically risky. India provided just one EIA report, which was incomplete (INHPC, 2012).

The EIA recommended a pick-up barrage downstream of the dam site to provide daily storage and irrigation to a broad command area (Mirza, 1997). Bangladeshi experts are concerned about the diversion despite India's promise to keep upstream water flowing (Burchi, 2014). Bangladesh responded with a 2005 hydrological impact assessment. IWM predicts that the Tipaimukh dam at the Amalshid point between the Barak River and the Surma–Kushiyara–Meghna river system may limit monsoon input by 10% in June, 23% in July, 16% in August, and 15% in September. The Kushiyara River in Amalshid may plummet by more than a metre. Fenchuganj, Sherpur, and Markuli may lose 0.25, 0.15, and 0.1 m, respectively. The Surma River may dip 0.75–0.25 m between Kanairghat and Sylhet. During dry monsoon years, the dam may decrease the water supply to the Barak–Surma–Kushiyara River (IWM, 2005).

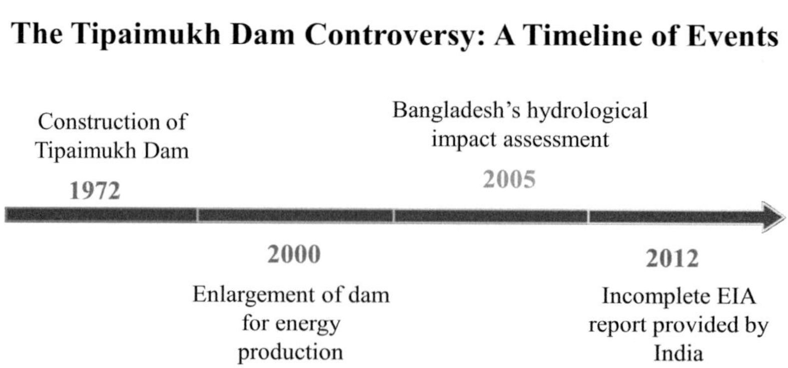

FIGURE 9.5 Timeline of the Tipaimukh dam controversy. (*Source:* Author's original illustration.)

THE FENI RIVER WATER-SHARING

India and Bangladesh are considering establishing bilateral water sharing on the Feni River. However, a 2019 MoU permits India to divert 1.82 cusecs of Feni River water for the Sabroom, Tripura water project. Bangladesh's Prime Minister Sheikh Hasina visited India on October 5, 2019, and signed the MoU. The agreement would improve the reliability of drinking water for over 7,000 homes in Sabroom. The Feni River, which connects India and Bangladesh, has been a source of conflict, and this agreement strengthens bilateral ties (Gupta, 2023).

THE KUSHIYARA RIVER WATER-SHARING

In 2022, India and Bangladesh signed the first temporary water-sharing agreement on the Kushiyara River in 25 years. This agreement tackles the fluctuating Barak River flow, favouring southern Assam in India and Sylhet, Bangladesh. This Memorandum of Understanding (MoU) permits Bangladesh to extract 153 cusecs of Kushiyara water to aid in producing Boro rice during the dry season (PIB, Govt. of India, 2022). Water will be delivered via the Sylhet Rahimpur Canal project to ensure agricultural supplies. This deal is part of a broader effort to strengthen cooperation in flood prevention, terrorism, and nuclear energy. The accord speaks well for the river water sharing between India and Bangladesh (India, Bangladesh Sign Pact on Sharing Waters of Kushiyara River, 2022).

KEY BARRIERS TO SUSTAINABLE TRANSBOUNDARY WATER COOPERATION

Despite decades of diplomatic engagement and multiple institutional mechanisms, transboundary water cooperation between India and Bangladesh continues to face substantial challenges (Figure 9.6). These barriers are not merely administrative but deeply embedded in legal ambiguities, political dynamics, institutional fragmentation, and a lack of climate-sensitive planning. The Ganga and Teesta river systems exemplify how such constraints collectively prevent the development of resilient, equitable, and adaptive transboundary governance frameworks. This section outlines the primary impediments to sustained cooperation, focusing on political, legal, institutional, and ecological dimensions.

POLITICAL FRAGMENTATION AND FEDERAL VETOES

A significant barrier to India–Bangladesh water cooperation lies in India's federal political structure, where water is a State List subject under Article 246 and Entry 17 of the Indian Constitution. This legal framework gives Indian states considerable autonomy over water management, including international rivers that pass through their territory. The Teesta dispute illustrates this vividly: although a near-final agreement was reached in 2011 at the bilateral level, it was stalled due to opposition from the Government of West Bengal, which objected to the proposed water-sharing formula, citing risks to local agriculture and irrigation security. This subnational veto

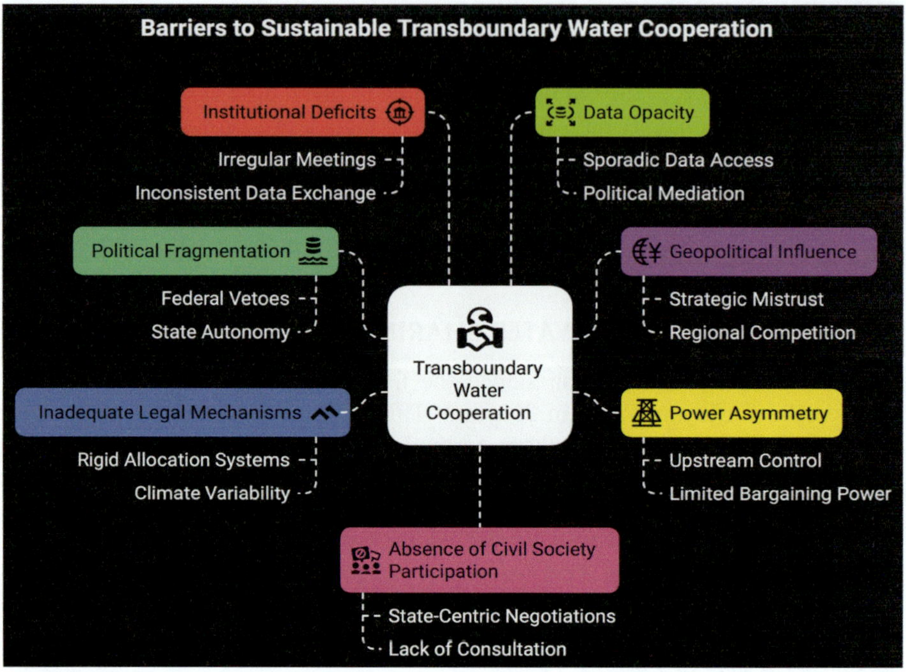

FIGURE 9.6 Barriers to sustainable transboundary water cooperation between India and Bangladesh. (*Source:* Author's original illustration.)

power creates a structural asymmetry in treaty implementation. While Bangladesh negotiates at the national level, India's internal political divisions allow state governments to effectively override national foreign policy decisions. Consequently, no agreement over Teesta has materialized in more than a decade, despite ongoing diplomatic assurances. The absence of a constitutional or legislative mechanism to harmonize state and central water priorities in international negotiations continues to weaken India's ability to fulfil cross-border commitments.

GEOPOLITICAL INFLUENCE AND STRATEGIC MISTRUST

India's hesitation in finalizing key transboundary water agreements, such as the Teesta Treaty, is partly shaped by China's expanding economic, infrastructural, and military presence in Bangladesh, especially in areas proximate to the Siliguri Corridor, a vital 22-km-wide land strip that connects India's north-eastern states to the mainland. This corridor has historically been considered India's strategic vulnerability or "Chicken's Neck," and any foreign encroachment near it triggers heightened security concerns in New Delhi. The revival of the Lalmonirhat airbase in northern Bangladesh with Chinese technical cooperation, only a short distance from this corridor, exemplifies the securitization of the region. Such developments have transformed India's water-sharing posture into one where geopolitical caution increasingly overrides purely hydrological or cooperative rationales. China's offer to finance the Teesta River Comprehensive Management and Restoration Project (TRCMRP) in Bangladesh has further compounded India's strategic unease. By presenting itself as an alternative hydrological partner, Beijing is not only asserting soft power in South Asia but also recalibrating Bangladesh's diplomatic alignment in a way that could diminish Indian influence.

India views this Chinese encroachment as an implicit challenge to its historical leadership in the region, interpreting Dhaka's engagement with Chinese-led water infrastructure as a strategic pivot. As a result, the water-sharing deadlock is no longer about mere allocation or state-level consensus; it has become entangled in a larger competition over regional influence and narrative control. India's hydro-diplomacy has entered a phase of securitized pragmatism, where legal frameworks for transboundary water cooperation are increasingly filtered through the lens of national security and territorial integrity. The delay in renegotiating the Ganga Water Treaty (due for renewal in 2026) further demonstrates India's recalibrated approach favouring cautious, bilateral, and politically flexible arrangements over binding long-term commitments, especially when the partner state is perceived to be drifting into China's strategic orbit. In this context, the failure to finalize the Teesta Treaty reflects not a lack of legal intent, but a deliberate deferral shaped by realpolitik where water is no longer a shared resource alone, but also a conduit of influence (The Economic Times, 2025).

POWER ASYMMETRY

The political economy of water cooperation in South Asia is strongly influenced by power asymmetry between riparian states. As the upper riparian and a regional hegemon, India controls the headwaters of most rivers flowing into Bangladesh, including the Ganga, Teesta, and several other tributaries. This upstream control allows India to pursue strategic infrastructure projects, such as barrages and dams, without prior consent or basin-wide consultation, as mandated under customary international law. Bangladesh, as a lower riparian state, remains highly vulnerable to upstream decisions and has limited bargaining power. While the Ganga Treaty (1996) offered some legal recognition of Bangladesh's dry-season water needs, the absence of similar frameworks for other rivers has created a perception of selective engagement and unilateralism. This imbalance undermines mutual trust, which is essential for long-term cooperation, especially in the context of increasing hydroclimatic uncertainty. Moreover, strategic concerns—such as India's growing focus on water security in its eastern and north-eastern border regions—have often superseded cooperative hydro-diplomacy (Gupta, 2016; Singh, 2018).

INADEQUATE LEGAL MECHANISMS FOR CLIMATE ADAPTATION

Existing legal instruments, particularly the Ganga Treaty, are ill-equipped to handle the accelerating impacts of climate change. The treaty relies on a rigid volumetric allocation system based on historical flow data, with little consideration for future hydrological variability (Kaura & Rani, 2020). There is no provision for adaptive allocation in response to monsoon failure, glacial melt surges, or seasonal droughts.

Teesta remains outside any formal legal framework, leaving Bangladesh's northern districts highly exposed to declining dry-season flows. Scientific studies confirm that the Teesta's seasonal variability is increasing, with climate-induced extremes threatening the sustainability of both ecosystems and livelihoods (Islam et al., 2021; ICIMOD, 2019). In the absence of treaty mechanisms that embed climate resilience, variability buffers, or basin-wide environmental planning, governance remains reactive rather than anticipatory.

INSTITUTIONAL DEFICITS AND WEAK ENFORCEMENT STRUCTURES

The JRC, though symbolically important, lacks the legal mandate, transparency, and operational authority to function as a robust cooperative institution. Meetings have often been irregular, data exchange inconsistent, and enforcement non-binding (Loucks & Van Beek, 2017). The JRC has no independent secretariat, no climate data modelling team, and no sub-basin monitoring cells that could provide real-time updates on flow, storage, or flood risks. Additionally, there is no basin-wide Integrated Water Resources Management (IWRM) plan that includes both countries. Without harmonized planning frameworks or joint investment in adaptive infrastructure—such as sediment management systems, early warning networks, or climate forecasting—river management remains fragmented and reactive. In recent years, extreme floods in the Brahmaputra basin and erratic flows in the Teesta have caught both governments unprepared, further revealing institutional fragility (Sharma, 2024).

DATA OPACITY AND ASYMMETRICAL INFORMATION FLOWS

Reliable hydrological data is critical to developing climate-resilient water-sharing mechanisms. However, one of the most persistent issues in India–Bangladesh river cooperation has been the lack of transparent and real-time data sharing. While there are bilateral agreements for some rivers, access to upstream flow data, sediment loads, water withdrawals, and ecological changes remains sporadic and politically mediated (Pramod & Eckstein, 2012). Bangladesh has repeatedly emphasized the need for joint hydro-meteorological stations, open access to discharge data during monsoon and lean periods, and joint impact assessments of proposed dams and barrages. However, institutional inertia and national security concerns have constrained full cooperation. The result is a persistent data asymmetry, where Bangladesh is often forced to negotiate with incomplete or outdated information.

ABSENCE OF CIVIL SOCIETY PARTICIPATION AND COMMUNITY ENGAGEMENT

Another key governance gap is the exclusion of local stakeholders, environmental experts, and civil society organizations from formal treaty-making and institutional monitoring processes. Transboundary water governance, especially under climate stress, requires multi-scalar participation to capture localized vulnerabilities and support decentralized adaptation measures. However, both India and Bangladesh have historically maintained state-centric, closed-door negotiations that leave out affected farming communities, fisherfolk, and indigenous groups living in riparian regions. This absence of participatory mechanisms has undermined treaty legitimacy and eroded grassroots support for water-sharing arrangements. In the case of the stalled Teesta agreement, for example, communities in Rangpur (Bangladesh) and Jalpaiguri (India) have expressed distrust of centralized decision-making, citing inadequate compensation for water diversion and lack of consultation during dam construction (Maja & Ayano, 2021). Without inclusive governance frameworks, long-term climate adaptation and conflict prevention remain difficult to achieve.

ENVIRONMENTAL VULNERABILITIES

Climate change and human interventions, such as dams and embankments, are changing river channels in India and Bangladesh, causing 117 rivers to dry up. This is a significant difficulty since the nations rely greatly on a reliable water supply for agriculture, industry, fisheries, and salinity management (Mahmood & Jia, 2016). The resulting changes in hydrological resources and their impacts are discussed next.

HYDROLOGICAL CHANGES

The Farakka Barrage has caused heated disputes, and Bharati and Jayakody (2011) show significant hydrological changes in the Ganges–Padma system. Despite variable precipitation, upstream diversion causes lower freshwater flow, which has downstream repercussions. India and Bangladesh face rising salinity, which disrupts navigation, fishing, and agriculture. The Barrage's effect is dramatic: reduced river velocity increases sedimentation, lowering navigability and worsening water shortages (Table 9.2). This loop reduces agricultural production and industrial efficiency and perpetuates poverty.

TABLE 9.2
Comparative Overview of Groundwater Depletion in the Ganges Basin: India v. Bangladesh

Aspect	India (Ganges Basin)	Bangladesh (GBM Delta)
Decline rate	2–4.5 cm/year storage loss, ~225 km^3 in 15 years	~1 km^3/year in some zones; north-eastern Bogra projected to double water table depth by 2060
Base-flow impact	Ganges summer flow down 59%, the river drying in summer	Recharge through surface hydrology helps, but impermeable areas show a net loss
Quality concerns	Largely quantity focused	Dual threats: depletion and arsenic migration
Policy response	Tech-level: Seeding of Rice on Beds (SRB); need crop policy shifts	Emphasis on modelling-based prediction, recycling, and safe-depth wells

Sources: Compiled by the authors from multiple sources including Handbook on Water Allocation in a Transboundary Context (2021), Wester et al. (2019), and IPCC (2023).

DECLINING GROUNDWATER LEVEL IN THE GANGES BASIN

Indian states and the southwestern part of Bangladesh have significant groundwater shortages and rely solely on the Ganges River system for agriculture and other operations. This region has just 4% of the nation's usable groundwater resources, estimated to be 1958 million cubic metres. The high demand for irrigation, limited rainfall, and excessive evapotranspiration exacerbate the issue. Overuse of groundwater, insufficient recharge, and decreased flow from the groundwater table to rivers have resulted in lower water levels in the Khulna division. Water tables in Khustia and Jessore districts have decreased from pre-diversion levels owing to excessive removal from hand and deep tube wells. Jalun, Kalpi, and other areas of Uttar Pradesh are also experiencing groundwater depletion.

OCCURRENCES OF FLOODING, SEDIMENTATION, AND RIVERBANK EROSION

Dams and embankments in the GBM river systems, constructed by Bangladesh and India, are intended to control flooding and help irrigation. However, their performance falls short of expectations, increasing flooding and riverbank erosion, notably after the Farakka Treaty (Faisal, 2002). Bangladesh, located between the Himalayas and the Bay of Bengal, suffers the burden of floods, with 80% of its territory susceptible to inundation (Table 9.3). Factors such as upstream water intake, hill flash

TABLE 9.3
Hydrological Hazards in the Ganges–Brahmaputra–Meghna (GBM) Basin

Category	India	Bangladesh
Major flood events	2022 Assam & North-East Floods: 9+ million affected; 113,000 ha crop loss. 2024 Northeast floods: severe in Tripura, Assam	1998 Flood: 30 million affected, 70% land submerged. 2024 flood: 502,500 displaced; 296,852 ha of crops damaged
Sediment load	Brahmaputra in Assam: ~540 million tonnes/year. Ganges near Farakka: ~440 million tonnes/year	Combined GBM: ~1 billion tonnes/year; only ~51% reaches sea, rest deposits on land and rivers
Sediment impact	Farakka Barrage: heavy siltation, upstream flooding, and riverbed rise	Delta clogging, increased floodplain waterlogging, and channel migration
Riverbank erosion	Brahmaputra: 868 km² lost (1912–1996); Majuli Island lost ~1.2–6.42 km²/year, 500,000+ displaced	Padma & Jamuna: ~180,000 ha lost (1970–2000); 10,000 ha eroded annually. 5,142 ha eroded (1993–2024)
Notable hotspots	Majuli Island, Rohmoria, Dibrugarh, Kaziranga, Palashbari	Harirampur (Padma): 2–12 km migration (1860–2009); Meghna & Padma plains: 159,000 ha eroded
Human impact	2,500 villages lost (1997–2008); frequent displacement in Assam & Bihar	1.6 million people displaced (1973–2010); recurring flood-induced migration

Sources: Compiled by the authors from multiple sources including Sweet (2024), UNICEF (2017), Wester et al. (2019), and IPCC (2023).

floods, and drainage congestion compound the issue, particularly when flood maxima coincide in all three rivers. The sedimentation of the GBM rivers leads to both land creation and flood aggravation. The proposed river interlinking may disturb silt deposition, affecting floodplain farming and coastal island building in Bangladesh (Chowdhury et al., 2022).

AGRICULTURE AND IRRIGATION

Post-Farakka troubles have spread to northwest and southwest Bangladesh, which relies on the Ganges and its tributaries for irrigation. It's reported that dry season surface water limitations aggravated hydrological and meteorological dryness. Since the mid-1970s, high-yielding variables (HYV) have increased irrigation requirements, particularly during the Boro season. Due to post-Farakka surface water restrictions, Bharati and Jayakody (2011) observed a shift to tube well irrigation. Many distributaries are dry since the lower Ganges water level has plummeted. According to Goes et al., 2021 agricultural loss has resulted from water limits in the G-K canals. River-linking projects may disrupt the delicate balance of water distribution by diverting surplus water to 'water-deficit regions. Misra et al. (2007) caution against the false idea of surplus water, which reduces surface and groundwater availability, resulting in waterlogging and agricultural output losses that endanger farmers' lives.

DRAINAGE CONGESTION

India and Bangladesh embank several transboundary rivers to avoid floods and boost agriculture. Both sides of the embankments choke drainage, although upstream is worse. India has barriers on the Kodalia and Isamati rivers, creating dry-season water shortages in Bangladesh. Several rivers have divided countries, and some have become borders. Since meandering rivers change channels often, new property ownership may be unclear. The Ganges and Kushiyara (a Barak branch in India) rivers also have geography concerns. Human embankments erode opposing banks, complicating matters. Nishat and Faisal (2000) differ on the Muhuri River bank erosion.

HUMAN MIGRATION

Environmental, political, religious, and ethnic factors influence voluntary or involuntary migration. Bangladesh's environmental disasters and internal and external migration statistics suggest that involuntary or forced migration has increased. Scholars argue that the Farakka-induced water constraint has resulted in massive migration from Khulna and Rajshahi (Ganges basin). The Ganges water dispute between Bangladesh and India may be used as an example of cooperation. Still, it is also an outstanding case study of an inter-state conflict in which two state actors rationally calculate their interests in a zero-sum scenario to obtain a single finite water resource. Environmental deterioration in a significant area of Bangladesh has forced river-dependent populations to relocate. According to Ahmed (2009), 67% of Khulna-Satkhira and Rajshahi migrants have relocated due to environmental instability. Winter hydrological dryness produced by the Farakka Barrage has aided regional migration.

TOWARDS CLIMATE-RESILIENT AND COOPERATIVE LEGAL FUTURES

The evolving climatic and geopolitical context across South Asia's transboundary river basins demands more than functional treaties and ad hoc institutional mechanisms. It necessitates a structural realignment of water diplomacy, legal design, and intergovernmental coordination to embed climate resilience, equity, and ecological sustainability into the very core of river governance. This chapter presents a forward-looking discussion of legal, institutional, and normative reforms that India and Bangladesh can consider in order to build durable, climate-informed, and cooperative water-sharing frameworks. Drawing from international legal standards and domestic policy innovations, the following discussion outlines actionable proposals and identifies key reform priorities.

INDIA'S NEED TO RATIFY THE UN WATERCOURSES CONVENTION

One of the foundational steps towards a legally consistent and globally aligned water governance posture would be for India to ratify the UN Convention on the Law of the Non-Navigational Uses of International Watercourses (1997). Though India participated in its drafting and abstained rather than opposed it during the vote, it remains a non-signatory, thereby depriving its transboundary water diplomacy of a clear legal benchmark. Ratification would majorly enhance India's legal credibility in transboundary water negotiations, including with Bangladesh, China, and Nepal. It will also allow for the formal adoption of equitable and reasonable utilization of resources and the obligation to cooperate as treaty obligations, not just aspirational norms, implying no significant harm. This might create legal momentum for regional hydro-cooperation frameworks in South Asia under customary international water law. Given the escalating ecological volatility in the Ganga–Brahmaputra–Meghna basin, India's adoption of multilateral legal instruments is a long-overdue necessity. It would also strengthen India's moral and legal standing in ongoing and future treaty negotiations with Bangladesh. Reforming transboundary water governance between India and Bangladesh requires more than technocratic fixes; it demands a structural and normative transformation rooted in climate reality, ecological interdependence, and legal clarity. As the expiration of the Ganga Treaty approaches in 2026 and the Teesta issue remains unresolved, both countries have an opportunity to reimagine water not as a contested commodity, but as a shared ecological and legal responsibility. Embedding climate resilience, cooperative sovereignty, and inclusive governance in future legal instruments will be central to building sustainable peace and environmental justice in the region.

MAINSTREAMING CLIMATE RESILIENCE THROUGH LEGAL OBLIGATIONS

The need to integrate climate resilience into transboundary legal arrangements is both urgent and non-negotiable. Current agreements like the Ganga Treaty and informal understandings on the Teesta are silent on climate variability, environmental shocks, or sustainability obligations. This omission weakens both treaties' ability to accommodate the rising unpredictability in flow, rainfall, and glacial dynamics.

To address this, India and Bangladesh should:

- Include a "climate resilience clause" in the 2026 Ganga Treaty renewal that mandates periodic flow reassessments based on IPCC data and scientific projections.
- Define "minimum environmental flow thresholds" to be guaranteed during lean seasons to protect riverine ecosystems.
- Mandate joint climate impact assessments before large infrastructure projects.
- Establish a shared Climate-Water Resilience Task Force within the JRC for integrated monitoring and legal response.

Such provisions will not only future-proof the treaty but also institutionalize long-term ecological thinking into bilateral water diplomacy.

TRANSITIONING TOWARDS ADAPTIVE LEGAL FRAMEWORKS

The rigidity of existing treaties, particularly the Ganga Water Sharing Treaty (1996), underscores the need to move from static volumetric allocations to flexible, climate-responsive legal frameworks. Unlike fixed allocation formulas based on historical flow data, adaptive frameworks allow for real-time adjustments based on actual hydrological and meteorological conditions. This can be operationalized through mechanisms such as flexible entitlements, seasonal flow bands, and minimum ecological flow guarantees, drawing inspiration from climate-informed treaty models such as the Columbia River Treaty (between Canada and USA) and the Mekong Agreement. These instruments incorporate climate projections, flow variability, and environmental flows into legal commitments (McCaffrey, 2020). As India and Bangladesh prepare to renegotiate the Ganga treaty post-2026, incorporating such flexibility would allow for better risk-sharing during drought years, enhanced ecological protection, and equitable flow distribution during stress periods.

INSTITUTIONAL REFORMS: THE ROLE AND LIMITS OF THE JOINT RIVERS COMMISSION

The India–Bangladesh JRC, established in 1972, was intended to serve as a permanent bilateral institution to ensure cooperation over 54 shared rivers. While the JRC has facilitated some data exchange and joint studies, its legal authority is limited, and its effectiveness has diminished over time due to infrequent meetings, lack of enforcement mechanisms, and absence of a basin-wide planning mandate. As climate-related risks increase and water disputes become more frequent, the institutional framework around the JRC requires significant reform. Currently, there is no IWRM approach, no climate data integration in water allocations, and no legal framework mandating adaptive policy revisions. The legal and institutional frameworks governing the Ganga and Teesta rivers are insufficiently equipped to address the complex, transboundary challenges posed by climate change. Existing agreements are rigid, incomplete, or absent altogether. Without legal innovations such as climate-adaptive treaties, dynamic allocation mechanisms, and empowered institutional bodies, India and Bangladesh may continue to face recurring hydro-political tensions and environmental insecurity in the coming decades.

Reinvigorating the India–Bangladesh JRC is critical to ensuring sustained institutional dialogue and evidence-based decision-making. The JRC, in its current form, lacks autonomous authority, scientific integration, and procedural transparency.

A reformed JRC should include:

- A permanent secretariat with interdisciplinary staff, including climate scientists, hydrologists, and legal experts.
- Real-time data-sharing protocols supported by joint hydro-meteorological observatories.
- Dispute resolution panels with legally binding jurisdiction.
- Community advisory boards representing vulnerable stakeholders from both upstream and downstream regions.

These institutional upgrades would elevate the JRC from a consultative body to a functional basin authority, capable of operationalizing treaty commitments and managing cross-border ecological risks proactively.

COMPARATIVE ANALYSIS WITH OTHER GLOBAL CASES OF TRANSBOUNDARY WATER MANAGEMENT UNDER CLIMATE STRESS

Transboundary water management is a primary global concern, compounded by the effects of climate change. Comparative research uncovers various worldwide examples of novel governance methods that have been effectively adopted. Various regional agreements and conventions across the globe address crucial concerns related to natural resource conservation and water management. These agreements promote cross-border collaboration for sustainable resource utilization, from the African Convention on the Conservation of Nature and Natural Resources to efforts such as the Danube River Protection Commission in Europe to accords in the Americas and Asia.

The African Convention on the Conservation of Nature and Natural Resources, signed in Algiers in 1968, aims to preserve the continent's natural heritage (Takang, 2014). This comprehensive framework covers various issues, including land and soil conservation, water management, species and genetic diversity preservation, and local community rights. The agreement emphasizes sustainable development, including avoiding land degradation and the long-term protection of natural resources (IUCN Environmental Law Programme, 2015). It aims to guarantee that Africa's natural resources are available for current and future generations by encouraging soil conservation, erosion avoidance, and sustainable agriculture methods (Erinosho, 2013).

In Europe, the International Commission for the Protection of the Danube River (ICPDR) was founded in 1994 to provide long-term water management and pollution control in the Danube River basin. The committee brings together riparian states and organizes efforts to protect the Danube River's resources (Chapman et al., 2016). Its emphasis on pollution control, accident risk reduction, and flood management demonstrates a commitment to using river resources sustainably while minimising environmental effects. Furthermore, the Rhine River Agreement among European nations, which focuses on pollution reduction and flood management, demonstrates how various national interests may be aligned via comprehensive treaties and shared environmental and economic incentives (Havinga, 2020).

In the Americas, accords such as the Framework Agreement for Integrated Water Resource Management (2004) and the La Plata Basin Treaty (1969) promote cross-border cooperation for sustainable water management (Elias, 2024). The Colorado River Basin Agreement between the United States and Mexico includes measures for drought mitigation and water sharing that are adjusted based on prevailing hydrological circumstances, showing flexibility in treaty frameworks to climatic variability (Grigg, 2023).

In Asia, collaborative accords such as the agreement on the Framework for Water Resources Management in the Euphrates-Tigris Basin (2000) and the founding of the Mekong River Commission (1995) seek to handle transboundary water issues (Altinbilek, 2004). These accords promote riparian states' cooperation in sustainably managing shared water resources. They aim to reduce climate change risks and provide equitable access to water for all stakeholders in the area by encouraging information sharing, technological transfer, and collaborative research (Gerlak & Haefner, 2017).

These studies give valuable insights into how flexible, adaptive management and cooperative frameworks might be built to deal with the uncertainties caused by climate change. These agreements encourage international cooperation in sharing water resources, intending to balance development and conservation (Sainz-Borgo, 2011). They aim to safeguard the long-term viability of water resources and ecosystems in their areas by encouraging collaborative efforts in resource management and infrastructure development.

ADOPTION OF BEST PRACTICES IN THE INDIA–BANGLADESH CONTEXT

Several essential lessons can be drawn from global examples to strengthen transboundary water governance between India and Bangladesh, particularly in light of increasing climate uncertainties (Figure 9.7). One of the most significant approaches is adaptive management, which involves embedding flexibility into water-sharing agreements through provisions for periodic revisions based on updated hydrological and meteorological data. This method helps both nations respond more effectively to the unpredictability and evolving nature of climate change. Equally important is the establishment of joint monitoring and data-sharing mechanisms, which foster transparency, trust, and timely decision-making. A successful model of this can be observed in the Mekong River Commission, where real-time data exchange supports

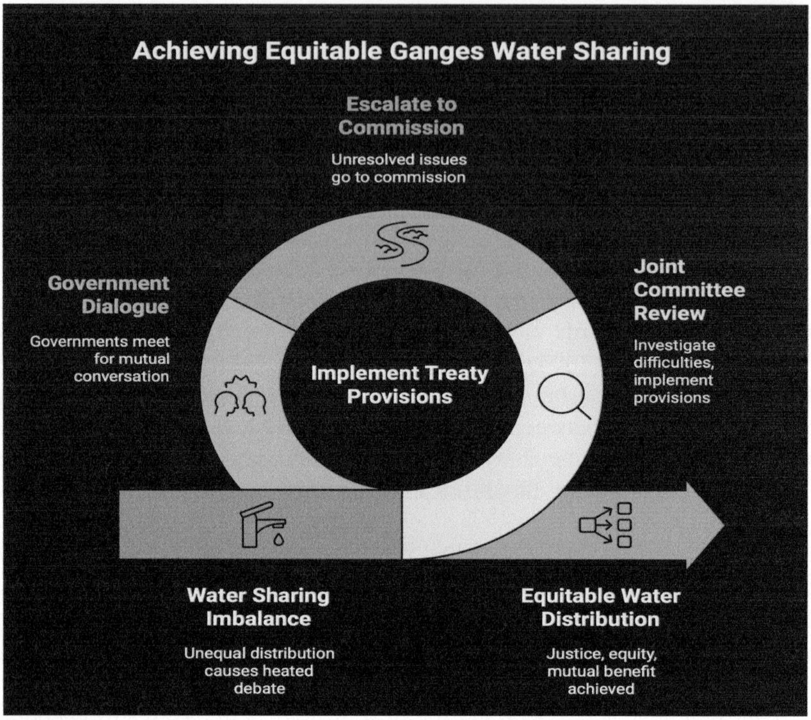

FIGURE 9.7 Factors towards achieving equitable water sharing. (*Source:* Author's original illustration.)

proactive and coordinated management across multiple countries. Furthermore, agreements that incorporate environmental and socio-economic objectives beyond mere volumetric water allocations tend to achieve greater long-term sustainability. The Danube River's integrated management framework is a prime example, as it balances ecological conservation with demands related to navigation and hydroelectric energy production (Humpesch, 1996). Additionally, embedding robust conflict resolution mechanisms within treaties is crucial for addressing disagreements constructively and preventing escalation.

The Nile Basin Initiative demonstrates how such frameworks can promote peaceful cooperation even among nations with contentious water claims (Mbaku, 2022). By adopting these global best practices, adaptive governance, collaborative institutions, integrated development goals, and structured dispute resolution, India and Bangladesh can build a more resilient and equitable transboundary water governance framework that is better equipped to manage the challenges of climate change and ensure sustainable water use for both nations. By incorporating these best practices, India and Bangladesh may create a more resilient framework for managing their shared water resources in changing climatic circumstances, resulting in more sustainable and equitable water governance.

CONCLUSION

The fragile future of South Asia's shared rivers, especially the Ganga and Teesta, sits at the intersection of ecological precarity, legal insufficiency, and geopolitical complexity. As this chapter has demonstrated, the existing legal and institutional frameworks governing transboundary water cooperation between India and Bangladesh are inadequate in both scope and design to address the climate-induced challenges that now define the region's hydrological realities. The Ganga Water Sharing Treaty of 1996, though a diplomatic milestone in its time, is now structurally misaligned with present and projected environmental risks. The prolonged failure to formalize the Teesta River agreement, further complicated by subnational vetoes and strategic mistrust, illustrates how domestic politics and geopolitical asymmetries continue to thwart meaningful cooperation. Meanwhile, accelerating glacial melt, erratic monsoons, and intensified floods and droughts threaten to destabilize livelihoods, ecosystems, and cross-border relations across the basin. In light of these developments, this research argues and supports paper has argued for a paradigm shift in transboundary water governance—one that centres climate resilience, legal adaptability, ecosystem protection, and institutional reform. Such a shift must be grounded in both international legal norms, particularly those enshrined in the UN Watercourses Convention (1997), and region-specific governance tools like a restructured JRC with binding authority and scientific expertise. The upcoming expiration of the Ganga Treaty in 2026 presents a rare but critical opportunity for India and Bangladesh to reimagine their water-sharing commitments through a sustainability lens. The inclusion of a climate adaptation and environmental flow clause in the renewed treaty would not only enhance ecological security but also institutionalize a forward-looking ethos in bilateral hydro-diplomacy.

Ultimately, the governance of transboundary rivers in the age of climate change must transcend extractive and unilateral models of sovereignty. Water, as this chapter contends, is not merely a strategic resource but a shared ecological and legal responsibility. Its management must reflect principles of equity, transparency, and intergenerational justice. For South Asia, a region historically shaped by water and increasingly threatened by its volatility, sustainable peace may well depend on how justly, cooperatively, and resiliently it governs its rivers. In conclusion, achieving sustainable transboundary water governance between India and Bangladesh amidst the challenges of climate change necessitates a comprehensive strategy that emphasizes cooperation, compliance with global agreements, holistic management, resilience to climate impacts, and openness. By focussing on these important areas, both nations can strengthen their ability to effectively manage shared water resources, guaranteeing the welfare of their people and the long-term viability of their river systems.

REFERENCES

Ahmed, I. (2009) Environmental refugees and environmental distress migration as a security challenge for India and Bangladesh. In: Brauch HS, Oswald Spring U, et. al. (eds), Facing Global Environmental Change: Environmental, Human, Energy, Food, Health and Water Security Concepts. Springer, Berlin Heidelberg, pp. 295–308.

Altinbilek, D. (2004). Development and management of the Euphrates–Tigris basin. International Journal of Water Resources Development, 20(1), 15–33. https://doi.org/10.1080/07900620310001635584

Arfanuzzaman, M. (2025). Transboundary water cooperation and joint river basin management are pivotal for climate resilient development in South Asia. World Development Perspectives, 38, 100681. https://doi.org/10.1016/j.wdp.2025.100681.

Bandyopadhyay, J., & Perveen, S. (2008) The interlinking of Indian rivers: Questions on the scientific, economic and environmental dimension of the project. In: Mirza MMQ, Ahmed AU, and Ahmad QK (eds), Interlinking Rivers in India: Issues and Concern. Taylor and Francis, Abingdon, pp. 53–76.

Baten, M. A., & Titumir, R. A. M. (2015, November 24). Environmental challenges of transboundary water resources management: The case of Bangladesh. Sustainable Water Resources Management. https://doi.org/10.1007/s40899-015-0037-0.

Bharati, L., & Jayakody, P. (2011). Hydrological impacts of inflow and land-use changes in the Gorai River catchment, Bangladesh. Water International, 36(3), 357–369.

Bhatt, M., Patel, R., Gleason, K., & Pandya, M. (2020, February 28). Disaster Risk Reduction and Climate Change Adaptation in South Asia. *Oxford Research Encyclopedia of Natural Hazard Science.* Retrieved July 2, 2025, from https://oxfordre.com/naturalhazardscience.

Bolch, T., et al. (2019). Status and change of the cryosphere in the extended Hindu Kush Himalaya region. In: Wester, P., Mishra, A., Mukherji, A., and Shrestha, A. (eds), The Hindu Kush Himalaya Assessment. Springer, Cham. https://doi.org/10.1007/978-3-319-92288-1_7.

Burchi, S. (2014, April 3). Flavia Rocha Loures and Alistair Rieu-Clarke (eds), The UN Watercourses Convention in force: Strengthening international law for transboundary water management. Water International, 39(3), 397–400. https://doi.org/10.1080/02508060.2014.903107.

Chapman, D. V., Bradley, C., Gettel, G. M., Hatvani, I. G., Hein, T., Kovács, J., et al. (2016). Developments in water quality monitoring and management in large river catchments using the Danube River as an example. Environmental Science & Policy, 64, 141–154.

Chowdhury, M. A., Hasan, M. K., & Islam, S. L. U. (2022, May). Climate change adaptation in Bangladesh: Current practices, challenges and the way forward. The Journal of Climate Change and Health, 6, 100108. https://doi.org/10.1016/j.joclim.2021.100108.

Douville, H., Raghavan, K., & Renwick, J., et al. (2023, July 6). Weather and Climate Extreme Events in a Changing Climate. Cambridge University Press. https://doi.org/10.1017/9781009157896.013.

Elias, G. (2024). Governance of the binational basin of the Bermejo and Grande de Tarija rivers between Argentina and Bolivia (1995–2019). Water International, 49(3–4), 327–334. https://doi.org/10.1080/02508060.2024.2319431

Erinosho, B. T. (2013, October). The revised African convention on the conservation of nature and natural resources: Prospects for a comprehensive treaty for the management of Africa's natural resources. African Journal of International and Comparative Law, 21(3), 378–397. https://doi.org/10.3366/ajicl.2013.0069.

Faisal, I. M. (2002, June). Managing common waters in the Ganges-Brahmaputra-Meghna region: Looking ahead. SAIS Review, 22(2), 309–327. https://doi.org/10.1353/sais.2002.0032.

Gerlak, A. K., & Haefner, A. (2017, September 20). Riparianization of the Mekong River Commission. Water International, 42(7), 893–902. https://doi.org/10.1080/02508060.2017.1376267.

Goes, B. J. M., Clark, A. K., & Bashar, K. (2021). Water allocation strategies for meeting dry-season water requirements for Ganges Kobadak Irrigation Project in Bangladesh. International Journal of Water Resources Development, 37(2), 300–320. https://www.tandfonline.com/doi/abs/10.1080/07900627.2020.1763265

Grigg, N. S. (2023, July 3). Colorado River Basin: Conflict management under hydrologic stress and institutional gridlock. International Journal of River Basin Management, 1–9. https://doi.org/10.1080/15715124.2023.2229802.

Gupta, J. (2016, July 2). The watercourses convention, hydro-hegemony and transboundary water issues. The International Spectator, 51(3), 118–131. https://doi.org/10.1080/03932729.2016.1198558.

Gupta, K. (2023, December 28). Resolving Teesta River conflicts: Harnessing institutional solutions for cross-border water cooperation. Journal of Asian and African Studies. https://doi.org/10.1177/00219096231218436.

Handbook on Water Allocation in a Transboundary Context | UNECE. (2021, December 1). https://unece.org/environment-policy/publications/handbook-water-allocation-transboundary-context.

Havinga, H. (2020, June 26). Towards sustainable river management of the Dutch Rhine River. Water, 12(6), 1827. https://doi.org/10.3390/w12061827.

Humpesch, U. H. (1996, September 18). Case study – The River Danube in Austria. River Systems, 10(1–4), 239–266. https://doi.org/10.1127/lr/10/1996/239.

India, Bangladesh Sign Pact on Sharing Waters of Kushiyara River. (2022, September 6). The Diplomat. https://thediplomat.com/2022/09/india-bangladesh-sign-pact-on-sharing-waters-of-kushiyara-river.

INHPC (Indian National Hydroelectric Power Corporation Ltd) (2012). http://www.nhpcindia.com.

IPCC, 2019: Climate Change and Land: an IPCC special report on climate change, desertification, land degradation, sustainable land management, food security, and greenhouse gas fluxes in terrestrial ecosystems. Accessed from https://www.ipcc.ch/site/assets/uploads/2019/11/SRCCL-Full-Report-Compiled-191128.pdf.

IPCC, 2023: Sections. In: Climate Change 2023: Synthesis Report. Contribution of Working Groups I, II and III to the Sixth Assessment Report of the Intergovernmental Panel on Climate Change [Core Writing Team, H. Lee and J. Romero (eds.)]. IPCC, Geneva, Switzerland, pp. 35–115. doi: 10.59327/IPCC/AR6-9789291691647.

Islam, M. N., Tamanna, S., Noman, M., et al. (2021, January 1). Climate Change Impact and Adaptation Strategies in Bangladesh to Strengthen Regional Cooperation. Springer Climate. https://doi.org/10.1007/978-3-030-71950-0_15.

IUCN Environmental Law Programme (2015). Draft International Covenant on Environment and Development – Implementing Sustainability – Gland, Switzerland: IUCN. xxxvi + 201 pp. dx.doi.org/10.2305/IUCN.CH.2015.EPLP.31/rev4.en

IWM (Institute of Water Modelling) (2005). Hydrological impact study of Tipaimukh dam of India on Bangladesh. Unpublished research report, Dhaka.

Kaura, V., & Rani, M. (2020, March). India's neighbourhood policy during 2014–2019: Political context and policy outcomes. Indian Journal of Public Administration, 66(1), 10–27. https://doi.org/10.1177/0019556120906072.

Loucks, D. P., & Van Beek, E. (2017, January 1). Water Resources Planning and Management: An Overview. Springer eBooks. https://doi.org/10.1007/978-3-319-44234-1_1.

Mahmood, R., & Jia, S. (2016, June 9). Assessment of impacts of climate change on the water resources of the transboundary Jhelum River Basin of Pakistan and India. Water, 8(6), 246. https://doi.org/10.3390/w8060246.

Maja, M. M., & Ayano, S. F. (2021, March 16). The Impact of Population Growth on Natural Resources and Farmers' Capacity to Adapt to Climate Change in Low-Income Countries. *Earth Systems and Environment*. https://doi.org/10.1007/s41748-021-00209-6.

Mbaku, J. M. (2022, March). The grand Ethiopian renaissance dam and the Nile basin conflict. Georgetown Journal of International Affairs, 23(1), 84–91. https://doi.org/10.1353/gia.2022.0014.

McCaffrey, S. (2001). The contribution of the UN Convention on the law of the non-navigational uses of international watercourses. International Journal of Global Environmental Issues, 1(3–4), 250–263. Retrieved from https://www.internationalwaterlaw.org/bibliography/IJGEI/03ijgenvl2001v1n34mccaffrey.pdf

McCaffrey, S. C. (2019). The Law of International Watercourses. Oxford University Press.

Mirza, M. M. Q., Ahmed, A. U., & Ahmad, Q. K. (eds) (2008) Interlinking of Rivers in India: Issues and Concerns. CRC Press/Balkema, Leiden.

Nishat, A., & Faisal, I. M. (2000). An assessment of the institutional mechanisms for water negotiations in the Ganges-Brahmaputra-Meghna system. International Negotiations, 5, 289–310.

NITI Aayog (2021). Report of the Committee constituted for formulation of strategy for Flood Management Works in entire country and River Management Activities and works related to Border Areas (2021–26).

Norouzi, N., & Ataei, E. (2022, October 1). Environmental economic diplomacy approach in international law of transboundary water sharing. International Journal of Politics and Security, 4(3), 221–250. https://doi.org/10.53451/ijps.883796.

Obaidullah, M., & Howlader, M. R. (2025). Assessing geopolitical and socio-economic consequences of India-Bangladesh water disputes. Discover Global Society, 3(1). https://doi.org/10.1007/s44282-025-00137-0.

Maja & Ayano, 2021, K., & Eckstein, G. (2012, December). Transboundary water resources management: A multidisciplinary approach. International Journal of Water Resources Development, 28(4), 693–695. https://doi.org/10.1080/07900627.2012.713907.

Pasricha, A. (2013) Problems and prospects of water management as a blue diplomacy for green economy development in Central Asia. Suraksha Chintan, 6(6), 122–125.

PIB, Government of India. (2022, September 28). Cabinet approves Memorandum of Understanding between India and Bangladesh on withdrawal of up to 153 cusecs of water each by India and Bangladesh from common border river Kushiyara [Press release]. Press Information Bureau. https://www.pib.gov.in/PressReleasePage.aspx?PRID=1862941

Rahman, K. S., Islam, Z., Navera, U. K., & Ludwig, F. (2019, February 27). A critical review of the Ganges Water Sharing arrangement. Water Policy, 21(2), 259–276. https://doi.org/10.2166/wp.2019.164.

Rashid, H. (2012) Teesta water issue: A few hard facts. The Daily Star. https://www.thedailystar.net/news-detail-217791

Rasul, G. (2015, March 24). Water for growth and development in the Ganges, Brahmaputra, and Meghna basins: An economic perspective. International Journal of River Basin Management, 13(3), 387–400. https://doi.org/10.1080/15715124.2015.1012518.

Sainz-Borgo, J. C. (2013). Transboundary water management in Venezuela. In Managing Transboundary Waters of Latin America (pp. 133–154). Routledge. https://www.taylorfrancis.com/chapters/edit/10.4324/9780203720509-8/transboundary-water-management-venezuela-juan-carlos-sainz-borgo

Sharma, A., Patel, P. L., & Sharma, P. J. (2025). Changes in the spatial variability of extreme climate characteristics across the Sabarmati River basin: Past and future. Gondwana Research. https://doi.org/10.1016/j.gr.2025.03.016.

Sharma, S. (2024, February 23). Depleting groundwater: Why India needs to rethink many agri practices. India Today. https://welllabs.org/depleting-groundwater-rethinking-agri-practices/

Singh, R. (2018). Perspectives on India-Bangladesh water dispute: The Farakka Barrage. Asian Themes in Social Sciences Research, 1(1), 14–20. https://doi.org/10.33094/journal.139.2018.11.14.20.

Sweet, S. A. (2024, October). Transboundary water management and flood in Bangladesh: Implications of dam releases by India. International Journal of Research and Innovation in Social Science, 8(10), 2293–2397. https://doi.org/10.47772/IJRISS.2024.8100197.

Takang, J. M. (2014). From Algiers to Maputo: The role of the African Convention on the Conservation of Nature and Natural Resources in the Harmonization of Conservation Policy in Africa. Journal of International Wildlife Law & Policy, 17(4), 168. https://doi.org/10.1080/13880292.2014.957034

Thankappan, N. (2024, January 1). Reducing the Risks of Transboundary Climate Change Impacts in India and Bangladesh: Options for Cooperation. *Disaster Risk Reduction*. https://doi.org/10.1007/978-981-99-9474-8_5.

The Economic Times (2025, June 27), India gets to test Bangladesh's nerve when they wanted desi chicken neck. Retrieved July 1, 2025, from https://economictimes.indiatimes.com/ganga-water-treaty-renewal-india-bangladesh-dispute-chicken-neck-siliguri-corridor/articleshow/122114306.cms.

Thomas, K. A. (2017). The Ganges water treaty: 20 years of cooperation on India's terms. Water Policy, 19(4), 724–740.

UNICEF. (2017) Flooding Affects Millions in Bangladesh, India and Nepal. UNICEF. Retrieved July 1, 2025, from https://www.unicef.org/stories/flooding-affects-millions-bangladesh-india-and-nepal

UN-Water. (2021). United Nations World Water Development Report 2021: Valuing water. UNESCO. https://digitallibrary.un.org/record/3905489?ln=en

Uprety, K., & Salman, S. M. A. (2011). Legal aspects of sharing and management of transboundary waters in South Asia: preventing conflicts and promoting cooperation. Hydrological Sciences Journal, 56(4), 641–661. https://doi.org/10.1080/02626667.2011.576252

Varady, R. G., Albrecht, T. R., Modak, S., Wilder, M., & Gerlak, A. K. (2023, February 3). Transboundary water governance scholarship: A critical review. Environments. https://doi.org/10.3390/environments10020027.

Wester, P., Mishra, A., Mukherji, A., & Shrestha, A. B. (2019). The Hindu Kush Himalaya Assessment: Mountains, Climate Change, Sustainability and People. Springer International Publishing. https://doi.org/10.1007/978-3-319-92288-1.

Winsemius, H. C., Jongman, B., Veldkamp, T. I. E., Hallegatte, S., Bangalore, M., & Ward, P. J. (2015). Disaster Risk, Climate Change, and Poverty: Assessing the Global Exposure of Poor People to Floods and Droughts (Policy Research Working Paper No. 7480). World Bank. https://documents.worldbank.org/curated/en/965831468189531165/pdf/WPS7480.pdf.

World Bank. (2021). Glaciers of the Himalayas: Climate Change, Black Carbon, and Regional Resilience. World Bank Publications. https://documents1.worldbank.org/curated/en/976841622778070962/pdf/Glaciers-of-the-Himalayas-Climate-Change-Black-Carbon-and-Regional-Resilience.pdf.

Xenarios, S., Laldjebaev, M., Schmidt-Vogt, D., Buurman, J., & Araral, E. (2021). Powering the uplands: Controversies of developing hydropower in upstream Central and Mainland Southeast Asia. In Handbook of Sustainable Politics and Economics of Natural Resources. Edward Elgar Publishing, pp. 152–185.

Yan, F., Wang, W., & Jiang, L. (2019, October 8). Dilemmas in and Pathways to Transboundary Water Cooperation between China and India on the Yaluzangbu-Brahmaputra River. Water. https://doi.org/10.3390/w11102096.

10 Water, Climate, and Sustainability
The Case of Bhutan

Chhimi Dorji and Deki Pem Tamang

INTRODUCTION

Water constitutes the cornerstone of Bhutan's ecological, economic, and cultural identity. As a mountainous country situated in the Eastern Himalayas, Bhutan is endowed with substantial water resources comprising glacial lakes, snow-fed rivers, monsoon catchments, and high-altitude wetlands. These hydrological systems underpin biodiversity conservation, agricultural production, energy generation, and spiritual traditions. Bhutan ranks among the highest globally in per capita water availability, with an annual estimate of over 94,500 m^3 per person. However, the mere abundance of water at the national scale masks severe challenges related to spatial and seasonal variability, climate-induced alterations, and increasing water-related risks. Climate change is manifesting visibly across Bhutan's hydrological landscape. Regional warming, particularly pronounced at higher altitudes and during winter months, is contributing to rapid glacial retreat, shifts in river seasonality, erratic monsoon behaviour, and the proliferation of glacial lakes. Between 1980 and 2010, Bhutan's total glacier area decreased by over 23% (Bajracharya et al., 2014), while the number of glacial bodies increased due to fragmentation. This dynamic shift has direct implications for downstream water security and hazard exposure. For example, moraine-dammed glacial lakes, formed as a consequence of accelerated melt, pose a significant risk of glacial lake outburst floods (GLOFs). Historical events such as the 1994 Lugge Tsho GLOF in Lunana underscore the catastrophic potential of these hazards, threatening lives, infrastructure, and ecosystems downstream.

Bhutan's hydrology is inherently seasonal, with approximately 76% of annual river discharge occurring during the monsoon months (June–September). This pattern leads to an overabundance of water in summer and acute scarcity during the dry winter period. Furthermore, the observed trend of declining dry-season flows in several river basins, coupled with increasing instances of spring sources drying up, signals the emerging reality of local water insecurity despite Bhutan's seemingly abundant resources. Climate projections further suggest warmer and wetter monsoons, drier winters, and heightened interannual variability, compounding the challenges of water resource planning. These hydrological shifts occur within a socio-economic context that is increasingly dynamic. Bhutan's economy is heavily reliant on hydropower, which accounts for more than 14% of GDP and constitutes the principal source of export earnings through bilateral trade with India (National

DOI: 10.1201/9781003599692-10

Accounts Statistics, 2016). Agriculture remains the primary livelihood source for over 60% of the population, yet it is predominantly rain-fed and highly vulnerable to climate variability. Rapid urbanization, projected to encompass over half the population by 2030, has placed additional stress on municipal water systems. Even in urban areas located near perennial rivers, water supply remains intermittent due to inadequate infrastructure and distribution inefficiencies.

Institutionally, Bhutan has demonstrated commendable leadership in environmental governance. The Water Act of 2011, the National Water Policy (2003), and the establishment of the National Environment Commission (NEC) have provided a strong legislative and institutional framework for water management. However, implementation remains fragmented across agencies, and local capacity is often limited. Integrated Water Resource Management (IWRM), although formally endorsed, still requires operationalization at the basin and community levels. The complex topography and dispersed settlement patterns further complicate centralized water governance, reinforcing the need for localized and adaptive solutions. Scientific research on Bhutan's water resources has expanded in recent decades, leveraging advances in remote sensing, GIS, and climate modelling. Studies by Krusic et al. (2015), Chhetri (2022), and Sonessa et al. (2019) have provided important insights into temperature variability, rainfall forecasting, and hydrological sensitivity to temperature change. Nonetheless, significant gaps persist. Long-term, high-resolution data on glacier mass balance, streamflow variability, and wetland hydrology remain limited. Moreover, there is a paucity of community-level studies that assess the differential impacts of climate change across diverse ecological and socio-economic settings.

In recognition of these evolving threats, Bhutan launched its first National Adaptation Plan (NAP) in 2023, identifying water as the highest-priority sector. The plan lays out a 15-year roadmap with an investment estimate of USD 14 billion. Flagship projects such as the Advancing Climate Resilience of the Water Sector (ACREWAS) are being implemented in climate-vulnerable districts, combining infrastructure upgrades with integrated irrigation and drinking water systems. Complementary initiatives such as NAPA-III have focused on ecosystem-based adaptation through wetland conservation, reforestation, and climate-resilient agriculture. Together, these efforts reflect Bhutan's commitment to balancing development with environmental sustainability. Notwithstanding these advances, Bhutan faces a set of emerging challenges that demand a more integrated, data-informed, and inclusive approach to water governance. Increasing hydrological variability has implications not only for energy and agriculture but also for disaster preparedness, public health, and regional cooperation. Bhutan's status as a climate-resilient yet vulnerable Himalayan state places it in a unique position to contribute meaningfully to global discussions on water security in high mountain regions.

SIGNIFICANCE AND NOVELTY OF THE STUDY

This chapter offers a comprehensive and multidisciplinary assessment of Bhutan's water resources in the context of climate change. It synthesizes hydrological data, policy frameworks, and field-based evidence to provide an in-depth analysis of vulnerabilities, emerging risks, and adaptation opportunities. While previous studies have focused on either glacier dynamics or sectoral modelling, this study distinguishes itself by adopting a systems-based perspective that connects upstream–downstream linkages, integrates ecological and socio-economic dimensions, and identifies both physical and institutional vulnerabilities. The novelty of this work lies in its emphasis on seasonal water insecurity despite national-scale abundance, its treatment of glacial lakes as dynamic risk indicators, and its contextualization of water governance within Bhutan's evolving development trajectory. Furthermore, it presents case studies that illuminate practical responses to climate risks and underscores the importance of community-based and ecosystem-based adaptation. As such, the study contributes to both academic knowledge and policy development, offering insights applicable to other mountainous developing countries confronting similar challenges.

RESEARCH QUESTIONS

To guide the inquiry and structure the subsequent analysis, the chapter addresses the following research questions:

1. How is climate change affecting the hydrology of Bhutan, particularly in relation to glacier retreat, glacial lake expansion, river flow variability, and water-related hazards?
2. What are the sectoral implications of these hydrological changes for agriculture, hydropower generation, and urban and rural water supply?
3. How effective are existing water governance institutions, policies, and adaptation strategies in addressing emerging water-related challenges in Bhutan?
4. What are the critical data, capacity, and infrastructure gaps that must be addressed to improve water security and resilience in Bhutan?
5. How can Bhutan's integrated and proactive approach inform regional and global efforts to manage mountain water resources in the face of climate change?

PREVIOUS STUDIES

A considerable volume of scientific research has been conducted on Bhutan's water resources, encompassing a wide range of thematic areas. These include water supply and sanitation systems, water quality, water balance, irrigation practices, institutional frameworks, glaciers, water availability, climate change impacts, groundwater resources, hydro-economics, and hydropower potential. Notably, Tariq et al. (2021) have provided a comprehensive overview of Bhutan's internal and external water-related issues, while Biswas and Bisht (2016) have explored transboundary water management challenges in the broader South Asian context. Similarly, Gain and Giupponi (2014) have examined water scarcity concerns in the lower Brahmaputra basin, highlighting Bhutan's exposure to future water stress. Research on Bhutan's hydroclimatic systems has also advanced significantly. Krusic et al. (2015)

reconstructed over six centuries of summer temperature variability, establishing a relationship between solar activity and decadal temperature fluctuations. Numerous studies have applied remote sensing and Geographic Information System (GIS) tools to assess water quality, monitor glacial lakes, evaluate GLOF)risks, and model water balance.

Participatory approaches to integrated water resources management have been proposed for Bhutan's river basins, including those by Saha et al. (2018). In the irrigation sector, Gurung (2017) has introduced innovative methodologies such as role-playing simulations, companion modelling, and multi-agent systems to address decision-making and system optimization. Satellite-based datasets, such as the Tropical Rainfall Measuring Mission (TRMM, 2010) Multi-satellite Precipitation Analysis (TMPA, 2010), have been used to model run-off in ungauged catchments like the Wangchhu River. Materna combined Global Navigation Satellite Systems (GNSS) and GRACE satellite data to analyze land deformation caused by excessive hydrological loading. In addressing the impacts of climate variability, Wangdi et al. (2019) conducted experimental studies using 100% throughfall exclusion and ambient control plots to assess drought-induced stress on Bhutan's montane ecosystems. Chhetri (2022) utilized advanced machine learning techniques, including linear regression, multilayer perceptron (MLP), convolutional neural networks (CNN), long short-term memory (LSTM), gated recurrent units (GRU), and bidirectional LSTM (BLSTM), to forecast monthly rainfall patterns. Additionally, Sonessa et al. (2019) using the Variable Infiltration Capacity (VIC) model, found that temperature variations significantly influence run-off in high-altitude regions exceeding 5,000 m above sea level (MASL). Tenzin and Bhaskar (2021) applied hydrodynamic modelling to simulate flash flood events, further contributing to the understanding of Bhutan's evolving hydrological risks.

WATER RESOURCE AND HYDROLOGICAL PROFILE OF BHUTAN

Glaciers Of Bhutan

According to recent analyses of Landsat satellite imagery, Bhutan was home to 885 clean glaciers and 50 debris-covered glaciers in 2010, collectively covering an estimated 658.1 km^2 (642 + 16.1 km^2). This represents approximately 1.6% of Bhutan's total land area, as illustrated in Figure 10.1 and Table 10.1. These glaciers are distributed across a wide altitudinal range, from around 7,230 MASL at their highest to approximately 4,050 MASL at their lowest margins. In terms of river basin distribution, the Punatshangchhu basin hosts the largest concentration, with 466 glaciers occupying a combined area of 361.6 ± 9.3 km^2. The Wangchhu basin contains 58 glaciers covering 32.8 ± 0.7 km^2, while the Mangdechhu basin supports 287 glaciers with a glaciated area of 179.7 ± 3.05 km^2. The most extensive coverage is found in the Drangmechhu basin, with 124 glaciers spanning 780.6 ± 16.1 km^2. Notably, the largest individual glacier in Bhutan has been identified as G090161E 28125N (coordinates: 90°16′1″E, 28°12′5″N), situated within the Punatshangchhu basin, with an area of approximately 36 km^2. The figure highlights the significance of Bhutan's glaciers as critical hydrological reserves, feeding the country's major river basins and sustaining

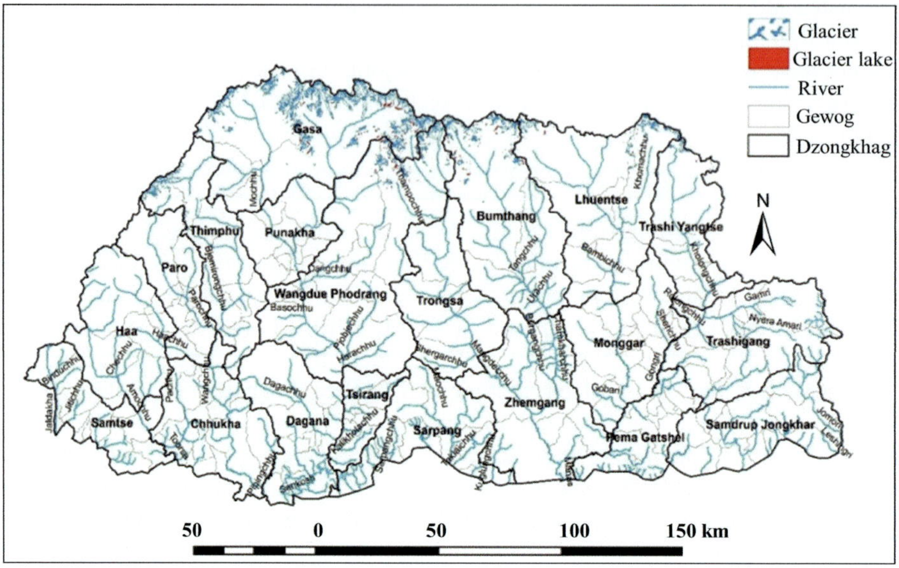

FIGURE 10.1 Glacier distribution in Bhutan. (*Source:* Prepared by authors.)

TABLE 10.1
Number and Areas of Glaciers

Basin	Sub-basin	Number of Glaciers		Glacier Area (km²)			Largest Glacier
		CI	DC	CI	DC	Total	km²
Wangchhu	Thimphuchhu	19	0	5.2 + 0.1	0.0 + 0.0	5.2 ± 0.1	1.5
	Parochhu	37	2	22.2 + 0.4	5.4 + 0.0	27.6 ± 0.5	7.2
	Subtotal	56	2	27.4 + 0.6	5.4 + 0.0	32.8 ± 0.7	7.2
Punatshangchhu	Mochhu	180	15	103.8 + 2.9	9.9 + 0.1	113.7 ± 3.1	10.4
	Phochu	256	15	201.1 + 4.9	46.9 + 0.2	247.9 ± 6.1	36.6
	Subtotal	436	30	304.8 + 7.9	56.8 + 0.4	361.6 ± 9.3	36.6
Mangdechhu	Mangdechhu	150	5	91.9 + 2.5	15.8 + 0.1	107.8 ± 3.05	20.3
	Chamkharchhu	125	7	62.1 + 1.5	9.8 ± 0.1	71.9 ± 1.8	15.6
Drangmechhu	Kurichhu	96	6	57.8 + 1.3	3.6 ± 0.0	61.4 ± 1.4	12.5
	Drangmechhu	22	0	6.7 + 0.1	0.0 ± 0.0	6.7 ± 0.0	2.0
	Subtotal	393	18	218.5 + 5.4	29.2 ± 0.3	247.7 ± 6.2	20.3
Total		885	50	550.7 + 13.8	91.4 ± 0.7	642.1 ± 16.1	36.6

Source: After Bajracharya et al. (2014).

downstream ecosystems and livelihoods. However, their spatial distribution and elevation also make them particularly vulnerable to the impacts of climate change, including accelerated melting and the formation of potentially hazardous glacial lakes.

Glaciers are very good indicators of climate change as they react quickly to changes in temperature and precipitation and provide direct visual evidence when their size changes (NSIDC, 2015). Glaciers change in terms of their thickness, area, and mouth location, which are directly proportional to the change in the atmosphere around them. Glaciers are also important sources of freshwater. Bajracharya et al. (2014) had looked at the change of glacier area from 1980 to 2010 in Bhutan. It was found that there had been a decrease in glacier area from 837.6 ± 0.7 km^2 to 642 ± 16.1 km^2 which is a decrease of –23.3 ± 0.9 % with a decadal trend of –6.39 ± 1.6% that is statistically significant at a P-value of 0.017 (Figure 10.2). The change in glacier areas in the Lunana area is pictorially shown in Figure 10.2. The number of glaciers, on the other hand, had increased from 771 in 1980 to 885 in 2010, which is an increase of +14.8%. The increase in glacier numbers is attributed to the shrinking and fragmentation of glaciers. The study confirmed that the glacier area in Bhutan is changing rapidly. The first in situ mass balance study in Bhutan on the Gangjula glacier in the Phunatshangchhu by Tshering and Fujita (2015) from 2003 to 2014 found that there is a negative mass balance of –1,550 ± 290 kg/(m^2 year) which indicates that the glaciers in Bhutan are also losing mass.

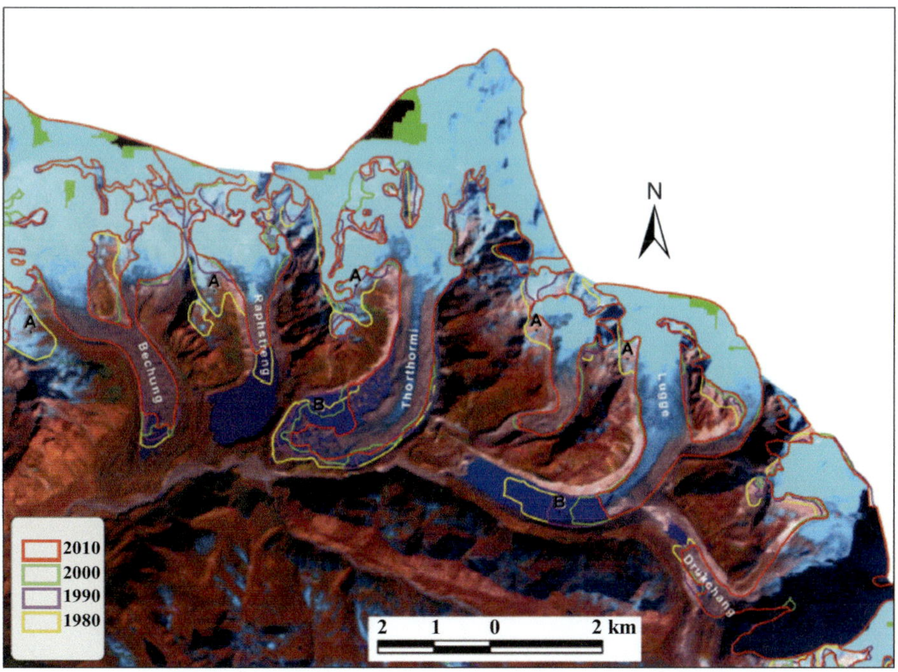

FIGURE 10.2 Decadal change in glacier area in Lunana. (*Source:* Bajracharya et al., 2014.)

Glacial Lakes in Bhutan

When glaciers melt, they form water bodies which later become glacier lakes. Lakes are formed when the rate of melting is more than the rate of snow accumulation on the glacier mass. Glacier lakes pose a serious threat to downstream communities as they have the potential to burst out in the form of GLOF. GLOFs occur when the moraine dams/supporting the mouth/outlet of the glacier lakes suddenly burst out due to the hydrostatic pressure of higher water levels in the lake itself or due to melting of the frozen compacted moraine dams. Understanding of glacier lakes is therefore crucial for planning and development of any water resources projects. One of the first efforts of its kind in Bhutan, a study by the International Centre for Integrated Mountain Development (ICIMOD) and the Department of Geology and Mines (DGM), Bhutan, in 2001 had identified that there are some 2674 lakes across the country. Of the lakes identified, only 562 lakes were associated with glaciers. From the long list, the study had identified that 24 lakes are potentially dangerous lakes, as shown in Figure 10.3. The potentially dangerous lakes were identified based on the condition of lakes, dams, associated mother glaciers, and topographic features around the lakes and glaciers.

A team from Japan and DGM did a field assessment of the Potentially Dangerous Glacier Lakes (PDGL) that was identified by the ICIMOD study from 2009 to 2012. The assessment found that further fieldwork and surveys are required to be done, along with a comprehensive assessment of topography, to validate the study results. A concrete recommendation by the team was on the establishment of early warning systems in the Mangde-Chamkharchhu basins. The Japan Aerospace Exploration

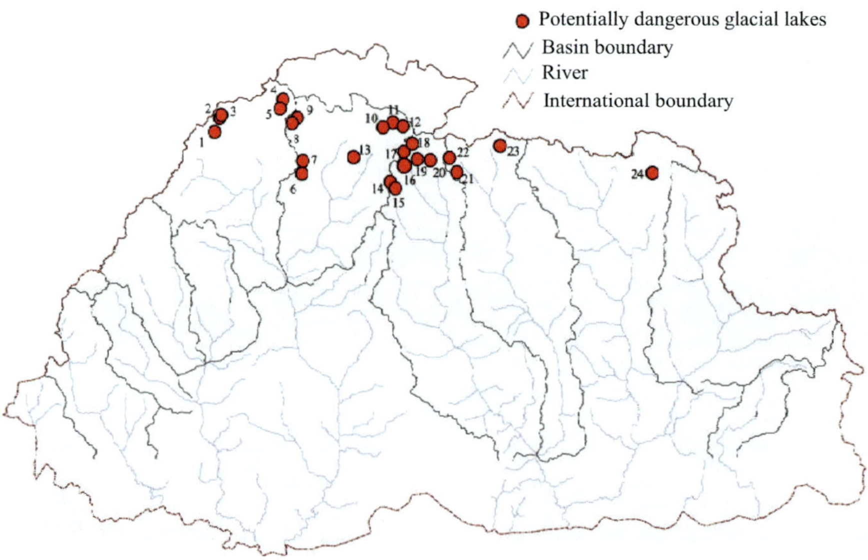

FIGURE 10.3 Potentially dangerous glacial lakes in Bhutan. (*Source:* Maharjan et al., 2018.)

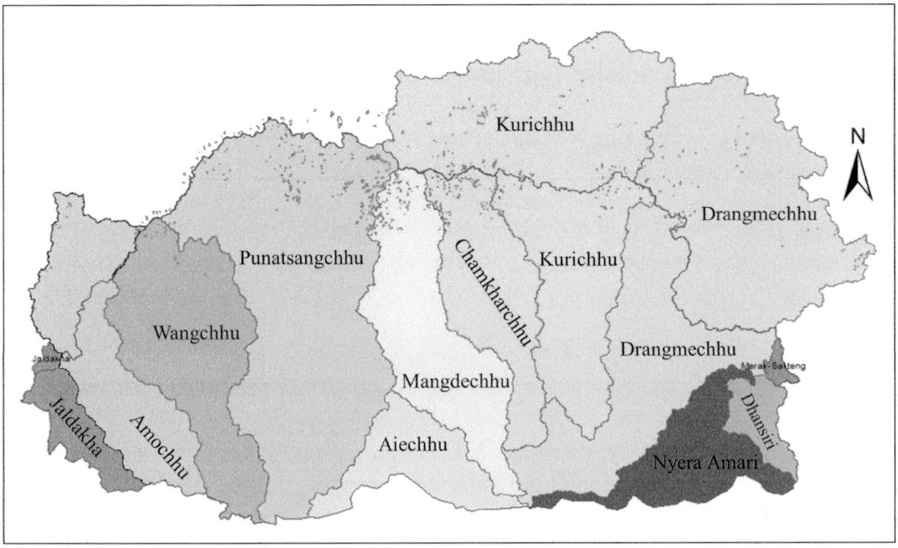

FIGURE 10.4 Glacial Lakes of Bhutan. (*Source:* JAXA, 2014.)

Agency (JAXA) made an inventory of glacial lakes in Bhutan using the Advanced Land Observing Satellite (ALOS, nicknamed "Daichi"), Version 12.03. According to the database, there are 733 glacial lakes in the Bhutan Himalaya, with a total area of about 82.5 km^2, as shown in Figure 10.4. An assessment of the melting of glacier lakes was conducted by the JAXA team in 2008. It was found that the glacier lake areas in the Lunana area are rapidly increasing in number, as shown in Table 10.2. This indicates that more glaciers are melting and forming glacier lakes.

High Altitude Wetlands

While there are no extensive studies done on freshwater lakes or wetlands, the Ugyen Wangchuck Institute for Conservation and Environment and WWF Bhutan have completed an inventory of high-altitude wetlands through satellite imagery analysis and field surveys. High-altitude wetlands had been defined as a combination of marshes, supra-snow lakes, and supraglacial and glacial lakes that are above 3,000 MASL

TABLE 10.2
Glacial Lakes Surface Area Change in Lunana

Areas	Bechung	Raphstreng	Thorthormi	Lugge
Area on Dec. 27, 1993 (ha)	1	127	0	118
Area on Nov. 9, 1994 (ha)	4	130	41	96
Area on Dec. 26, 2007 (ha)	18	126	88	127

Source: JAXA (2014).

TABLE 10.3
High Altitude Wetland Type and Area Statistics

Wetland Type	# Lake	Area Subtotal (m^2)	Average Area (m^2)	Largest Lake (m^2)	Smallest Lake (m^2)
Supra Snow Lake	110	52,327.0	475.7	4758.8	36.2
Supraglacial Lake	495	28,554,801.3	57,686.5	1,517,436.4	133.6
Glacial lake	637	23,230,604.6	36,468.8	878,311.5	114.7
Lake	1722	49,973,272.8	29,020.5	868,048.9	34.6
Marsh	63	497,334.4	7894.2	63,811.1	126.1

Source: (Wangdi and Norbu, 2010.)

According to the inventory, there are 3,027 high altitude lakes covering an area of 102 km^2, which is 0.26% of the total area of Bhutan and 0.61% of the high altitudes, forming an important water resource for the country. It was also found that about 52% of total wetlands in Bhutan are found in the Mangdechhu and Chamkharchhu river systems. The Chamkharchhu river system, with 655 high-altitude lakes, has the highest number of lakes. The most common lakes are lakes in alpine meadows with 1722 numbers covering an area of 499.7 km^2. This is followed by glacier lakes with 637 numbers; the least is 63 marshes with an area of 49.7 km^2 as shown in Table 10.3. The high-altitude wetlands are within the range of 3098 to 5546 MASL. The distribution is highest in the range of 3900 to 5300 MASL as shown in Figure 10.5.

Water Resources of Bhutan

In total, Bhutan has some 3182 rivers and rivulets within its territory which is based on a 90×90 m SRTM DEM on a 1:1,500,000 Map with a total length of 4741 km. However, only 280 streams and rivers have substantial flows and are delineated on maps.

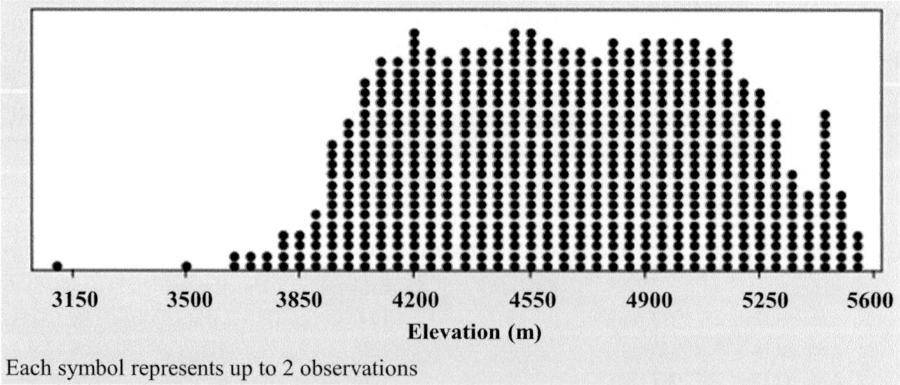

FIGURE 10.5 Frequency of wetlands and lakes distribution along elevation in Bhutan. (*Source:* Wangdi and Norbu, 2010.)

TABLE 10.4
Basins and Their Flows

Management Basin	Area (km²)	River Basins	Area (km²)	Annual Flow (MCM)
Amochhu	3,252	Jaldakha	942	2,715.64
		Amochhu	2,310	6,659.36
Wangchhu	4,596	Wangchhu	4,596	5,209.06
Punatshangchhu	11,582	Punatshangchhu	9,645	19,129.79
		Aiechhu	1,937	6,989.14
Mangdechhu	7,380	Mangdechhu	7,380	11,797.24
Drangmechhu	11,584	Drangmechhu	8,457	13,569.14
		Nyera Amari	2,348	3,383.89
		Jomori	642	925.24
		Merak-Sakteng	137	197.44
Total	38,394	Total	38,394	70,576.01
		Population		746,773
		Per capita water available		94,508.04 m³/annum
		Flow		2,238.0 m³/s

Source: NECS (2016).

According to the analysis done, the average flow in Bhutan is 2,238 m³/s, while the total annual water volume is 70,576.02 MCM (million cubic metre). This indicates that the per capita water available is 94,500 m³/capita/annum, considering a projected population of 746,773 in 2015. The details of the average flow availability and 80% dependable flow for each basin and district are in given Table 10.4. Based on hydrological analysis, there are five main hydrological basins and five minor hydrological basins in Bhutan. However, for the purpose of management, five management units are considered as specified in Table 10.4.

River Basins of Bhutan

Based on management purpose, the river basins of Bhutan are the Amochhu, Wangchhu, Punatshangchhu, Chamkharchhu, and Drangmechhu. The minor basins are Jaldakha, considered as part of Amochhu, Aiechhu as part of Punatshangchhu, and Jomori, Merak-Sakteng chhu, and Nyeramari as part of Drangmechhu management basins. The basins are shown in Figure 10.6. The distribution of districts to respective basins is shown in Table 10.5.

Amochhu

Amochhu is the smallest of the main basins in the far south-west of Bhutan and covers an area of 3,252 km². The transboundary area contributing to Amochhu is about 1,917.0 km². The main river is Amochhu which originates in China and flows through the districts of Samtse, Haa and Chhukha. The 358 km long river exits next to Phuentsholing city and then flows into India, where it is known as Torsa. The

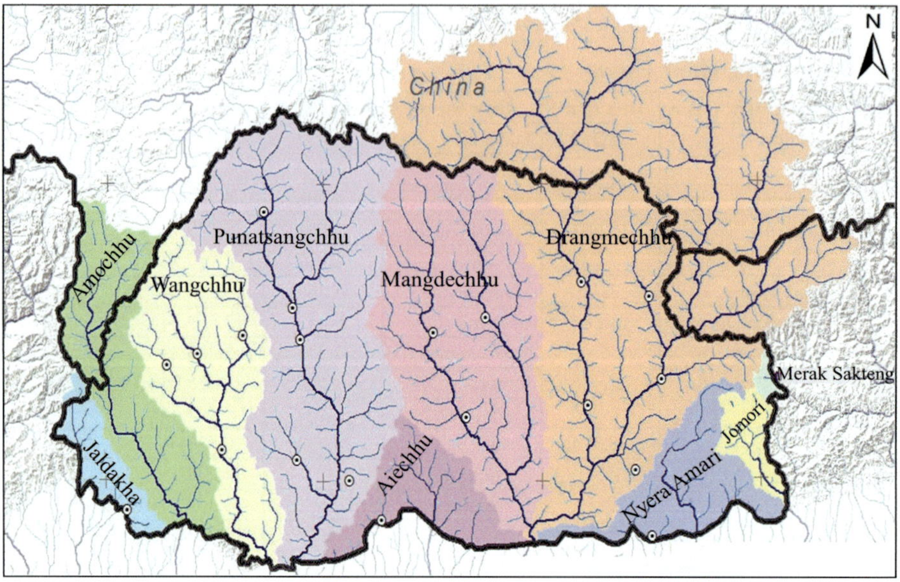

FIGURE 10.6 The river basins of Bhutan. (*Source:* NIWRM, 2016.)

Jaldakha river basin, with its tributaries, is also considered part of the Amochhu although they do not join within Bhutan. There are also a few other tributaries such as Bindu Khola and Tendu River. The total annual flow generated in the basin is 9,375.07 MCM. There are 32 streams in the basin with a total river length of 645 km.

Wangchhu

The Wangchhu basin is 4,596 km^2 and covers the districts of Haa, Paro, Thimphu, Chhukha, and a very small portion of Dagana. The Wangchhu is considered one of the most important basins in the country, with the highest population and economic activities. The main rivers in the Wangchhu are Haachhu, Parochhu, Thimphuchhu, and Pippingchhu. The longest river is Parochhu with about 173 km length. The overall length of rivers in the basin is about 553 km with 18 tributaries. The total annual flow in the basin is 5,209.06 MCM. The Wangchhu River flows into the plains of Jalpaiguri district in India where it is known as the Raidak.

Punatshangchhu

With the highest concentration of glaciers and glacier lakes in the country, Punatshangchhu is the riskiest river basin in Bhutan covering an area of 9,645 km^2. It is also the location of the largest hydropower projects in the country with Punatshangchhu I and Punatshangchhu II hydropower projects. Punatshangchhu basin covers the districts of Gasa, Punakha, Wangdue Phodrang, Tsirang, and Dagana. The Phochhu river is 222 km long; the total length of the 83 streams in the

TABLE 10.5

Distribution of Districts in Basins

Districts	Wang chhu	Punatshangchhu	Mangdechhu	Drangmechhu	Amochhu	Aiechhu	Nyera Ama Chhu
Thimphu	78.5%	21.5%					
Gasa		100.0%					
Punakha		100.0%					
Wangdue Phodrang		74.3%	26.1%				
Tsirang		100.0%					
Dagana	5.0%	94.0%				0.8%	
Trongsa			100.0%				
Bumthang			100.0%				
Zhemgang			87.2%	12.8%			
Lhuentse				100.0%			
Monggar				100.0%			
Trashi Yangtse				100.0%			
Trashigang				58.1%			41.9%
Pema Gatshel				69.2%		30.8%	
Chhukha	42.2%				57.9%		
Haa	55.4%				44.4%		
Paro	100.0%						
Samdrup Jongkhar							100.0%
Samtse					100.0%		
Sarpang						100.0%	
Total area (km^2)	**4,596.0**	**9,645.0**	**7,380.0**	**8,457.0**	**3,252.0**	**1,937.0**	**3,127.0**

Source: NIWRM (2016).

basin is about 1,343 km, which is inclusive of the Aiechhu basin rivers. The annual flow of the Punatshangchhu basin is 19,129.79 MCM. It consists of the rivers of Phochhu, MoChhu, Dangchhu, and Dagachhu.

Aiechhu basin consists of parts of Zhemgang and Dagana districts, while the whole of the Sarbang district is within Aiechhu. In a hydrological sense, Aiechhu is not part of the Punatshangchhu, but for management purposes it is clubbed with the Punatshangchhu basin. The total area is 1,937 km^2, most of which is the floodplains in the southern bordering the great Indian plains. The annual flow in the Aiechhu basin is 6,989.14 MCM.

Mangdechhu

Mangdechhu basin is 7,380 km^2 covering the districts of Bumthang, Trongsa Zhemgang, and parts of Wangdue Phodrang with an annual flow of 11,797.24 MCM. The main rivers are Nikachhu, Durchhu, Tangchhu, Urachhu, Chamkharchhu, and Mangdechhu. Chamkharchhu is the longest river in the basin with about 174 km length. There are some 58 rivers and rivulets in the basin with a total length of 789 km.

Drangmechhu

Drangmechhu is the largest basin in Bhutan with an area of 8,457 km^2 within Bhutan. It has two transboundary tributaries of Kurichhu from China and Gongri from China and India in the North with an annual flow of 15,684.70 MCM. The international area contributing to Kurichhu is 5,510 km^2 and Gongri is 6,676.9 km^2. The other tributaries are Khomachhu, Sherichhu, Kholongchhu, and Gamri. The basin covers the districts of Mongar, Lhuentse, Trashi Yangtse, Trashigang, and parts of Pema Gatshel with a total annual flow of 13,569.14 MCM. Kurichhu is the longest river in the Drangmechhu basin with about 187 km length. The total length of rivers in the basin is 1,411 km, from the sum of 89 streams and rivers in the combined basin of Drangmechhu. In addition to the main Drangmechhu basin, the Nyera Amari basin is also considered part of the Drangmechhu management unit. The combination of Nyera Amari chhu with Jomori (also called Dhansari) and Merak-Sakteng basin has an area of 3,127 km^2 with an annual flow of 4,506.57 MCM. Part of the district of Trashigang and the whole of Samdrup Jongkhar fall within the basin.

Rainfall Patterns and Seasonality

Historical hydrological and climatic data had been collected by the Department of Hydro-Met Services (DHMS) under the Ministry of Economic Affairs of Bhutan (MOEA) since 1985 (now called the National Center for Hydrology and Meteorology, NCHM). However, reliable and consistent data were available only from mid-1990s. The average temperature is minimum in January (10.1°C) and maximum in August (22.2°C). Overall, the annual average temperature is 17.2°C. The annual rainfall for the year is least in the month of December (6.37 mm), followed by January (11.52 mm) and November (9.54 mm). The annual total rainfall averaged over the country is 1,797.6 mm.

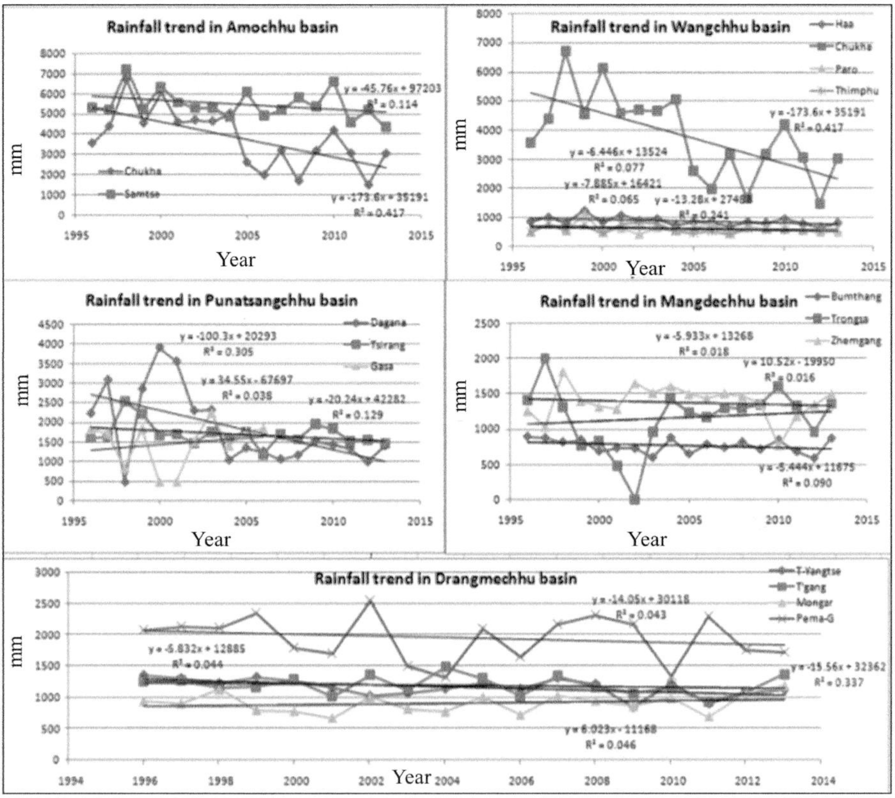

FIGURE 10.7 Trend of total annual rainfall in Bhutan. (*Source:* NIWRM, 2016.)

Trend of Total Annual Rainfall

The trend of total annual rainfall across Bhutan is given in Figure 10.7. The trends of almost all the stations are showing a negative slope, indicating a decrease in total rainfall. However, R^2 is close to zero, indicating that the decrease is not significant and consistent. The only stations that are showing a positive slope are Gasa, Trongsa, and Mongar, which are only 3 of the 20 stations.

Total Monthly Precipitation

The average total monthly precipitation in 20 Class A meteorological stations across Bhutan (1996–2013 data) was used to find out the months with the least precipitation. The months with the lowest precipitation are December (0.4%), November (0.5%), January (0.6%), and February (1.3%), starting with the minimum in order. The maximum rainfall is in the months of July (23.2%) and June (18.9%), as shown in Figure 10.8.

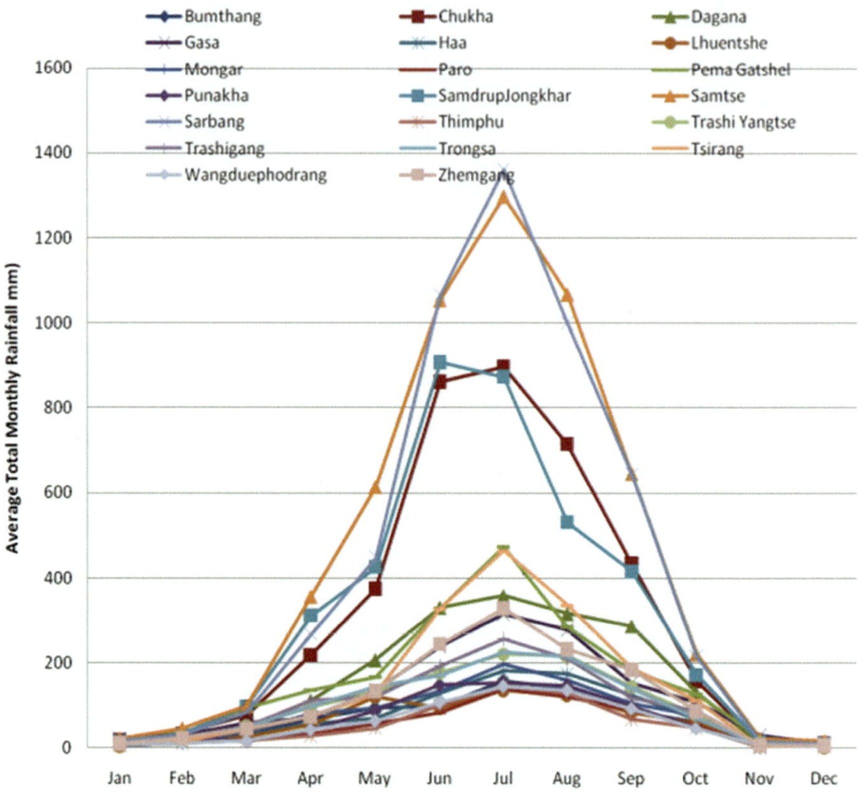

FIGURE 10.8 Average total monthly precipitation across Bhutan. (*Source*: NCHM, 1996–2013.)

IMPORTANCE OF WATER TO BHUTAN'S SOCIETY, CULTURE, AND ECONOMY

Society and Culture

Bhutan is a place of social and cultural diversity with local dialects, social norms, local beliefs, and livelihood practices varying across the country. This may be attributed to the high mountains, deep gorges, and fast flowing rivers that have isolated communities. The limited communication and interaction with other communities have led to an evolution of livelihood practices, dialects, religious beliefs, traditions, and culture that are unique to the ecological zones. The higher Himalayas are inhabited by nomads who are dependent on livestock and dairy products. Those living along the mountain slopes practise dryland agriculture, depending primarily on wheat, barley, maize, and horticulture. Some communities practise terraced rainfed as well as irrigated paddy cultivation. The fertile valleys of the inner Himalayas and southern foothills are inhabited by communities that are dependent on paddy and horticultural production. Such livelihood practices are still prevalent in the rural communities that make up an estimated 69% of the country's population; day-to-day livelihood practices still revolve around subsistence farming and dependence on local natural resources and environmental services, especially water supply and forest products. The majority of the population practise Buddhism, followed by a smaller section of Hindus and Christians. Buddhist reverence for living beings and beliefs in local deities contribute immensely to the preservation and protection of nature and natural resources. Both urban and rural communities maintain strong links with ancestral traditions, culture, and religious beliefs that promote respect for nature. Today, Bhutan strongly pursues the preservation of these cultural heritages in which water plays a major role.

Bhutan is a welfare state with free education, health services, and agricultural subsidies. The population is expected to grow from 745,000 to some 886,500 by 2030. Although the average registered household size is about 4.6, most of the family members explore livelihood opportunities in urban areas. About 30% of the population resides in urban areas and this is expected to reach 50% by 2020. Some 13% of the population lives in Thimphu alone. Given the difficulties of farming on the mountain slopes, more of the educated younger generation migrate to urban areas in search of employment, leaving behind the older members of the family. Migration from rural areas has resulted in a decline in the agricultural workforce and has translated into rapid urbanization, unemployment, and associated social issues. Half of the population is under 22 years old. The child (under 5 years old) mortality rate is 6.16%.

Economy

As of 2023, Bhutan's economy has shown steady recovery and growth over time. According to the National Statistics Bureau (NSB), the country's Gross Domestic Product (GDP) reached Nu. 249.39 billion (approximately USD 3.02 billion) in 2023, reflecting a growth rate of 4.88%. This marks a notable improvement from the economic slowdown experienced in 2013, when growth had decelerated to 2.07%. Bhutan's GDP per capita has risen significantly from about Nu. 142,983.67 (USD 2,300) in 2013 to between USD 3,500 and USD 3,900 in 2023, depending on the data source (World Bank (2024), FRED (2025), and Trading Economics (2025)). When adjusted for purchasing power parity (PPP), Bhutan's GDP per capita in 2023 stood at approximately USD 14,645. The overall increase in per capita income and stable growth reflects Bhutan's gradual economic diversification and resilience following the challenges faced in earlier years.

The World Bank has classified Bhutan as a lower-middle-income country. Bhutan's economy is based primarily on agriculture and forestry, providing livelihoods for about 69% of its population. Agriculture mainly consists of subsistence farming and animal husbandry. The share in GDP for key sub-sectors related to water is given in Table 10.6. The country has embarked on increasing investments in hydropower development for export to India. With over 2000 MW of hydropower generation capacity and another few thousand megawatts of hydropower under construction, the country's economy is commonly regarded as hydropower driven.

TABLE 10.6
Share of Sub-sectors in GDP

Sub-sector	% of Share in GDP
Agriculture	9.18
Livestock	4.11
Forestry (and logging)	2.89
Water supply	0.03
Electricity	14.15
Manufacturing	8.51

Source: National Accounts Statistics (2016).

High-value low-impact tourism is also a major source of foreign currency earnings. Industries are mainly small-scale cottage based, but there are breweries, mines, and a chipboard industry.

Through the bilateral India–Bhutan hydropower cooperation agreement, most of the hydropower investments come from India and are made available in the mixed grant-loan mode. Bhutan's foreign debt with India for hydropower has surpassed the value of its GDP. Hydropower-driven economic development, urbanization, and a lack of investment in agriculture have made Bhutan dependent on cereal imports. Improving food self-sufficiency through agricultural development and irrigation is a key water-related objective of the Royal Government of Bhutan (RGOB).

WATER-RELATED ISSUES IN BHUTAN

This chapter gives an overview of water-related issues identified through surveys of water issues at Gewog levels and technical assessments of climate and water resources. Details are provided in the stand-alone documents supporting the plan, while the results and main conclusions are briefly presented in the following sections.

Multi-sectoral Coordination

Many different agencies are involved in or concerned with the management of water resources. So far, the approach to management has largely been sector-based, but efforts for better coordination have been made. The Water Policy (2003), Water Act (2011) and Regulations (2014) paved the way for adopting IWRM. The Water Act of 2011 is explicit in empowering the NEC as the apex body for all matters related to the water resources of Bhutan. Hence, it has the responsibility of coordinating the efforts of competent authorities and line agencies in managing the water resources of the country. The current level of coordination is not adequate to fulfil the requirements of the Water Act and its regulations. Water-related agencies pursue their own agendas and priorities for the use and management of water. Defined mechanisms need to be put in place to coordinate and enhance cooperation among water-related agencies/authorities and other environmental organizations. Efforts have been made by the government to foster coordination through the establishment of a water board consisting of director-level officials from various agencies. However, the board does not meet regularly, and it is not clear how it fits within the overall coordination role and mandated functions of the NECS.

A new Department of Water is formed under the NEC, which is responsible for strengthening water resource management as per the provisions of the Water Policy, Water Act, and the Water Regulation. The department also has the mandate to implement the NIWRMP as the overall institution for the water sector coordination and capacity building of players in the sector to plan, implement, and monitor water management in a coordinated manner.

Spatial and Seasonal Water Availability

Bhutan has a high per capita availability of water when assessed at the level of basins and districts. However, most of the abundantly available water resources vary immensely over time. The climate of Bhutan is distinctly seasonal, with monsoon rains occurring between April and October. Other months of the year have low or no precipitation. Given the rugged terrain, there is a high proportion of run-off into small streams and rivulets. These discharge into tributaries and rivers that are characterized by steep gradients and narrow stream beds that have very little storage. As a result, surface water is quickly evacuated from the river basins. Whatever water is available during the lean winter season is the result of subsurface base flow and this is marginal. Furthermore, access to water stored in mountain lakes is very difficult. Bhutan's high per capita water availability thus must be considered in the light of there being much water at the wrong time and the wrong place. The ensuing shortages can be mitigated by retaining water through watershed conservation measures and maintaining forest and/or ground cover to intercept rainfall, hold the water, and promote infiltration. Erosion gullies, which make the water run faster down the slopes, should be controlled.

Limited Managed Water Storage

Water shortage is particularly predominant in the winter season and in higher elevations. This may be attributed to a lack of intervention to harness the abundant water from monsoon rains. Managing the excess of water during the monsoon to provide a supply during the lean season must be explored. Apart from watershed conservation, water outflow can also be slowed down artificially in the form of storage structures. Higher on the hillsides, many small rainwater harvesting structures can store small amounts of water for domestic use. Lower down the hillsides, check dams may be somewhat larger and store water for domestic and irrigation uses and perhaps for micro-hydropower generation. Large reservoirs could be built in the tributaries and/or main rivers, but in these locations, water is not scarce. Such reservoirs could level out the strongly seasonal flow fluctuations for the generation of hydropower, but would contribute little to domestic use and agriculture. Currently, there are no retention structures that can be managed to store water for provision during times of need.

WATER AVAILABILITY VERSUS ACCESSIBILITY

Bhutan's water balance does not show any water scarcity at the national, basin, or district level when considering time periods of one month. But this does not exclude the possibility that within the districts there can still be areas with temporary water shortages. The primarily rugged terrain and altitudinal variations over small distances make certain places have too much water, while adjacent areas experience a shortage. Most of the communities that are located along slopes depend on smaller streams, springs, and lakes for drinking and irrigation. The abundant water is largely available in the form of major rivers and tributaries flowing in low-lying river valleys and deep gorges. For example, some housing blocks in Thimphu city located near the perennial Thimphu River have a piped water supply that is only operational for a few hours every other day. This shows that the occupants of the housing block have limited access to water, even though the river is flowing right past them. Likewise, a local water source may temporarily run dry, leaving irrigated fields without a supply of water, even while large volumes of water run down a stream slightly further away. This aspect is depicted in Figure 10.9. These examples show the following three aspects:

- District-level water balances have been assessed, but they have little relevance at the household or farm level. Household or farm-level studies are possible. However, these are best undertaken on a case-by-case basis.
- Water may be available but not accessible in the absence of investment in equipment (storage reservoirs, pumps) and operations (diesel or electricity cost of pumping, maintenance). It is not a problem of water shortage, but a problem of water management. This aspect is compounded by the low

FIGURE 10.9 Schematic presentation of water availability. (*Source:* NIWRM, 2016.)

density of the rural population living in highly dispersed settlements on hill slopes.
- Managing water at the local level cannot be done from the centre. This is particularly true in Bhutan because of its large altitudinal variations and difficult terrain.

Demand Management

When confronted with (potential) water scarcity, one often looks at supply-side solutions like augmenting the water supply or accessibility in the form of putting up retention reservoirs, tapping new (remote) water sources, or using groundwater. However, one may also try to contain or reduce the demand for water in order to reduce water stress. Examples of demand management measures can be grouped into two categories, as given below.

Higher Agricultural Productivity

To enhance agricultural resilience and productivity in the face of climate change and other challenges, several adaptive practices can be adopted. Timely land preparation using power tillers and the prompt transplantation of paddy ensure optimal use of the growing season and improve yields. Diversifying crops and selecting improved varieties such as those with shorter growing seasons, high-yielding potential, lower water requirements, and resistance to pests and diseases can further enhance food security. Implementing the System of Rice Intensification (SRI) improves water use efficiency and crop productivity. Effective drainage systems help mitigate the risks of waterlogging caused by excessive rainfall. Additionally, reducing pre- and post-harvest losses through improved handling, storage, and processing methods minimizes food waste. Lastly, Integrated Pest Management (IPM) offers a sustainable approach to controlling pests and diseases by combining biological, cultural, mechanical, and chemical tools in a way that minimizes environmental and health risks.

Higher Water Use Efficiency

Improving water use efficiency in agriculture is critical to ensuring sustainable water management and climate resilience. The use of mulch or plastic sheets helps reduce evaporation losses from the soil surface, conserving moisture for crop use. Optimized irrigation scheduling prevents over-irrigation, ensuring that water is applied only when necessary and in appropriate amounts. Better puddling of paddy fields minimizes seepage losses by enhancing soil compaction. Similarly, the use of piped or lined supply channels significantly reduces water loss during conveyance. Transitioning from traditional surface irrigation methods, such as basin or furrow systems, to more efficient technologies like sprinklers or drip irrigation can drastically reduce water wastage and improve delivery precision. Incorporating plant residues into the soil improves its organic matter content and enhances its ability to retain moisture. Additionally, introducing water pricing mechanisms can promote the judicious use of water by encouraging farmers to adopt more efficient practices and technologies.

Water as a Common Pool Resource

Water is a common pool resource used by everyone to meet personal needs but also to support various economic activities such as irrigation, hydropower generation, commercial car wash, chipboard industries, and water bottling. Because access is open to all, competition for use can arise, often leading to overexploitation or conflict. This situation reflects what is termed the "tragedy of the commons," where individuals seek to maximize their personal gain from a shared resource rather than cooperate for collective benefit (Ostrom, 1993). However, around the world, well-designed collective management systems have successfully governed common resources for centuries, for instance, the traditional rainwater harvesting structures in Rajasthan and community-managed grazing lands in parts of Africa.

In Bhutan, similar challenges are emerging as water resources are degrading in both quality and quantity, mainly due to increasing demand and the discharge of effluents into rivers and streams. According to the National Integrated Water Resources Management Plan (Asian Development Bank & Royal Government of Bhutan, 2016), weak institutional mechanisms and limited enforcement have contributed to this decline. Since water is state property, effective monitoring and regulation remain difficult, and the capacity of the National Environment Commission Secretariat (NECS) and relevant line agencies needs to be strengthened to regularly update water quality standards and monitor compliance.

At the local level, the establishment and empowerment of Water Users Associations (WUAs) can play a crucial role in managing water as a common pool resource. The Water Act of Bhutan (2011) and its Rules and Regulations (2014) recognize the importance of such community-based institutions in promoting equitable distribution, sustainable use, and conflict resolution related to water. Strengthening these local governance systems, alongside national oversight, will be key to ensuring the sustainable management of Bhutan's water resources.

Unreliable or Insufficient Water Supply and Sanitation

The most important use of water resources is for human drinking, and it has been given the highest priority in the Water Act. Potable water is usually tapped from wells, springs, and small rivulets on the hill slopes upstream of dwelling areas. These sources are not always reliable, and therefore alternative sources are often sought at relatively great expense. Drinking water for larger cities is mainly abstracted from perennial streams where the supply is more secure throughout the year. Bhutan has made considerable progress in terms of coverage for both rural and urban water supply and sanitation in the last two to three decades. Improving public health by reducing the incidence of water-related illnesses through the provision of safe drinking water and improved sanitation facilities is actively being pursued. According to a UN/WHO country report for 2014, the water supply infrastructure in both rural and urban areas needs to be upgraded. Data from this report indicate that more than 50% of the urban population has intermittent water supply. In rural areas, only about 69% of rural water supply schemes are reportedly functional. Delivery of water supply services in rural areas is based on source protection specified in water safety plans, but there are no treatment systems in place. Rural sanitation facilities consist predominantly of pit toilets, followed by pour flush toilets and others. Except in core areas of major cities, septic tanks are the most common sewage treatment facility in urban areas. Although construction regulations require septic tanks to be equipped with soak pits, there are many substandard systems leading to the discharge of effluents directly into the environment.

Unreliable or Insufficient Irrigation Water Supply

Agriculture is the biggest consumer of water through irrigation. Most fields are fragmented and scattered on the hillsides, where water is tapped from springs and rivulets. Local and/or temporary shortages of irrigation water occur frequently. Tapping alternative sources is generally expensive because of the topography. There are some larger irrigation schemes in the southern foothills. The National Irrigation Master Plan has been elaborated in order to increase food production and self-sufficiency.

Impact of Climate Change

Rainfall is expected to increase in the future, particularly in the south of the country, and specifically during the monsoon period when water is already abundant. This will increase run-off as well as discharge and sediment load in the rivers during summer. The sediments will settle along the river when the flow rate reduces towards the lean season, which may increase flood hazards during the next monsoon. Greater variability and uncertainty are also expected. There would be a greater chance of extremely high river discharge during the wet season, calling for new design procedures and standards for water works. The traditional approach for determining design floods based on observed return periods would grossly underestimate the future river discharge and levels, as shown in Figure 10.10. As can be seen, by the middle of the century (2045), the flow for a return period of 50 years may increase by around 26% in Wangchhu, about 7% in Punatshangchhu, and up to 38% in Mangdechhu and Drangmechhu. There also is an expected trend towards lesser river flows in winter, although no change in minimum flows. However, it should be noted that the climate change assessments were constrained by a shortage of data. Available data have limited spatial representation and often cover less than the minimum requirement of 30 years of observations. Hence, the projections become less realistic when magnifying resolution from regional into smaller areas such as districts and Gewogs.

In climate models, future decades are to have drier and warmer winter months and wetter and warmer summer months. Under such a warming climate, river flows will become more erratic as rainfall is rapidly converted to run-off instead of being stored as ice. In turn, a combination of steep topography and the projected increase in summer rainfall and river flows in the coming decades poses major water-related hazards. Accelerated glacier melting is already increasing the number and size of glacial lakes and the risk of catastrophic flooding if these lakes are breached. In South Asia, half of the natural disaster events have been due to floods and landslides associated with extreme weather events. Climate change is predicted to make such extreme weather events more destructive, in particular through the intensified Asian

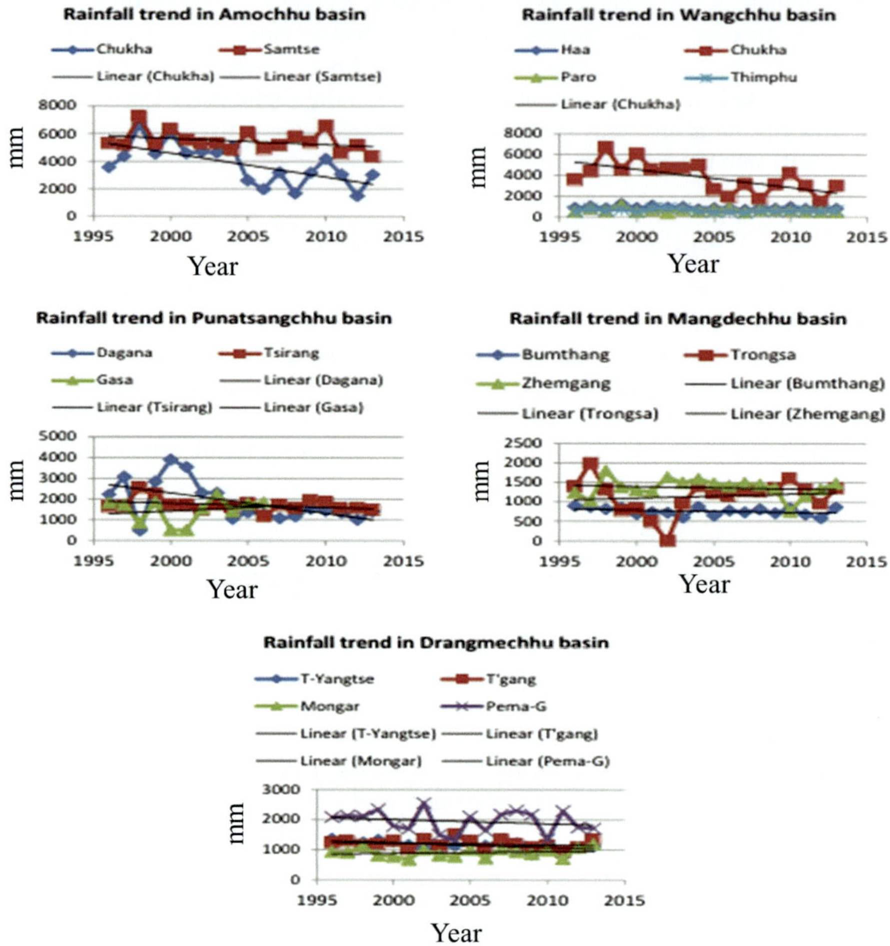

FIGURE 10.10 Recent rainfall trends in different basins. (*Source:* NIWRM, 2016.)

monsoon system. The Himalayan region in particular has become even more vulnerable to natural disasters spawned by melting glaciers, which form high-altitude lakes that can suddenly breach and cause catastrophic flooding downstream.

Water Sources Drying Up

There are frequent reports of water sources drying up. This appears to be at odds with predictions of increased rainfall caused by climate change. Investigation has shown that rainfall has, in fact, declined over the last two decades in a number of stations spread over the country, without a discernible spatial trend (Figure 10.10). Plotting rainfall figures over longer time periods tends, however, to conceal this observation and shows an increasing trend instead. This shows how delicate it is to draw conclusions in relation to climate and climate change when data remains limited.

SUSTAINABILITY: CLIMATE-RESILIENT INFRASTRUCTURE

NAP and ACREWAS Project: Bhutan is taking significant steps to enhance climate resilience through strategic infrastructure development, guided by its first NAP, launched in 2023. The NAP prioritizes water as the most vulnerable of seven key sectors and sets a comprehensive framework for adaptation over the next 15 years, with an estimated investment of USD 14 billion. A cornerstone of this plan is the ACREWAS project, which focuses on building climate-proof drinking water systems in Gasa and Punakha, and integrated irrigation and drinking water systems in Tsirang, collectively benefiting over 37,000 people. Similar interventions in Paro, Dagana, and Tsirang are expected to support an additional 36,000 individuals. These initiatives include the construction and upgrade of climate-resilient infrastructure such as water supply systems, dams, and irrigation networks that are engineered to withstand extreme weather events, floods, and landslides. Recognizing the vital role of hydropower in Bhutan's economy, adaptation strategies are being implemented to address changing hydrological patterns caused by erratic river flows and glacial melt. Key measures include enhancing river conveyance capacity, enforcing no-build zones along riverbanks, and improving flood risk management through strengthened river systems and protective infrastructure. These efforts collectively aim to secure water and energy resources while safeguarding communities against the growing impacts of climate change.

Watershed Management and Reforestation

Ecosystem-Based Adaptation: Bhutan has adopted an ecosystem-based approach to climate adaptation, placing strong emphasis on restoring degraded landscapes, reforestation, wetland conservation, and sustainable land use practices. The country's third National Adaptation Programme of Action (NAPA-III), implemented between 2017 and 2023, applied an integrated landscape management strategy that enhanced biological corridors, promoted climate-resilient agriculture, and supported the livelihoods of vulnerable communities. Key interventions included tree planting, watershed protection, and wetland restoration—nature-based solutions that play a vital role in regulating water availability, reducing soil erosion, and improving groundwater recharge. Additionally, Bhutan's commitment to REDD+ initiatives and the extensive network of protected areas contribute significantly to safeguarding ecosystem services that are crucial for ensuring water security and strengthening resilience to climate change impacts.

Enhancing Water Efficiency and Storage

Water Source Mapping and Modernization: Bhutan is proactively strengthening its water resource management in response to increasing climate pressures. The government has undertaken extensive mapping and monitoring of water sources, identifying that out of 7,399 water sources, 69 have already dried up and 1,586 are at risk. This data-driven approach enables targeted interventions to safeguard at-risk communities. To promote sustainable water use, Bhutan is investing in water use efficiency through improved irrigation systems, the adoption of climate-smart agricultural practices, and the development of modern water infrastructure, all of which help minimize losses and ensure a reliable water supply for both farming and household needs. Adaptation measures also include the promotion of natural infiltration and water buffering techniques, along with the construction of emergency water storage and controlled-release systems to mitigate the impacts of GLOFs. Furthermore, the government is exploring public–private partnerships as a means to bolster investment and innovation in water resource management and infrastructure development. These comprehensive efforts reflect Bhutan's commitment to building climate resilience and securing water sustainability for the future.

Regional and Global Cooperation

Transboundary and International Collaboration: Bhutan recognizes the importance of regional and global partnerships in addressing the shared challenges of climate change, particularly in the water and energy sectors. The country is currently engaged in negotiations with India and the state of Assam to expand hydropower exports and explore the potential for transboundary water supply for agricultural purposes. These efforts underscore the value of regional cooperation in enhancing water and energy security. At the global level, Bhutan actively participates in international climate adaptation initiatives and receives financial and technical support from key partners such as the Green Climate Fund, the Global Environment Facility, and the United Nations Development Programme (UNDP) to advance water sector adaptation projects. The country's adaptation strategies are closely aligned with its Nationally Determined Contribution (NDC) under the Paris Agreement, while the NAP process integrates insights, best practices, and resources from these international engagements. Moreover, Bhutan prioritizes knowledge sharing, capacity building, and joint research with international partners to strengthen its climate adaptation efforts. Collaborative initiatives in early warning systems and climate-resilient planning help build national expertise and ensure a proactive response to climate risks. Through such transboundary and international collaborations, Bhutan is reinforcing its commitment to sustainable development and regional resilience.

CONCLUSION AND RECOMMENDATIONS

Bhutan's water resources are both a blessing and a challenge. The nation's rivers, glaciers, and high-altitude wetlands are fundamental to its ecological integrity, economic development, and cultural heritage. However, these resources are under increasing threat from climate change, as evidenced by rising temperatures, shifting precipitation patterns, and accelerated glacial melt. These changes have led to greater variability in water availability, increased frequency and severity of hazards such as floods and GLOFs, and heightened risks of water scarcity during dry periods. The impacts of these changes are far-reaching. Agriculture, which underpins rural livelihoods, is highly sensitive to water availability and timing. Hydropower, the mainstay of Bhutan's economy, depends on reliable river flows, which are becoming more unpredictable. Communities, particularly those in vulnerable areas, face growing risks to their health, safety, and food security. Despite Bhutan's high per capita water availability, the country's unique topography, climate sensitivity, and socio-economic context make it vulnerable to water-related shocks.

Recommendations

Strengthen Climate-Resilient Infrastructure: Invest in robust infrastructure such as early warning systems for GLOFs and floods, resilient irrigation networks, and water storage facilities to buffer against seasonal variability and extreme events.

Enhance Watershed and Ecosystem Management: Prioritize reforestation, wetland conservation, and integrated watershed management to stabilize catchments, regulate flows, and maintain ecosystem services.

Expand Data Collection and Research: Improve hydrological, glaciological, and meteorological monitoring networks to generate high-resolution, long-term data for informed decision-making and adaptive management.

Promote Water Use Efficiency: Encourage adoption of water-saving technologies and practices in agriculture, municipal supply, and industry to reduce demand and increase resilience to shortages.

Foster Regional and Transboundary Cooperation: Engage proactively with neighbouring countries to manage shared river basins, exchange data, and coordinate disaster risk reduction efforts.

Empower Communities and Build Local Capacity: Strengthen community-based water management, support traditional knowledge systems, and provide education and resources for local adaptation initiatives.

KEY FINDINGS

Bhutan faces a complex set of water-related climate challenges, despite having one of the highest per capita water availabilities in the region. This abundance, however, is unevenly distributed across seasons and geographic regions, resulting in acute water shortages during dry periods. Compounding this issue is the rapid retreat of glaciers over the past 30 years; the country's glacier area has shrunk by more than 23%, while the number and size of glacial lakes have expanded significantly. This trend increases the risk of GLOFs, posing a severe threat to downstream communities. The country is also witnessing a rise in the frequency and intensity of water-related hazards, including floods, GLOFs, and droughts, which endanger lives, livelihoods, and critical infrastructure. Agriculture and hydropower, two key pillars of Bhutan's economy, are particularly vulnerable to changing water availability, with direct consequences for national food and energy security. In response, Bhutan has initiated several climate adaptation measures, notably the NAP 2023 and targeted interventions to reduce GLOF risks. While these efforts mark significant progress, implementation remains a challenge due to financial, technical, and institutional constraints. Additionally, there is limited research on wetlands and high-altitude lakes, despite their critical role in Bhutan's hydrology. A deeper understanding of these ecosystems is essential for developing informed and effective water management strategies in the face of climate change.

Gaps

Despite Bhutan's proactive efforts in climate adaptation and water resource management, several critical gaps hinder the effectiveness and scalability of its initiatives. One of the key challenges is the lack of high-resolution and long-term data on river flows, glacier dynamics, and wetland hydrology. This data deficiency constrains the ability to develop accurate climate models, assess risks, and plan for future scenarios with precision. Additionally, most existing research and monitoring efforts are conducted at national or river basin levels, with limited focus on localized impact assessments. As a result, the specific vulnerabilities and adaptation needs of individual communities are often overlooked, reducing the effectiveness of targeted interventions. The ecosystem services provided by high-altitude wetlands—such as water regulation, biodiversity support, and climate resilience—also remain under-researched, limiting the integration of these ecosystems into broader adaptation strategies. Transboundary water management presents another area of concern, with limited research and regional cooperation to address the downstream implications of upstream developments in neighbouring countries. Furthermore, there is a lack of detailed assessments on how water-related hazards affect different socio-economic groups, particularly women and marginalized communities. This absence of disaggregated data impedes the development of inclusive, equitable, and socially responsive adaptation measures. Addressing these gaps is essential for building a more resilient and adaptive water management system in Bhutan.

FUTURE RESEARCH

To strengthen Bhutan's climate resilience and ensure sustainable water resource management, a range of focused research areas must be pursued. First, enhanced glacier and glacial lake monitoring is essential. This involves the development and deployment of advanced remote sensing technologies and field-based monitoring systems to track glacier mass balance, glacial lake evolution, and associated hazards in near real-time, enabling timely risk assessment and response. Second, community-based adaptation studies are needed to evaluate the effectiveness of local strategies, including traditional water management practices and grassroots initiatives. Understanding what works at the community level can help design scalable and culturally appropriate interventions. Third, there is a pressing need for comprehensive research into wetland hydrology and ecosystem services, particularly in high-altitude regions. These studies should explore the ecological functions of wetlands and their potential to buffer climate impacts and support water security. Fourth, integrated water–energy–food nexus analyses can provide insights into how water availability affects hydropower generation and agricultural productivity, especially under varying climate change scenarios. This holistic approach is crucial for informed decision-making and cross-sectoral planning. Fifth, transboundary water governance research should explore the potential for regional collaboration on shared river basins. Key areas include data sharing mechanisms, joint disaster risk reduction efforts, and harmonization of water management policies across borders. Finally, socio-economic vulnerability and resilience mapping are vital to ensure no one is left behind. Developing spatially explicit models to identify household- and community-level vulnerabilities will support more targeted and equitable policy responses and investment decisions. These research priorities will provide the evidence base needed to shape adaptive, inclusive, and forward-looking water governance in Bhutan.

REFERENCES

Bajracharya, Samjwal Ratna, Sudan Bikash Maharjan, and Finu Shrestha, 2014. The status and decadal change of glaciers in Bhutan from the 1980s to 2010 based on satellite data. Annals of Glaciology, 55(66). doi: 10.3189/2014AoG66A125

Biswas, A. K., & Bisht, A. S. (2016). Unruly waters: A glimpse into Bhutan's water security challenges. In Water governance and sustainability (pp. 45–67). Springer.

Brown, A. O., & Matlock, M. D. (2011). A review of water scarcity indices and methodologies (White Paper #106). The Sustainability Consortium, University of Arkansas. https://sustainabilityconsortium.org/download/a-review-of-water-scarcity-indices-and-methodologies/

Chhetri, T. B. (2022). Climate change impacts on hydrology and water resources of Himalayan basins: A review. Current World Environment, 17(1), 32–43. https://doi.org/10.12944/CWE.17.1.4

Deutsche Welle. (2012, January 23). Bhutan grapples with climate challenges. https://www.dw.com/en/bhutan-grapples-with-climate-challenges/a-17151430

FRED. (2025). Gross Domestic Product per Capita for Bhutan. https://fred.stlouisfed.org/series/PCAGDPBTA648NWDB

Gain, A. K., & Giupponi, C. (2014). Assessing water policy alternatives in the context of basin-wide integrated water management: A literature review. Water Resources Management, 28(14), 4567–4585.

Govindaraju, R. S., & Das, S. (2016). Event and continuous modeling with HEC-HMS [Technical paper]. ResearchGate. https://www.researchgate.net/publication/245289459_Event_and_continuous_modeling_with_HEC_HMS

Gurung, T. B. (2017). Companion modelling for irrigation management in Bhutan: Role-playing simulations and multi-agent systems. Agricultural Systems, 156, 1–12.

Huffman, G. J., Bolvin, D. T., Nelkin, E. J., Wolff, D. B., Adler, R. F., Gu, G., Hong, Y., Bowman, K. P., & Stocker, E. F. (2010). The TRMM Multisatellite Precipitation Analysis (TMPA). In R. F. Adler, G. J. Huffman, & D. T. Bolvin (Eds.), Satellite rainfall applications for surface hydrology (pp. 3B42–3B52). NASA Goddard Space Flight Center.

International Association of Dredging Companies. (2016). Water availability. http://www.iadclexicon.org/water-availability/

Japan Aerospace Exploration Agency (JAXA). (2008). EORC topics: 2008/04/02. http://www.eorc.jaxa.jp/en/imgdata/topics/2008/tp080402.html

Japan Aerospace Exploration Agency (JAXA). (2012). Glacial Lake Inventory of Bhutan using ALOS data. https://www.eorc.jaxa.jp/ALOS/en/bhutan_gli/

JAXA, 2014. https://www.eorc.jaxa.jp/en/earthview/2008/tp080402.html

Kharat, R. S., & Mundra, A. (2016). Water issues in Bhutan: Internal disputes and external tensions. In Water issues in Himalayan South Asia: Internal challenges, disputes and cooperation (pp. 123–145). Springer.

Komori, J., Koike, T., Yamanokuchi, T., & Tshering, P. (2012). Chapter 2: GLOF events in the Bhutan Himalayas. In Final Report of SATREPS Project. Graduate School of Environmental Studies, Nagoya University, Japan.

Krusic, P. J., et al. (2015). Six hundred thirty-eight years of summer temperature in the Bhutan's eastern Himalaya, reconstructed from tree rings. Geophysical Research Letters, 42(7), 2354–2362. https://doi.org/10.1002/2015GL063566

Maharjan, S. B., Mool, P. K., Lizong, W., Xiao, G., Shrestha, F., Shrestha, R. B., … & Baral, P. (2018). The status of glacial lakes in the Hindu Kush Himalaya-ICIMOD Research Report 2018/1 (2018).
Ministry of Agriculture (MOA). (2008). Integrated River Basin Management (IRBMP), Wang River Basin: Master Plan Document. Social Forestry Division (SFD), Department of Forests (DOF), Royal Government of Bhutan (RGOB).
Ministry of Agriculture and Forests (MOAF). (2012). Statistical Yearbook of Bhutan 2013: Data from Bhutan RNR Statistics. Policy and Planning Division, MOAF, Royal Government of Bhutan. https://www.nsb.gov.bt/publications/statistical-yearbook/
Mouse River Enhanced Flood Protection Project. (2015). Preliminary Engineering Report and project documents. http://mouseriverplan.com
National Adaptation Programme of Action Phase II (NAPA II). (2015). Addressing the risk of climate-induced disasters through enhanced national and local capacity for effective actions. Funded by LDCF through GEF and UNDP, Royal Government of Bhutan. https://weadapt.org/wp-content/uploads/2023/05/terminal_evaluation_bhutan_napa_2.pdf
National Snow and Ice Data Center (NSIDC). (2015). NSIDC homepage and cryosphere information portal. Retrieved from https://www.nsidc.org
National Statistics Bureau (2016) National Accounts Statistics 2016. Thimphu: Royal Government of Bhutan. https://www.nsb.gov.bt/publications/statistical-yearbook/
Nayar, A. (2009). Climate: When the ice melts. Nature, 461(7267), 1042–1046. https://doi.org/10.1038/4611042a
NCHM (1996–2013). https://www.nchm.gov.bt/home/pageMenu/801
NEC S. (2016). *Bhutan State of the Environment Report 2016*. Thimphu: National Environment Commission Retrieved from http://www.nec.gov.bt/necs/wp-content/uploads/2020/10/Bhutan-State-of-Environment-Report-2016.pdf
NECS, ADB and TA (April, 2016). Wangchhu Basin Management Plan. *Adapting to climate change through IWRM*. https://www.moenr.gov.bt/wp-content/uploads/2024/04/Wangchhu-River-Basin-Mangement-Plan.pdf
Nepal, S., & Shrestha, A. B. (2015). Impact of climate change on the hydrological regime of the Indus, Ganges and Brahmaputra River basins: A review of the literature. International Journal of Water Resources Development, 31(2), 201–218. https://doi.org/10.1080/07900627.2015.1030494
NSB, 2005. Settlement Map of Bhutan. National Statistical Bureau, Thimphu, Bhutan.
Ostrom, E. (1993). Design principles in long-enduring irrigation institutions. Water Resources Research, *29*(7), 1907–1912. doi:https://doi.org/10.1029/92WR02991
Pilli-Sihvola, K., Phuntsho Namgyal, and Chhimi Dorji, 2014. Socio-Economic Study on Improved Hydro-Meteorological Services in the Kingdom of Bhutan. Report Prepared for the Strengthening Hydro-Meteorological Services for Bhutan (SHSB) project. Finnish Meteorological Institute and Department of Hydro-Met Services, Bhutan, 66 pp.
Saha, S., et al. (2018). Participatory integrated water resources management in Bhutan: Lessons from the Wangchhu basin. Journal of Hydrology: Regional Studies, 18, 112–130.
Sakthivadivel, R. (2006). Water balance studies and hydrological modeling for IWRM. In P. P. Mollinga, A. Dixit, & K. Athukorala (Eds.), Integrated water resources management: South Asian experiences (pp. 189–217). SaciWATERs. http://ndl.ethernet.edu.et/bitstream/123456789/40990/1/9.pdf
Snow and Glacier Division (SGD), DHMS, MOEA. (2015). Glacier and snow monitoring reports. Department of Hydro-Meteorology Services, Ministry of Economic Affairs, Royal Government of Bhutan. https://www.nchm.gov.bt
Sonessa, A. S., et al. (2019). Hydrological sensitivity to temperature change in the Himalayan region. Journal of Hydrology, 578, 123456. (Note: Exact DOI unavailable in accessible records; verify through institutional access or contact corresponding author.)

Stroud Water Research Center. (2015, October 30). Clean water essential to Bhutan's happiness. https://stroudcenter.org/press/2015/clean-water-essential-to-bhutans-happiness.shtm

Tariq, M. A. U. R., Wangchuk, K., & Muttil, N. (2021). A critical review of water resources and their management in Bhutan. Hydrology, 8(1), 31.

Tenzin, U., & Bhaskar, N. V. (2021). Hydrodynamic modeling of flash floods in Bhutanese river basins. Natural Hazards, 107(2), 1456–1478.

Trading Economics. (2025). Bhutan GDP per capita. https://tradingeconomics.com/bhutan/gdp-per-capita

Tshering, P., & Fujita, K. (2016). First in situ record of decadal glacier mass balance (2003–2014) from the Bhutan Himalaya. Annals of Glaciology, 57(71), 289–294. https://doi.org/10.3189/2016AoG71A036.

United Nations. (2015). Water scarcity. International Decade for Action "Water for Life" 2005-2015. http://www.un.org/waterforlifedecade/scarcity.shtml

U.S. Army Corps of Engineers, Hydrologic Engineering Center. (2016). HEC-HMS: Hydrologic Modeling System. https://www.hec.usace.army.mil/software/hec-hms/

U.S. Geological Survey. (2016). Wetland definitions. In History of wetlands in the conterminous United States (Water-Supply Paper 2425). https://water.usgs.gov/nwsum/WSP2425/definitions.html

Wangdi, N., et al. (2019). Experimental assessment of drought stress on montane ecosystems in Bhutan using throughfall exclusion. Forest Ecology and Management, 445, 45–56.

Wangdi, S. N., & Norbu, N. (2010). Inventory of High Altitude Wetlands in Bhutan. Environmental Science, Geography. https://www.semanticscholar.org/paper/INVENTORY-OF-HIGH-ALTITUDE-WETLANDS-IN-BHUTAN-Wangdi-Norbu/3e161fe92fc7640647b1c0cf95e73cc4c50709ea

World Bank. (2015). Modernizing weather, water, and climate services: A road map for Bhutan (Water Partnership Program Report No. 98534). https://www.gfdrr.org/sites/default/files/publication/Bhutan%20Roadmap%202015.pdf

World Bank. (2024). GDP per capita (current US$) - Bhutan. https://data.worldbank.org/indicator/NY.GDP.PCAP.CD?locations=BT

11 The Cryosphere and Delta

Navigating Transboundary Dynamics of Ganga River Basin

Radhika Mulay and Vijay Paranjpye

INTRODUCTION

Historically, the Cryosphere and Delta regions of the Ganga River system remained largely unexplored due to the treacherous terrain of the Hindu Kush Himalayas and the wilderness of the Sundarbans. Aside from a few mythological stories and anecdotes about *rishis* (sages) who sought solitude in the highlands and stories of a few traders and pastoralists who ventured through the Himalayas passes, early documented explorations began in the 1500s with a Spanish missionary at the Mughal court and later by the British East India Company. In order to improve the topographic understanding of the region, the Survey of India organised a systematic programme to map the Hindu Kush Himalaya in the mid-19th century (Chatterjee & Bishop, 2025). Although the Delta region has been inhabited since the 1200s, it was rapidly explored by the British for land revenue purposes, leading to the establishment of human settlements practicing paddy cultivation and fishing in the saline territories of the Bay of Bengal (Eaton, 1990; Ghosh et al., 2015). Furthermore, different nomenclature for rivers across territories complicated the understanding of their interconnectedness. Therefore, the present-day understanding of the origin and mouth of the Ganga River system is relatively young and still evolving. From an Indian perspective, the Ganga River is believed to originate at Gomukh in the Gangotri glacier, flowing through Uttarakhand, Uttar Pradesh, Bihar, Jharkhand, and West Bengal, and finally meets the Bay of Bengal at Gangasagar island, about 100 km south of Kolkata. This perspective overlooks the historical shifts and avulsions that have shaped the river's transboundary identity, shared by multiple riparian countries.

Consequently, there is comparatively limited research on the interlinkages between the Cryosphere and Delta regions with the GRB. Most studies focus on the mainstream of the Ganga from Haridwar to Kolkata, within India, generating disproportionate data, research, and planning on this section, while neglecting the upstream and downstream interconnections. Existing literature points to various studies both in the Himalayas and the Delta; however, they are often conducted

DOI: 10.1201/9781003599692-11

independently, ignoring the interlinkages. Thus, one of the aims of this chapter is to explore the Cryosphere and Delta together to emphasize upstream and downstream connections, especially in light of climate uncertainties. The study synthesizes quantitative and qualitative research from cartographic maps, archival records, academic studies, government reports, and ethnographic accounts, which were gathered and triangulated.[1] Primary data from field visits in the Hindu Kush Himalaya and Sundarbans, as well as multiple mountaineering expeditions, research studies, and training programmes from 2010 to 2024, have informed the study's ecological, demographic, and sociocultural contexts. The objective is to document the evolution of the heads and tails of GRB, their interlinkages, their role in hydrological, ecological, sociocultural, and political contexts, and future scenarios.

The chapter argues that these interlinkages are often overlooked due to political borders separating the Hindu Kush Himalayas in the Tibet Autonomous Region (China), Nepal, India, and the Delta region in eastern India and Bangladesh. The introduction outlines the background, rationale, and methodology. Subsequent sections profile the study area, detail historical river avulsions, and discuss the significance of the Cryosphere and the Delta region within the larger GRB. Sections on ecological diversity, sociocultural dynamics, and geopolitical issues culminate in providing recommendations for multilateral cooperation and the precautionary principle for sustainable water resources management. The chapter concludes by advocating for an integrated approach to the GRB that incorporates the Cryosphere and Delta regions to ensure a resilient water future.

PROFILE OF THE STUDY AREA: A TRANSBOUNDARY RIVER BASIN

Within the larger Ganga-Brahmaputra-Meghna River basin (GBM),[2] the entire drainage or catchment area of Ganga River Basin (GRB) spans across four riparian countries, from the Himalaya mountains in the Tibet Autonomous Region (China), Nepal, and India, and empties out through India and Bangladesh into the Bay of Bengal, forming the heads and tails of GRB. Table 11.1 and Figure 11.1 demarcate the GRB and provide approximate estimates of the country-wise area and water-share

TABLE 11.1
Country-wise Area and Water Share Contribution in Ganga River Basin

Country	Area (km²)	Area (%)	Water-Share Contribution (%)
India	8,61,451	80	54
Nepal	1,47,100	13	40
Bangladesh	46,300	4	5
Tibet	33,000	3	1
Total	10,87,852	100	100

Source: Aryal and Rajkarnikar (2011); Central Water Commission & Regional Remote Sensing Centre-West (2012); Frenken (2012).

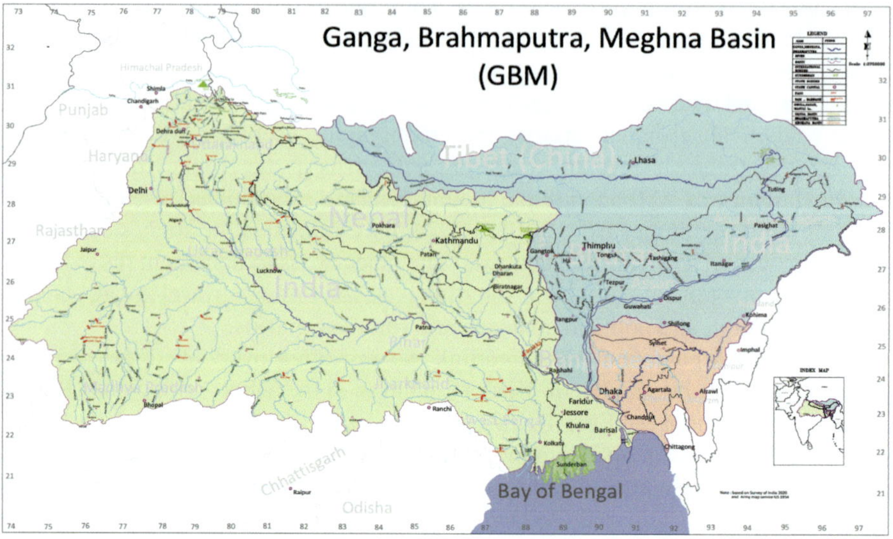

FIGURE 11.1 Map of the entire Ganga River Basin spread across Tibet Autonomous Region (China), Nepal, India, and Bangladesh. *Note:* Cryosphere or the headwaters of Ganga are in the north in higher elevations in the Hindu Kush Himalayan region and Delta region is in the south-eastern part of the highlighted region near the Bay of Bengal. (*Source:* Sadoff et al., 2013.)

contribution in GRB, which occupies an area of about 10,87,300 km^2 (Central Water Commission & Regional Remote Sensing Centre-West, 2012). Based on water-share contribution, India and Nepal are almost equal shareholders, and Bangladesh and Tibet, although they have comparatively less share, are part of the river basin and, therefore, need to be acknowledged in the river basin plans.

THE HEADWATERS OF GANGA RIVER BASIN

The Ganga River, as an entity, comes into existence at Devprayag in the Tehri Garhwal district of Uttarakhand, where two principal tributaries, the Bhagirathi and Alaknanda rivers, merge. The Bhagirathi originates at Gomukh at Gangotri glacier, considered the traditional and cultural origin of the Ganga,[3] while Alaknanda arises at the terminus of Satopanth and Bhagirathi-Khadak glaciers near the Nanda Devi.[4] Their confluences are collectively known as the *Pancha Prayag*, or the 'five sacred confluences', one of which is Devprayag. After this confluence, Ganga is joined by several left-bank tributaries, including the Ramganga, Gomati, Mahakali (Sarada), Ghaghara (Karnali), Gandaki (Narayani), Sapta Kosi, and Mahananda. The Mahakali (Sarada) originates in the Kalapani region and marks the Indo-Nepal border. Gandaki rises in Tibet near Nepal, while Sapta Kosi is formed by seven tributaries out of which Tamur Kosi originates from the Kangchenjunga ridgeline on the border of Nepal-Sikkim border,[5] and Arun River from Tibet. All the glaciers and rivers south of Mt. Everest drain into the Sapta Kosi River.

According to the Central Water Commission of India, Ganga's total yield is about 525 BCM (Billion Cubic Metres). The Nepal Water and Energy Commission Secretariat (WECS) reports that nearly 225 BCM flows from left-bank tributaries in Nepal and the Tibet Autonomous Region (China), accounting for approximately 40–42% of Ganga's average annual flow, and around 60-70% of the dry-season flow when the flows from right-bank tributary Yamuna and her sub-tributaries are negligible (Aryal & Rajkarnikar, 2011; Frenken, 2012). On the right bank, the Ganga is fed by the Yamuna, originating at the Yamunotri glacier near the Banderpunch peaks, and her sub-tributaries, including the Chambal, Betwa, and others. A major tectonic event around 5,000 to 6,000 years ago redirected the Yamuna River to join the Ganga at present-day Prayagraj (Allahabad)[6] contributing about 97 BCM annually (Hoiberg, 2000; Jain et al., 2007; Valdiya et al., 2016). Thus, Ganga's headwaters extend beyond Gangotri, encompassing regions from the Banderpunch mountains (origin of the Yamuna River) in the west to the Kangchenjunga ridgeline in the east, with all of Nepal lying within GRB, divided into ten sub-basins (Figure 11.2). These headwaters lie in the Cryosphere (frozen part of the earth, including ice, snow, and permafrost) of the Hindu Kush Himalayan region, the Third Pole, and form the heads of GRB.

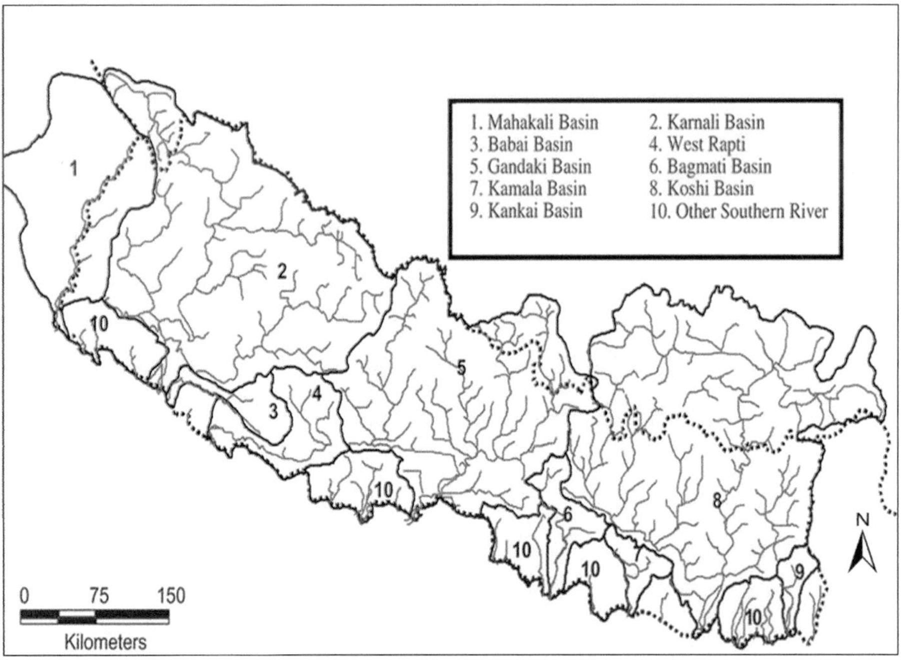

FIGURE 11.2 Major tributaries of Ganga River through Nepal territory. The map indicates that the entire geographical area of Nepal lies in the Ganga River Basin. (*Source:* After Gautam & Acharya, 2012.)

THE TAILS OF GANGA RIVER BASIN

The right-bank and left-bank tributaries of the Ganga River contribute to the mainstream until she reaches the northern limits of the Murshidabad district in West Bengal, India, where the dispersion or attrition of the mainstream begins. Near the Farakka Barrage, high sediment load and a gentle riverbed gradient cause sluggish water flow. The first distributary to branch off is the Bhagirathi-Hooghly, marking the point where the mainstream splits into the Bhagirathi-Hooghly and the Padma River, which flows into Bangladesh. Unlike the Yamuna avulsion, which occurred about 5,000 to 6,000 years ago, the river avulsions near the Ganga are relatively recent. A 15th-century Bengali literary reference suggests that the Bhagirathi-Hooghly was the main channel for the Ganga entering the Bay of Bengal near Gangasagar Island, historically known as Adi Ganga (Mukherjee, 2016).

Following the major earthquake of 1505 CE, the Ganga began migrating eastward, with shifting sands pushing the shoreline southward, and leading to eastward progradation river's tail-end, as well as that of the Brahmaputra, independently. This eastward progradation occurred in periodic stages G1, G2, and G3 (Figure 11.3) and resulted in increased flow in the Padma River, which gained significance due to the rising water discharge, while the Bhagirathi-Hooghly began to desiccate and required revival through the artificial Feeder Canal (Allison et al., 2002; Sarker et al., 2013).

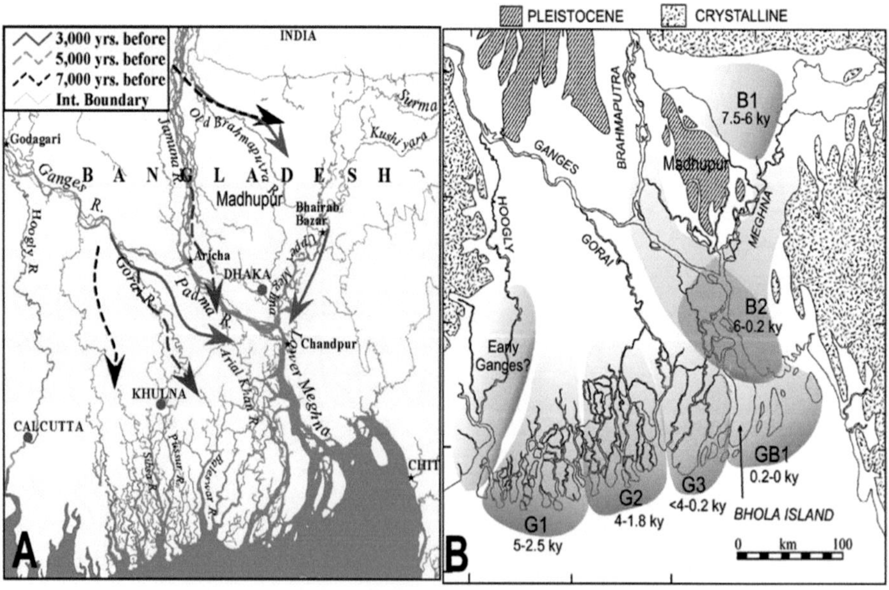

FIGURE 11.3 (A) Map of the Ganga-Brahmaputra Delta with (B) changing phases showing the movement of river channels: Ganga as G1, G2, and G3; Brahmaputra as B1 and B2; and combined Ganga-Brahmaputra as GB1. (*Source:* After Sarker et al., 2013.)

Subsequently, the Teesta River and the Brahmaputra rivers also experienced drastic river avulsions due to the 1782 earthquake and 1787 megafloods. The Teesta River changed its flow to join the Brahmaputra,[7] leading to the formation of an anabranch[8] named Jamuna, while the original old Brahmaputra channel underwent westward migration which eventually merged with the Ganga (Padma) near Rajbari district in Bangladesh. The confluence of the Ganga (Padma), Brahmaputra, and Meghna (Barak) formed the GBM basin, flowing into the Bay of Bengal near Bhola Island. The final discharge point of the GBM extends from Gangasagar Island in India to Bhola Island in Bangladesh, creating a complex series of eight major and multiple minor distributaries, spanning approximately 350 km along the southern coastline of the Bengal Delta, from east to west. The series of changes and avulsions in the tail-end reaches of the GBM river system created one of the most dynamic Delta configurations, covering 105,641 km^2, which forms the tails of the Ganga, with two-thirds in Bangladesh and one-third in India.

On a geological timescale, it is crucial to recognize that these changes in river courses and channels are not isolated events; rather, they are ongoing processes that continuously alter the drainage patterns and flow dynamics of the river system. A distinctive feature of the Delta is that, in addition to the eight major mouths, there are literally hundreds of smaller rivers that directly feed into the Bay of Bengal through a network of continuously evolving creeks and estuaries. In the headwaters, heavy rainfall caused four major rivers in Uttarakhand's headwaters, Kosi, Gaula, Nandhaur, and Dabka, to alter their courses in October 2021 (Azad, 2021).

Another significant example of ongoing avulsion is the Kosi River avulsion. After cutting through the Himalayas with steep gradients, the Sapta Kosi River enters the low-gradient Indian plains, where water velocity drops, depositing millions of tons of silt annually. This build-up raises the riverbed above the surrounding land, forcing the river to carve new channels. This process, spanning millennia, has created the world's largest alluvial 'mega-fan' (Chakraborty et al., 2010). Factors like intense rainfall, damaged drainage systems, and wetland loss due to anthropogenic activities continue to shape the morphology of this river system and result in severe annual floods in Bihar and Uttar Pradesh. These avulsions, anabranching, braiding, twisting, and migration of channels influence the current morphology of the Ganga River system and are vital for understanding the transboundary dynamics. According to this understanding, a schematic transboundary line diagram from the headwaters to the tails—from Cryosphere to the Delta—suggests that the total length of the Ganga adds up to 2,616 km (Figure 11.4).[9]

CRYOSPHERE: SIGNIFICANT CONTRIBUTOR TO THE HYDROLOGICAL BUDGET

As elaborated in the previous sections, the headwaters of the Ganga are situated in the Cryosphere (areas above the snowline, roughly above 4000–5000 m in altitude) of Hindu Kush Himalayas, primarily encompassing the transboundary regions of the states of Uttarakhand, small parts of Himachal Pradesh and Sikkim in India, parts of Nepal, and a significant portion of the Tibet Autonomous Region in China. Of the

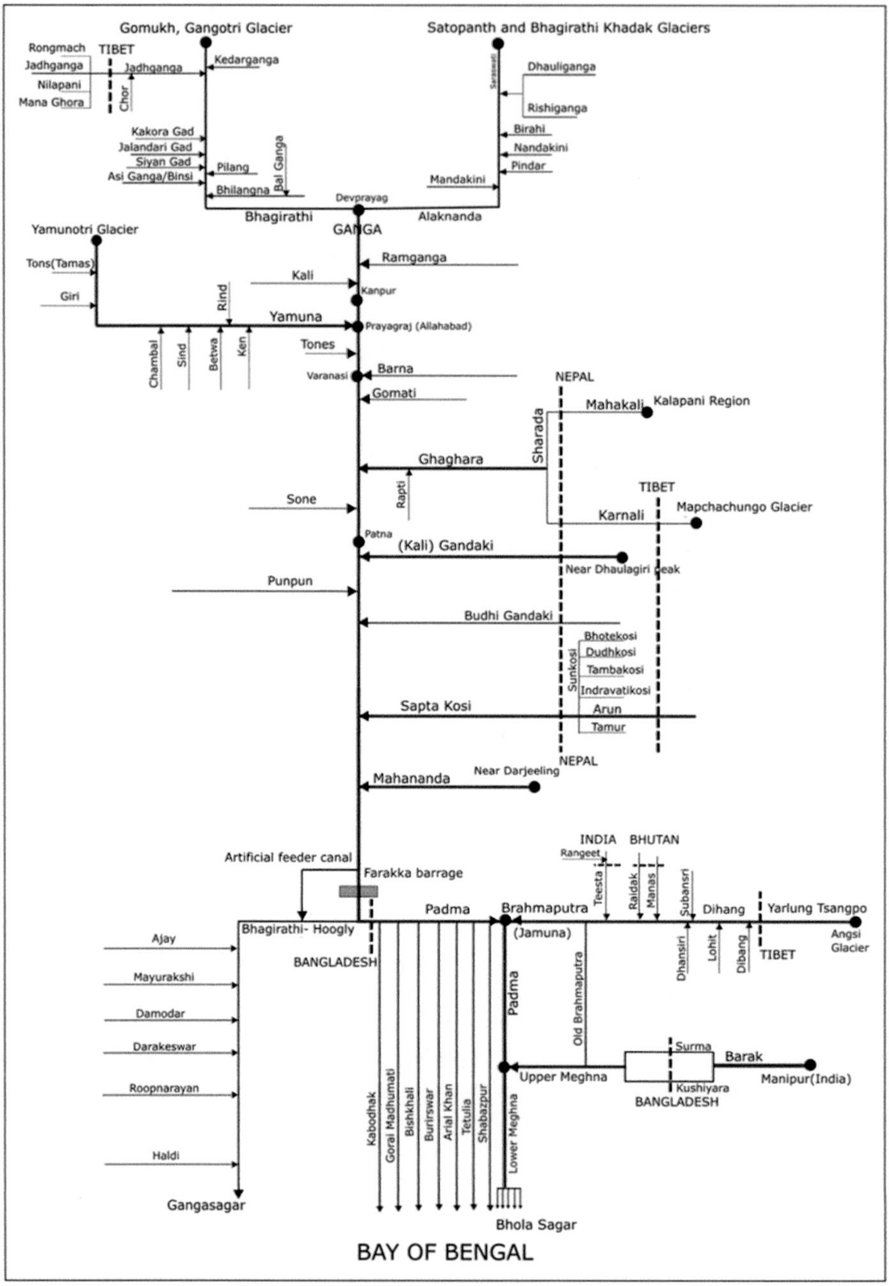

FIGURE 11.4 Transborder line diagram of Ganga River illustrating the transboundary nature of the tributaries and distributaries in Tibet Autonomous Region (China), Nepal, India, Bhutan, and Bangladesh. (*Source:* Prepared by authors.)

three distinct water sources of the GRB—monsoon rainfall, groundwater reserves, and Cryosphere—the millennial stock of freshwater stored in the Cryosphere contributes nearly 9–10% to the hydrological budget of the Ganga River Basin, primarily through ice melt and snowmelt[10] (Miller et al., 2012; Siderius et al., 2013). Based on a study conducted by Immerzeel et al. (2010), through the integration of an empirical degree-day snowmelt run-off model with a global hydrological modelling framework, out of the 9–10% estimate, around 40% is derived from ice melt and 60% from snowmelt.

This snow and ice melt emerges from the various components of the Cryosphere including around 6,237 to 7,963 glaciers with a total ice reserve of about 793.53 km^3, about 4707 glacial lakes with a catchment area of 247,109 km^2, as well as snowbound areas on land slopes, peaks, rocky outcrops, valleys, and gorges (Space Applications Centre (ISRO), 2011; Bajracharya, 2011; Rao et al., 2021).[11] Permafrost, ground that remains frozen at temperatures of 0°C or lower for a minimum period of two consecutive years, forms a very crucial component of the Cryosphere, but as it cannot be visually seen, and since its dynamics are unknown, it is the least understood and neglected part of the Cryosphere. Subsequently, thermokarst lakes, a special category of high-altitude lakes formed in topographic depressions when permafrost melts, also remain less understood and documented within the Cryosphere of GRB.

While rainfall is the dominant source of water flow in the Ganga, during the summer months, when rainfall is minimal, the contribution from snow and ice melt becomes crucial for maintaining the minimum environmental flows (E-flows) during the lean periods. It is generally accepted that while annual precipitation may increase in the future, the occurrence of extreme point rainfall events (EPREs) will also rise. As a result, even if total annual precipitation increases, the number of rainy days may decrease, leading to longer periods of drought. As noted earlier, approximately 40–42% of the water in the Ganga River comes from glacial-fed rivers in Nepal. The Cryosphere, the largest storehouse of freshwater in Asia, thus plays a vital role in regulating annual flows, particularly during the eight months from October to May. Glacial meltwater plays a significant role in sustaining river flow, especially during dry years and late summer, once the seasonal snow packs have melted. In the GRB at Rishikesh, the contribution of glacial ice melt contributes an average of ~32% to stream discharge, peaking at ~40% during summer and monsoon (Maurya et al., 2011). A study especially focusing on the Bhagirathi River basin highlights that the monthly proportions of snowmelt vary from around 42.9% in the monsoon season to about 91.9% in the summer months (Rai et al., 2019). Therefore, the Cryosphere, specifically glaciers, act as buffers, regulating stream flow and providing insurance against periods of low flow (Sharma & Condappa, 2013; Singh et al., 2011). These lean flows are essential for supporting unique biodiversity, livestock, and pastoral communities in the high-altitude regions as well as crucial for sustenance in the floodplains and the Delta region in the middle and downstream regime.

THE MEGA-DELTA OF GANGA RIVER BASIN

In the tail-end reaches, there are nearly 700 rivers within the Delta region, with a combined length of around 24,000 km influenced by four major geological phenomena:

(a) gravitational forces, (b) tidal back-lift from the sea that causes the ingress of seawater, (c) long-term plate tectonic shifts (still active in this region), and (d) sediment deposition (bed load or silt load). The Delta is not the largest solely because it drains water from two of the largest river sub-basins in Asia, Ganga and Brahmaputra, but because these rivers transport the highest volumes of suspended sediments and bed load from the youngest, tallest, and most unstable mountain ranges, which are fragile and brecciated. Around 75% to 80% of the Delta has been formed by sediment deposition over the past 100,000 years, creating the Upper-Fan Delta, Fluvio-tidal Delta, and Tidal Delta (Figure 11.5). It is estimated that approximately 2.4 billion tons of sediment are transported annually, with a portion deposited in the Delta and a significant amount carried into the Bay of Bengal (Banglapedia, 2021). This makes this Delta the largest sediment dispersal system in the world. Such extensive deposition has led to the creation of a transboundary shallow submarine shelf that extends several kilometers into the Bay of Bengal shared between India and Bangladesh territories.

This Delta plays a crucial role in the functioning of the Ganga River system and broader ecological and hydrological health of the region. The sediment deposited in the Delta naturally regulates the river flows, especially during the monsoon

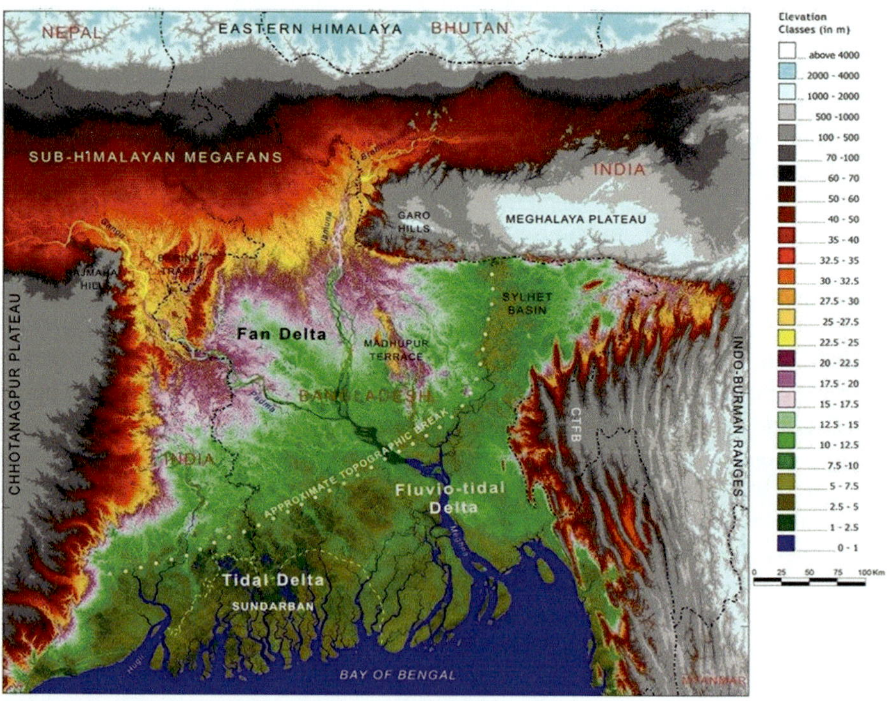

FIGURE 11.5 Physiographic setting of the Delta of GRB. Subsidence due to tectonic movement, accretion due to sedimentation accompanied by tidal movement, and sea level rise are constantly altering the entire Delta region. (*Source*: After Bandyopadhyay, 2019.)

season. The intricate network of rivers, tidal creeks, and mangrove forests acts as a natural buffer, absorbing and slowing the flow of water, thus reducing the severity of floods downstream. Additionally, the groundwater in the Delta contributes to the flow of water during dry periods, ensuring the maintenance of the river's base flow, especially during times when glacier meltwater or monsoon rainfall is insufficient. Beyond its importance in sediment deposition and flood regulation, the Delta also plays a critical role in coastal protection, climate regulation, and supporting biodiversity. It serves as a crucial buffer zone that connects the freshwater ecosystems of the river with the coastal and marine environments of the Bay of Bengal. Moreover, the Delta's resources are integral to the well-being of millions of people in the region, contributing to both their livelihoods and cultural heritage in India as well as Bangladesh. These ecological and sociocultural dimensions are explored in the following sections.

ECOLOGICAL DIVERSITY ACROSS UNIQUE NICHES

These morphological characteristics and environmental conditions in the Cryosphere and the Delta region determine the ecological biodiversity shaped through evolutionary processes of mutation and adaptation that have influenced microbial communities, flora, and fauna over millennia. This biological diversity spans various sub-ecosystems, including glacial ecosystems teeming with microbial life, glacial lakes, wetlands, river and spring systems, meadows, grasslands, forests, floodplains, agricultural landscapes, and the Delta ecosystem, each with its own ecological niche. These ecosystems do not have distinct boundaries but instead form interfaces where multiple sub-ecosystems overlap. For example, the edges of glacial lakes in the Gangotri region or the forests near Gomukh demonstrate such transitions. High-altitude meadows like Nandanban also serve as transitional zones between different sub-ecosystems. Ganga's self-purifying quality is attributed to several factors: low temperatures non-conducive for bacterial growth at high altitudes, high levels of dissolved oxygen (7–13 mg/l), low human population and settlements, the absence of industrial activity in its upper reaches, and most notably the natural presence of bacteriophages (bacteria-eaters)—microorganisms that destroy harmful bacteria like *Escherichia coli*. These bacteriophages, which are released when permafrost melts, continuously replenish the Ganga and other Himalayan glacial tributaries, making it a crucial subject for further research. It is important to note that these bacteriophages can only deal with bacterial pollution but remain inactive against chemical, industrial, plastic, and other kinds of non-organic pollutants.

The faunal diversity in the region shows remarkable adaptability to extreme physical and climatic conditions. In the highlands, species like the Bharal (blue Himalayan sheep) thrive in low temperatures, scarce water, and limited nutrients, while in the Delta, animals adapt to heavy sedimentation and saline waters. A notable example is the Ganga River Dolphin (*Platanista gangetica*), a nearly blind aquatic mammal adapted to the turbid, muddy waters of the river. The nutrients in the Ganga support diverse species in the Delta, including the endangered Bengal tiger, various fish species, and migratory birds. In the Delta region, the Sundarbans,

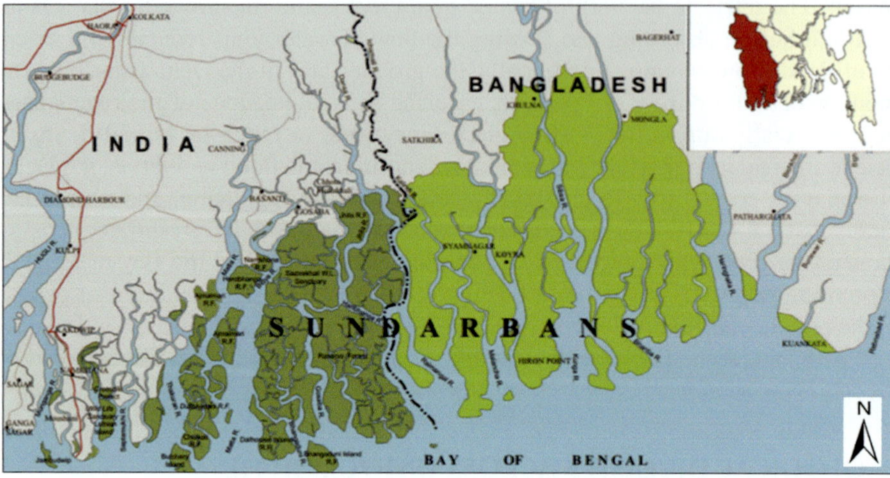

FIGURE 11.6 This map shows the location of Sundarbans Mangrove Forest situated approximately 60% in Bangladesh and the other 40% in India. (*Source:* After Baltazar, 2015.)

named after the Sundari mangroves (*Heritiera fomes*), spanning about 26,000 km^2 [12] (60% in Bangladesh and 40% in India), host a rich diversity of aquatic and terrestrial life at the junction of freshwater and saline waters (Figure 11.6). Therefore, the Himalayan Cryosphere and the expansive Delta are vital carbon sinks that contribute to weather patterns and rainfall regulation. The Delta's mangroves provide protection against coastal erosion, storm surges, and cyclones. Unique ecological niches, including the glaciers, glacial lakes, permafrost, spring ecosystems and mangrove forests, wetlands, and coastal ecosystems, thus, play a crucial role in regulating local and regional climates in the heads and tails of the Ganga.

SOCIOCULTURAL DYNAMICS

In addition to the unique biodiversity and ecosystem services, the headwaters and tail-end regions of the GRB have long been home to diverse communities that adapted to the challenging climatic conditions of the Cryosphere and Delta regions.

THE TRANSHUMANCE OF HIMALAYAS

The earliest human settlers in the mountainous areas were likely ascetics who, around the 8th century CE, ventured into these remote and hostile regions in search of solitude. Over time, other communities capable of enduring the harsh environment developed sustainable practices, allowing them to establish permanent settlements in this extreme terrain. Livelihoods in these areas were shaped by the Cryosphere's dynamics and prevailing weather conditions, leading to practices such as subsistence agriculture and pastoralism. Many communities settled in the middle and lower hill ranges, forming small villages or hamlets and adopting a hill-farming lifestyle

FIGURE 11.7 Images of pastoral practice, subsistence farming, and human settlements from field visits to the villages within the GRB in Himachal Pradesh and Uttarakhand regions in India and Nepal in the Cryosphere region of the Hindu Kush Himalaya. (*Source:* Prepared by authors; *Photo credits:* Swanand Kulkarni, Vijay Paranjpye, and Anagha Athavale (2017–2019).)

(Figure 11.7). With increasing interactions with the plains, Peninsular India, and Tibet, trading communities began to thrive. Tectonic movements, which caused river shifts and significant physical upheavals, also altered migration routes and influenced the settlement patterns of communities such as the Khasha, Kirata, Bhotia, Vanraji, Lepcha, Gurung, and others. Today, the Cryosphere region of the Ganga River supports a population of approximately 30 million people, with sparse density distributed across the Tibet, Nepal, Bhutan, and Indian territories.

THE BOAT PEOPLE OF THE DELTA

In the Delta region, the swampy, dense forests, marshy terrain, and brackish lagoons were initially uninhabitable and, therefore, less populated. However, since the colonial period, when Kolkata was chosen as the British capital due to its 'ecologically subsidized' nature, the demographics in the region have changed considerably.[13] Colonial officers undertook extensive clearing of mangrove forests for land revenue generation, converting the swampy marshes and forests into rice fields, which were then leased to local farmers. Over the past 200 years, a new community of farmers and fisherfolk gradually settled and adapted to life in the deltaic region. Today, approximately five million people reside around the Sundarbans, living in small village islands and frequently venturing into protected forest areas to collect timber, firewood, honey, beeswax, fish, and other resources. Their livelihoods are diverse, including freshwater and saline fishing, subsistence rice cultivation in uplands and

FIGURE 11.8 Images of fishing, boating practices, and human settlements from field visits to the Sundarbans Delta in Ganga River basin. (*Source:* Prepared by authors. *Photo credits:* Radhika Mulay, and Sukrit Sen (2018–2024).)

swampy areas, and, in some regions, a transition to commercial fishing, aquaculture, and prawn farming (Figure 11.8). These communities rely heavily on natural ecosystems, with their land and habitations periodically reshaped by flooding and the shifting courses of tributaries.

Having deeply naturalized and adapted to these unique conditions, the transhumance in the Cryosphere and the boat people in the Delta region can be considered 'ecosystem people'. This term reflects their nomadic existence, characterized by continual migration within the Himalayan highlands in response to shifting snowlines and between islands and creeks in the Delta region, which are transboundary in nature. These communities embody resilience and adaptability, shaped by their dynamic and challenging environments in the river basin. While their lifestyle has often been celebrated for its sustainability and harmony with natural cycles, these communities are highly vulnerable to rapidly changing climatic conditions in the river basin. The interconnectedness of the river basin's upstream and downstream areas exacerbates the impacts on both fragile ecosystems and the communities that rely on them. Increasing challenges—such as unpredictable climate, ecological pressures, limited access to education, healthcare, and essential services—and the unsustainable growth of tourism (tourism remains relatively low in the Delta compared to the upper Ganga region, primarily due to access difficulties and limited infrastructure) force youth to migrate in search of better opportunities in urban centres.

In Nepal, the Department of Labour and Employment issued over 2.7 million labour permits between 2008 and 2015. Similarly, by 2018, around 60,000 people had migrated from the Sundarbans Delta. During the 2024 field visit, the women in Kumirmari island village in the Sundarbans narrated their memories of loss and death, recalling the moment 'when the river turned black' (signifying death), during the devastating cyclone

Aila of 2009 and subsequent other disasters, which profoundly altered their lives and forced men and youth to migrate to the mainland in search for employment and stability. This ongoing mass migration has created 'ghost towns and villages' in the headwater and tail-end regions, where only women, children, and the elderly remain. Along with this demographic shift, the loss of traditional livelihoods and cultural practices raises concerns about the erosion of the region's identity and resilience.

ANTHROPOGENIC ACTIVITIES AND MAN-MADE DISASTERS IN THE CRYOSPHERE AND DELTA

The demography and lifestyles of these 'ecosystem people' have also been significantly altered due to increasing anthropogenic activities. Within India, infrastructure projects began in the early 1900s and accelerated post-independence, with major projects like the Farakka Barrage (1975) and Tehri Dam (1978) leading to over 1,547 structures by 2019, over 70–75% of them primarily located in the Cryosphere regions (Dam Safety Monitoring Directorate, 2019; WWF-India, South Asia Network on Dams, Rivers and People (SANDRP), 2019). Unlike India, the growth of dams and hydropower projects in Nepal Himalaya has been slow due to political instability and a lack of technical and financial capacity. In the Delta region, after the Farakka Barrage, several other barrages were constructed, including the Gazaldoba Barrage on the Teesta (1987) and the Duani Barrage in Lal Monirhat (1990). The proposed Tipaimukh dam on the Meghna River is under contention due to resistance from affected communities in Assam and Bangladesh.

During the last 75 years, water abstraction, dam construction, sand mining, and encroachments have reduced water discharges—E-flows in the Ganga River by almost 45% and 57% at Farakka Barrage and at Gangasagar Island, respectively (WWF-India, South Asia Network on Dams, Rivers and People (SANDRP), 2019). The water received by Bangladesh beyond the Farakka Barrage varies from 170 cumecs to less than 5 cumecs during the lean period, altering the deltaic morphology, water quality, and biodiversity. This leads to periods of water scarcity during lean months (October to June), dispelling the common misconception of 'mirage of water surpluses' in the Delta region. Untreated waste and agricultural run-off significantly impact water quality, biodiversity, and livelihoods across the entire river basin. Current pollution levels show a decline in Ganga's self-purification capacity, with high fecal coliform levels detected even at Gangotri glacier, ranging between 300 MPN/100 ml in winter and at times exceeding 500 MPN/100 ml in summer (Baghel et al., 2005; Chandra, 2017). High population density in the Delta (1,200 per km^2) further exacerbates water distribution issues due to water scarcity, saline water intrusion, and deteriorating water quality.

Increasing sedimentation due to reduction in water flows raises dead weight, operation, and maintenance costs for hydroelectric projects due to corrosion caused by sediments. In Nepal, in July 1993, the Kulekhani Project faced high sediment and debris inflow into the reservoir because of flooding, lowering efficiency, and production capacity. In addition, most river valley project reports lack crucial data on changes in the snowbound areas, glacier movement such as receding glaciers due to rise in temperature, and flood frequency, impacting water flow estimates upstream of dam sites, increasing chances of failures and damages.

CLIMATIC UNCERTAINTIES AND FUTURE SCENARIOS IN THE CRYOSPHERE AND THE DELTA

The *Assessment of Climate Change over the Indian Region* report by the Ministry of Earth Sciences revealed that the Himalayas are warming faster than the rest of India (1951–2018) (Krishnan et al., 2020). Over the past four decades, apart from the 'Karakoram anomaly', rising temperatures have caused significant shrinking of the Cryosphere, including glacial retreat and permafrost temperature increases. Rising EPREs, avalanches, landslides, and glacial lake outburst floods (GLOFs) exacerbate the effects of unchecked infrastructure development, converting natural phenomena into man-made disasters. The February 2021 Raini disaster, a result of the collapse of a hanging glacial cliff in the Nanda Devi region, led to an avalanche and deluge in Dhauliganga, Rishiganga, and Alaknanda rivers destroying two power projects, NTPC's Tapovan-Vishnugad hydropower project and the Rishiganga hydropower project, and loss of human life trapped in the hydropower tunnel. Snow thawing and glacial retreat destabilize high-altitude slopes, releasing hibernating microorganisms which may have unknown ecological consequences downstream (IPCC, 2019). About 40–42% of the Ganga's water is contributed by Nepali Rivers, necessitating research on bacteriophages in these glacial regions.

Intergovernmental Panel on Climate Change (IPCC) in the Summary for Policymakers: *Sixth Assessment Report on Impacts, Adaptation and Vulnerability*, 2023 projects that river run-off in the Ganga is likely to increase by ~33% by 2050 due to accelerated glacier melting, and extreme precipitation as per the RCP4.5 and 8.5 scenarios (IPCC, 2023). The increased run-off might initially appear positive, but the combined effects of sea level rise, land subsidence, and salinity intrusion can lead to a complex and potentially negative impact on water availability in the Delta region. Increasing disasters and inadequate prediction systems in the Cryosphere threaten water security for local communities in the hilly areas, as well as in the Delta. Rising sea levels accompanied by climate-induced catastrophic floods, droughts, thunderstorms, cyclones, land subsidence, tidal surges, and extreme events like tsunamis may submerge islands in highly disaster-prone lowlands in West Bengal and Bangladesh, displacing around 1.5 million people as 'climate refugees' (Gupta, 2018). Various plans, including the Bangladesh Climate Change Strategic Action Plan (BCCSAP) 2009 and the National Adaptation Program of Action (NAPA) 2009, aim to address these issues. However, it is crucial to note that the repercussions of these disasters will not be limited to political borders, and therefore a bilateral or multilateral approach between Bangladesh and India is essential.

EXISTING BILATERAL TREATIES AND GEOPOLITICAL ISSUES

Addressing these hydrological, climatic, ecological, and socio-economic complexities and uncertainties of the transboundary river basin requires strong bilateral and multilateral cooperation among all riparian countries. Water-related negotiations are inherently long-term, often leading to unexpected outcomes during changes in governance or shifts in international power dynamics. As such, water diplomacy—viewing water as a shared resource—requires sustained, long-term collaboration grounded in mutual respect and equal rights. Such cooperation is essential for developing effective, long-lasting solutions to water-related challenges and for implementing an Integrated Approach to River Basin Management. The following section outlines the current status of bilateral water treaties among riparian countries, highlighting areas where resolutions and opportunities for cooperation exist. In the Cryosphere region, India and Nepal have a long history of transboundary water cooperation, dating back to the British colonial period. As the lower riparian state, India relies on Nepal for critical needs such as flood management, irrigation, drinking water, hydropower, and navigation. With India's rapidly industrializing economy, the demand for hydropower is rising exponentially, and Nepal, with the potential to generate approximately 40,000 MW of hydropower, is a key partner.

Annual floods wreak havoc, emphasizing the need for effective cross-border flood-control mechanisms and collaboration. Nepal's growing population also necessitates higher agricultural output, making irrigation a critical concern. The fertile Indo-Gangetic plains of India and Nepal's Terai region depend on the Ganga for irrigation. Beyond these areas, sectors such as navigation, tourism, fisheries, and trade further highlight the interdependence between the two riparian nations. Bilateral cooperation in these areas is indispensable, detailed in Table 11.2.

TABLE 11.2
Brief Overview of the Bilateral Treaties between India and Nepal

Year	Name of the River/Treaty	Nature of the Treaty/Agreement
1920	Mahakali Treaty (Sarada)	During the British India Government, Nepal and India signed the first treaty: the construction of the Sarada Barrage on the Mahakali/Sarada River
1954	Kosi Treaty	To control flooding in the Kosi River in India and Nepal: construction of the Bhimnagar or Hanumannagar Barrage
1959	Gandak Treaty	Construction of a barrage at Bhainsa-lotan: for irrigation purposes in India and Nepal, and to build a hydropower project to supply electricity to Nepal
1991	Memorandum of Understanding for Tanakpur Barrage	The controversial Tanakpur Treaty was signed: construction of the Tanakpur Barrage on the Mahakali River
1996	Mahakali Treaty	The treaty also set forth the plan for the Pancheshwar Multi-Purpose Development Project: Tanakpur and Sarada Barrages to provide water and energy to Nepal; build a link road between Tanakpur Barrage to Mahendranagar in western Nepal for transportation

Source: Authors, after Dhungel and Butler (2016).

In addition to the treaties addressing large-scale water management projects, India and Nepal have initiated small-scale, localized studies and dialogues:

- Feasibility studies for multipurpose Karnali-Chisapani Project, the Sapta-Kosi High Dam Multipurpose Project (3300 MW), and the Sun-Kosi-Kamala Diversion Project
- Investigating navigational potential of Kosi, Gandaki, and Karnali Rivers
- Collaborative efforts for flood forecasting data exchange, and flooding and inundation management plans

Since the 1920s, 11 different joint committees have been established to facilitate dialogue and implement water-related projects between India and Nepal, including the Joint Standing Committee on Inundation Problems, the Nepal-India Joint Commission on Water Resources, and the Joint Standing Technical Committee. Despite these binding treaties, water dialogues remain strained due to fragmented and non-functional committees, a lack of data, poor transparency, flawed project designs, and inadequate infrastructure maintenance. These issues are exacerbated by the territorial dispute over the Mahakali River, which borders India (Uttarakhand) and Nepal. The contention centres around the ownership of the 372 km^2 Kalapani region, stemming from differing interpretations of the Mahakali River's source, designated under the 1815 Sugauli Treaty. Resolving this issue will pave the way for the implementation of the Indo-Nepal (water) Treaty, open the India-Kailash Mansarovar pilgrimage route, and revive the traditional trade route between Uttarakhand and Tibet via the Kalapani Valley. This would enhance bilateral cooperation and unlock the regional cultural, spiritual, and economic potential. Amid complex water and energy dynamics, the historic Power Trading Agreement (PTA) of 2014 and subsequent trading from 2021 marked a new beginning for bilateral energy cooperation. This cooperation must be extended to integrating the Cryosphere into river management plans and shifting focus from hydropower to irrigation and flood control, biodiversity, and sustainable economic activities like tourism, navigation, and fisheries.

India-Bangladesh: The partition of the erstwhile united Bengal Delta into West Bengal, with Calcutta as its capital, and East Pakistan (now Bangladesh), with Dhaka as its capital, was based solely on religious considerations, without any regard for geomorphological continuity. This division also fragmented the river basin. Following the Partition of India in 1947, a total of 54 sub-basins in the GBM River basin became transboundary rivers. Of these, only one treaty, the Ganga Water Sharing Treaty of 1996, was signed between India and Bangladesh. Despite the treaty, the arbitrary partitioning of land left unresolved water-sharing issues between the two countries and the fragmentation of private lands. While the 2015 India-Bangladesh Land Boundary Agreement tried to address the long-overdue land settlement process and citizenship, many challenges remain unresolved. Consequently, issues such as inundation, flooding, water-logging, saline water intrusion into freshwater systems, soil erosion, and the destruction of embankments persist without mechanisms for relief, rehabilitation, or conflict resolution. This results in severe socio-ecological impacts on the river system and surrounding communities.

Indo-China: A Memorandum of Understanding (MoU) on sharing flow data of the Yarlung Tsangpo was first signed between China and India in 2002 and renewed periodically in 2018. Supported by the India-China Expert Level Mechanism (ELM) on Transboundary Rivers, this MoU aims to provide advance flood warnings by requiring China to share rainfall, water level, and discharge data. It also mandates data sharing during non-monsoon periods when water levels approach danger levels. India, in turn, pays a mutually agreed sum and shares information on data utilization. However, experts view the agreement as redundant because China's selected gauging stations Nugesha, Yangcun, and Nuxia do not cover the high-rainfall areas of the south-eastern Himalayas. It is recommended to shift the stations to Gompo Ne and Mêdog in Tibet to enhance India's access to critical data for early flood warnings and to strengthen Indo-China hydro-political relations concerning the Brahmaputra River.

Indo-Bhutanese: Bhutan becomes part of the basin with the consideration of the larger GBM basin. India and Bhutan have established a Hydro-Meteorological and Flood-Forecasting Network comprising 32 stations across eight shared rivers: Puthimari, Pagladia, Sun Kosi, Manas, Raidak, Torsa, Aie, and Jaldhaka. Additionally, the two nations collaborate through a Joint Group of Experts on Flood Management and the Indo-Bhutan Joint Hydro-Power Project, which includes ventures like the 600 MW Kholongchhu project.

AVENUES FOR MULTILATERAL COOPERATION

As elaborated above, the key feature of the Cryosphere and Delta in GRB is its continuous geomorphological zone extending across political borders, from the southern Tibet Autonomous Region (China) to Nepal, India, Bhutan, and Bangladesh. While various bilateral agreements exist, there is a growing need for a multilateral negotiation on data and water sharing, flood control, and cooperative disaster management, especially considering the escalating risks posed by climate change in the highly vulnerable Cryosphere and Delta region. This requires multilateral cooperation and agreements that transcend competition over water resources for purely economic gain. An example of such a multilateral cooperation is the Kailash Sacred Landscape (KSL) Conservation and Development Initiative, a UNESCO Cultural Heritage Zone involving the origins of several South Asian rivers. Initiated by ICIMOD and supported by UNEP, Global Resource Information Database (GRID), and German Federal Ministry of Economic Cooperation and Development from 2012 to 2017, this collaborative programme included representatives from China, India, and Nepal, showcasing potential for collaborative governance.

Multiple such multilateral umbrella agreements and long-term strategic water management plans for the GRB are urgently needed. Recommended measures include:

- Strengthening and revising existing bilateral treaties for water sharing, sustainable agricultural and industrial practices across different countries,
- Engagement in multilateral research collaborations, cross-disciplinary and joint partnerships for studies such as glacier monitoring, sedimentation

impacts, etc., and leveraging international funding for development of a Centralized Data Repository for Cryosphere and Delta,
- Multilateral Cooperation for Disaster Risk Reduction and Relief and Rehabilitation Plan: implementation of early warning systems, flood forecasting, and emergency preparedness including coordination between upstream and downstream regions and involving local communities for monitoring and action is crucial,
- Transboundary collaborations for ecosystem restoration programmes for Cryosphere protection including reforestation and reduction of black carbon deposition in glacier zones in Tibet Autonomous Region (China), Nepal, India, and Bhutan and Delta conservation initiatives such as mangrove restoration and sediment management to combat erosion between India and Bangladesh, and
- Improve efficiency of existing dams and projects by revising old rule curves, reservoir operational schedules, install mechanized ultra-modern flood warning systems, mechanization of spillways and gates, and instrumentation for measuring flows and discharge rates.

As a signatory to the International Convention on the Protection and Use of Transboundary Watercourses and International Lakes, 1992, and the UN Watercourses Convention, 1997, the upper and lower riparian states are required to obtain prior consent from the involved parties before altering the river systems. Considering the transboundary nature of Ganga River system, as discussed in this chapter, comprehensive agreements for rivers flowing from Nepal into Indian territory and the 54 transboundary rivers in Delta region are recommended for sustainable implementation of transboundary projects such as 'Inland Waterways Development Plan' (IWDP) or 'Interlinking of Rivers Project' to guard the integrity and wholesomeness of the entire Delta.

CONCLUSION

Although the Cryosphere constitutes a small part of the GRB, critical changes at the margins have a negative 'last-mile' downstream impact on E-flows, water availability, and livelihoods. Unique phenomena such as permafrost thawing and microbial activity still remain a blind spot, but they will have an unprecedented impact due to climate change. Both the Cryosphere and Delta are dynamic and fragile regions with high levels of ecological uncertainty, yet they remain underrepresented in the river basin plans. Current administrative policies disregard the upstream-downstream linkages, particularly in infrastructure development across upper and middle reaches, neglecting critical implications for the Delta. Integrating the Cryosphere and Delta regions in GRB plans calls for the need to go beyond political borders, the historical legacy of diplomatic relations, and humility to recognize human limitations when dealing with the strength and magnanimity of such transboundary rivers. This can be achieved by adopting 'precautionary planning principles' to ensure systematic planning with foresight in the administrative processes to develop sustainably, over a long period of time, without compromising the integrity of the river system, and ensure preservation of the sociocultural fabric in these unique niches.

Immediate steps include fostering data sharing, joint water balance assessments, collaborative research programmes, multilateral disaster management, and preparedness programmes. Climate change must be embedded in all river planning strategies, recognizing the risks from altered rainfall patterns, glacial retreat, rising temperatures, and sea level rise, as all the riparian nations will face the impacts with varying degrees of intensity. While respecting administrative sovereignties is essential, this chapter advocates for a transboundary approach which emphasizes precautionary and collaborative river management strategies to navigate the complexities and dynamics of the Cryosphere and Delta within the GRB.

ACKNOWLEDGEMENTS

This chapter is based on insights gained from a comprehensive study of the headwaters and tails of the Ganga, conducted in collaboration with Chaitrali Kulkarni under the guidance of Prof. Vijay Paranjpye from 2019 to 2023 supported by INTACH-Delhi.

NOTES

1. This study refers to research studies and reports generated by organisations such as Wildlife Institute of India, Dehradun; India Water Resources Information System (GoI); Wadia Institute of Himalayan Geology; Central Water Commission, New Delhi; National Institute of Hydrology, Roorkee; National Mission for Clean Ganga; International Center for Integrated Mountain Development (ICIMOD), Nepal; Kathmandu University, Nepal; Water Conservation Foundation, Kathmandu; FAO AquaStat; Bangladesh Water Partnership, Dhakka; University of Dhaka, Bangladesh; Nepal Water Partnership; SANDRP; Observer Research Foundation, Kolkata; People's Science Institute, Dehradun; Barh Mukti Abhiyan; Yamuna Jiye Abhiyan; and many others.
2. The Ganga-Brahmaputra-Meghna (GBM) basin covers a total catchment area of about 17 million km^2 and is spread over five countries—divided between India (64%), China (18%), Nepal (9%), Bangladesh (7%) and Bhutan (3%)—and forms the third largest river basin in the world, after the Amazon and Congo in South America and Africa, respectively (FAO, 2011). Note: Bhutan is included when we consider the GBM basin.
3. Bhagirathi is joined by seven major tributaries: Kedar Ganga, Jadhganga, Kakora Gad, Jalandari Gad, Siyan Gad, Asi Ganga, and Bhilangana. Jadhganga River originates in Tibet, merging with three tributaries, Rongmach, Nilapani, and Mana Ghora, before entering India and joining Bhagirathi at Bhaironghati. Maps from the Survey of India are considered authoritative regarding their positions.

   Note: Scientific discussions of the Ganga River's origins lead to varying conclusions. Some scholars view the Alaknanda River as the source, given its larger volume and highest elevation compared to the Bhagirathi. Others consider the Tons River, a tributary of the Yamuna, as the furthest upstream point in the basin. Some even suggest the Brahmaputra River as a source within the larger Ganga-Brahmaputra-Meghna basin. However, historically and culturally, Gomukh at the Gangotri Glacier is recognized as the origin of the Bhagirathi River and the Ganga. The Ganga River Basin covers 1.32 million km^2, supporting around 43% of India's population, making it the primary river in the GBM basin (Central Water Commission & Regional Remote Sensing Centre-West, 2012).
4. Alaknanda is fed by seven major tributaries: Mandakini, Nandakini, Pindar, Dhauliganga, Vishnuganga, Rishiganga, and Birahi.

5. Parts of Sikkim are included in this section as water from the eastern side of the ridge-line of the Kangchenjunga peak flows into the Tamor River. Three rivers Yangma, Ghunsa, and Simbua join the Tamor River. The origin of the Simbua River is the Yalang glacier which lies at the foot of Mt. Kangchenjunga (personal communication with Dr. Chiranjibi Bhattaria, Nepal).
6. Prior to this, the Yamuna was part of the Ghaggar-Hakra paleochannels (Saraswati), which disappeared in the late Holocene period (Valdiya et al., 2016).
7. Note: Brahmaputra River literally means the son of Brahma, giving a masculine identity to the river unlike other rivers in the subcontinent.
8. An anabranch is defined as "a section of a river or stream that diverts from the main channel or stem of the watercourse and rejoins the main stem downstream" (North et al., 2007).
9. The Ganga River, measured from its mouth at Gangasagar Island near Kolkata, is about 2,525 km long (India-WRIS, 2021). However, after a series of river avulsions, its flow merges with the Brahmaputra as the Padma River, before reaching the Bay of Bengal near the Bhola Island. Considering this, the total length of the Ganga is approximately 2,616 km.
10. Snowmelt is water which forms a part of the annual hydrological cycle, whereas the ice melt forms that share of the stream which is a part of water which has accumulated over millennia.
11. There are variations in estimates in the number of glaciers, glacial lakes, areas covered by snow and glaciers, total ice reserves, annual release of water from the Cryosphere due to different criteria of size, length, seasons, area covered, and other parameters applied by different agencies.
12. Not to be confused with revenue land area (10,000 km^2) since it excludes aquatic/marine areas, which are normally not included in terrestrial catchment. According to IUCN and ZSI, 26,000 km^2 is the area under Sundarbans which includes 104 islands. This divergence in the area of almost 16,000 km^2 in area covered is because IUCN and ZSI include the 104 islands and the adjacent areas which support marine species.
13. Delta's unique waterscapes offered significant advantages, including opportunities for mercantile trade, defensibility, riverine transport, and natural solutions for drainage, sewerage, and sanitation in the developing city (Ghosh & Sen, 1987). Driven by the East India Company's primary interest in land revenue, the area was surveyed as early as 1769 by the Surveyor General. Systematic management of the forested delta began in the 1860s with the establishment of the Forest Department in Bengal Province.

REFERENCES

Allison, M., Khan, S., Goodbred, S., Kuehl, S. (2002). Stratigraphic evolution of the late Holocene Ganges–Brahmaputra lower delta plain. *Sedimentary Geology*, *155*(3–4), 317–342.

Aryal, R., Rajkarnikar, G. (2011). Water resources of Nepal in the context of climate change. Water and Energy Commission Secretariat (WECS), Government of Nepal. Available on https://awsassets.panda.org/downloads/water_resources_of_nepal_final_press_design.pdf

Azad, S. (2021). Rains change course of four major rivers in Uttarakhand. *The Times of India*. October 29. Available on https://timesofindia.indiatimes.com/city/dehradun/rains-change-course-of-four-major-rivers-in-uttarakhand/articleshow/87345378.cms

Baghel, V., Gopal, K., Dwivedi, S., Tripathi, R. (2005). Bacterial indicators of faecal contamination of the Gangetic River system right at its source. *Ecological Indicators*, *5*(1), 49–56.

Bajracharya, S. R., Maharjan, S. B., Shrestha, F., Bajracharya, O. R., Baidya, S. (2014). *Glacier status in Nepal and decadal change from 1980 to 2010 based on Landsat data* (Research Report 2014/2). International Centre for Integrated Mountain Development (ICIMOD), Kathmandu. Available on: https://lib.icimod.org/records/pgbnf-p3v73

Baltazar, C. (2015). Save the Sundarbans! *Environmental Synthesis and Communication*. Available on https://blogs.wellesley.edu/es39901fa15/2015/12/10/save-the-sundarbans-its-more-than-the-tigers-although-theyre-important-too/

Bandyopadhyay, S. (2019). Sundarban: A Review of Evolution and Geomorphology. *IWA Sundarban Joint Narrative*. World Bank. Available on https://documents.worldbank.org/curated/en/119121562735959426/pdf/Sundarban-A-Review-of-Evolution-and-Geomorphology.pdf

Banglapedia. (2021). River Systems in Bangladesh. *National Encyclopedia of Bangladesh. Available on* https://en.banglapedia.org/index.php/River_and_Drainage_System

Central Water Commission, Regional Remote Sensing Centre-West. (2012). *River Basin atlas of India*. Ministry of Water Resources, Government of India.

Chakraborty, T., Kar, R., Ghosh, P., Basu, S. (2010). Kosi megafan: Historical records, geomorphology, and the recent avulsion of the Kosi River. *Quaternary International*, *227*(2), 143–160.

Chandra, S. (2017). The ugly, unholy truth about Ganga. *Governance Now*. December 2. Available on https://www.governancenow.com/news/regular-story/the-ugly-unholy-truth-about-ganga

Chatterjee, S., Bishop, C. (2025). Himalayas. *Encyclopedia Britannica*. Available on: https://www.britannica.com/place/Himalayas

Dahal, S. H. (2018). China-Nepal-India Triangle: The Dark Side of Indo-Nepal Relations. Xlibris US. Available on: https://books.google.co.in/books/about/China_Nepal_India_Triangle.html?id=eS6xtwEACAAJ

Dam Safety Monitoring Directorate. (2019). National Register of Large Dams. Central Dam Safety Organization, Central Water Commission, Government of India.

Dhungel, D., Butler, C. (2016). Water resource cooperation between Nepal and India: What persists? In *Water Resources Cooperation between India and Nepal*. KW Publishers Pvt Ltd.

Eaton, R. (1990). Human settlement and colonization in the Sundarbans. *Agriculture and Human Values*, *7*, 6–16.

FAO. (2011). AQUASTAT Transboundary River Basin Overview – Ganges-Brahmaputra-Meghna. Food and Agriculture Organization of the United Nations (FAO). Rome, Italy.

Frenken, K. (Ed.). (2012). Irrigation in Southern and Eastern Asia in figures: AQUASTAT Survey – 2011 (FAO Water Reports No. 37). Food and Agriculture Organization of the United Nations. Available on: https://www.fao.org/4/i2809e/i2809e.pdf

Gautam, M., Acharya, K. (2012). Streamflow trends in Nepal. *Hydrological Sciences Journal*, *57*(2), 344–357.

Ghosh, A., Schmidt, S., Fickert, T., Nüsser, M. (2015). The Indian Sundarban mangrove forests: History, utilization, conservation strategies and local perception. *Diversity*, *7*(2), 149–169.

Ghosh, D., Sen, S. (1987). Ecological history of Calcutta's wetland conversion. *Environmental Conservation*, *14*(3), 219–226.

Gupta, J. (2018). Rising sea swamps island along Bengal coast. *India Climate Dialogue*. Available on: at: https://dialogue.earth/en/climate/rising-sea-swamps-island-along-bengal-coast/

Hoiberg, D. (Ed.). (2000). Students' Britannica India (Vols. 1-5). New Delhi, India: *Encyclopaedia Britannica (India)*. Available on: https://openlibrary.org/books/OL6816908M/Students%27_Britannica_India

Immerzeel, W., van Beek, L., Bierkens, M. (2010). Climate change will affect the Asian water towers. *Science*, *328*, 1382–1385.

India-WRIS. (2021). Ganga. *India* Water Resources Information System. Available on: https://indiawris.gov.in/wiki/doku.php?id=ganga

IPCC. (2019). IPCC Special Report on the Ocean and Cryosphere in a Changing Climate. Intergovernmental Panel on Climate Change. Available on: https://www.ipcc.ch/srocc/

IPCC. (2023). Climate Change 2023: Synthesis Report. Contribution of Working Groups I, II and III to the Sixth Assessment Report of the Intergovernmental Panel on Climate Change. Intergovernmental Panel on Climate Change. Available on: https://www.ipcc.ch/report/ar6/syr/h

Jain, S., Agarwal, P., Singh, V. (2007). *Hydrology and Water Resources of India* (Vol. 57). Water Science and Technology Library. Springer.

Krishnan, R., Sanjay, J., Gnanaseelan, C., Mujumdar, M., Kulkarni, A., Chakraborty, S. (Eds.). (2020). Assessment of Climate Change over the Indian Region. Report of the Ministry of Earth Sciences (MoES), Government of India.

Maurya, A., Shah, M., Deshpande, R., Bhardwaj, R., Prasad, A., Gupta, S. (2011). Hydrograph separation and precipitation source identification using stable water isotopes and conductivity: River Ganga at Himalayan foothills. *Hydrological Processes, 25*, 1521–1530.

Miller, J., Immerzeel, W., Rees, G. (2012). Climate change impacts on glacier hydrology and river discharge in the Hindu Kush–Himalayas. *Mountain Research and Development, 32*(4), 461–467.

Mukherjee, J. (2016). The Adi Ganga: A forgotten River in Bengal. *Economic & Political Weekly, 51*(8).

North, C., Nanson, G., Fagan, S. (2007). Recognition of the Sedimentary Architecture of Dryland Anabranching (Anastomosing) Rivers. *Journal of Sedimentary Research, 77*(11), 925–938.

Rai, S., Singh, D., Jacob, N., Rawat, Y., Arora, M., Kumar, B. (2019). Identifying contribution of snowmelt and glacier melt to the Bhagirathi River (Upper Ganga) near snout of the Gangotri Glacier using environmental isotopes. *CATENA, 173*, 339–351.

Rao, B., Gupta, A., Ruhi Maheshwari, S., Venkat Raju, P., Venkateshwar Rao, V. (2021). *Glacial Lake Atlas of Ganga River Basin*. National Remote Sensing Centre, ISRO, Hyderabad, India, pp. 1–194.

Sadoff, C., Harshadeep, N., Blackmore, D., Wu, X., O'Donnell, A., Jeuland, M., Lee, S., Whittington, D. (2013). Ten fundamental questions for water resources development in the Ganges: Myths and realities. *Water Policy, 15*(S1), 147–164.

Sarker, M., Atker, J., Rahman, M. (2013). Century-scale dynamics of the Bengal Delta and future development. *Proceedings of the International Conference on Water and Flood Management (ICWFM) (Dhaka, Bangladesh)*, 91–104.

Sharma, B., Condappa, D. (2013). Opportunities for harnessing the increased contribution of glacier and snowmelt flows in the Ganges basin. *Water Policy, 15*(S1), 9–25.

Siderius, C., Biemans, H., Wiltshire, A., Rao, S., Franssen, W., Kumar, P., Gosain, A., van Vliet, M., Collins, D. (2013). Snowmelt contributions to discharge of the Ganges. *Science of The Total Environment, 468–469*, S93–S101.

Singh, P., Kumar, A., Kishore, N. (2011). Meltwater storage and delaying characteristics of Gangotri Glacier (Indian Himalayas) during ablation season. *Hydrological Processes, 25*(2), 159–166.

Space Applications Centre, Indian Space Research Organisation (2011). *Snow and Glaciers of the Himalayas: Inventory and Monitoring (Discussion Paper II).* Space Applications Centre, ISRO, Ahmedabad, India. Available online: https://prod-qt-images.s3.amazonaws.com/indiawaterportal/import/sites/default/files/iwp2/Snow_and_Glaciers_of_the_Himalayas_Discussion_Paper_II_ISRO_SAC_2011.pdf

Valdiya, K., Bisht, R., Tandon, S., Sinha, R., Bhadra, B., Kar, A., Prabhakar, V., Chatterjee, R. (2016). Palaeochannels of North West India: Review and assessment: Report of the Expert Committee. Central Ground Water Board, Government of India.

WWF-India, South Asia Network on Dams, Rivers and People (SANDRP). (2019). *Rejuvenating Ganga: A Citizen's Report.* Available on https://sandrp.in/wp-content/uploads/2019/12/rejuvenating-ganga-citizen-report.pdf

12 Contested Waters and Clumsy Solutions

Plural Rationalities in Transboundary Water Governance in Nepal-India Borderlands

Udyan Devkota, Tek Raj Koirala, and Bibek Dhoj Thapa

INTRODUCTION

Water governance in transboundary river basins remains one of the contested and politically sensitive issues in international environmental politics. There are around 263 transboundary river basins across the globe, which cover 145 countries, and 45% of the land, serving approximately 40% of the world population (Rieu-Clarke et al., 2016). Governing these shared resources across national boundaries poses a significant challenge for regional cooperation, diplomacy, and sustainable development. Transboundary rivers all over the world have witnessed a rise in tensions over water sharing, competing interests, and unequal power relations often generated due to geopolitical rivalries and historical grievances. As climate change intensifies hydrological extremes such as floods and droughts in the era of Anthropocene, it has further complicated the governance (Biggs et al., 2013; Pahl-Wostl et al., 2008). South Asia stands out as a densely populated and impoverished region of the world, riddled with geopolitical rivalries and historical grievances, and climate change has raised the stakes especially high. Nepal and India, despite their close geographical, social, economic, cultural and religious ties, have struggled to govern their shared water resources, particularly rivers, cooperatively and equitably, often leading to diplomatic strain and bitterness (Gupta et al., 2021; Gyawali & Dixit, 1999).

Nepal-India borderlands are traversed by major river basins such as Koshi, Gandaki, Karnali, and Mahakali, along with hundreds of minor rivers and rivulets, which represent a microcosm of broader issues and dilemmas associated with transboundary resource governance. The seasonal changes in these rivers shape livelihoods, agriculture, migration, and disaster risks for the millions living in the region. These rivers are often subjected to the hydrological extremes of floods and drought. The major river basins, such as Koshi, Gandaki, and Mahakali, are governed by

DOI: 10.1201/9781003599692-12

contested bilateral treaties, focusing on state-centric technocratic solutions of building dams and embankments over innovative, integrated, and multi-stakeholder-led governance, which has resulted in the worsening livelihoods of borderland communities (Prakash et al., 2025). At the same time, hundreds of minor transboundary rivers and streams lack bilateral treaties and mechanism, which complicate water sharing, disaster management, and even fuel geopolitical friction (Karki, 2025; Ranjan, 2016). The main crux of the problem lies in that the existing bilateral treaties, such as Koshi and Gandak, prioritise hierarchical technocratic solutions, ignoring local realities, resilience, and accommodation of diverse actors and forces across borders, which further perpetuates the cycle of crisis. The local communities of the borderlands living along these rivers often have little to say in governing these rivers that shape their lives irrespective of their size.

This study explores a central question on how diverse rationalities, viz. hierarchical, individualist, egalitarian, and fatalist, shape the governance of transboundary rivers in Nepal-India borderlands. It applies the Cultural Theory or Theory of Plural Rationalities (Thompson et al., 1990) to understand how the river water is governed in three distinct rivers across a spectrum of formality and visibility: Koshi (a treaty-governed major river), Tinau (a medium-sized river), and Pandai (a marginal transboundary river). It seeks to investigate the plural nature of governance and the uneven experiences of local communities across the borderland in dealing with water-related risks and state interventions. By examining these questions, research seeks to contribute to the broader debate on governance and governance approaches to transboundary rivers in the era of Anthropocene.

The impetus for this research was provided by escalating failures of state-centric technocratic approaches on transboundary water governance to address cascading crises of climate change and growing socio-economic and political inequity, not only in Indo-Nepal borderlands but also beyond the region. The Indo-Nepal borderlands are home to more than 60 million people dependent on the river basin. Originating from the high Himalaya to the mid-Himalaya region, almost 6000 rivers flow from Nepal to India (Shrestha & Aryal, 2011). The region has witnessed rising flood frequency and severity due to heavier rainfall and rapid glacier retreat linked to climate change (Alfthan et al., 2018). Between 1954 and 2014, Nepal experienced 41 floods, killing 6500 people and affecting over 3.5 million (Gaire et al., 2015), while in Bihar, floods claimed 9500 lives between 1979 and 2017 (Gupta et al., 2021). While much attention has been paid to floods, drought-equally severe hydro-logical extreme rarely grabs similar attention from the media, experts, political leaders, civil society actors, and activists. With the increasing incidence of drought, which can be attributed to climate change, greater problems of sharing water between states and people in borderlands have emerged as a much greater challenge (Bhattarai et al., 2024). The case of Pandai is a glaring example of drought and the issue of water sharing, which has already witnessed conflict across communities between India and Nepal (Prasad, 2020). Despite having rich nature and water resources, people living across the Indo-Nepal borderlands have a higher incidence of poverty than the rest of the country, partly due to frequent floods, devastation and resulting in financial burden (Mishra, 2008). Yet the existing governance remains fragmented, fractured, and inadequate. The Koshi 1954 and Gandak 1959 agreements, and later amended after Nepali dissatisfaction, were designed as technical solutions for mainly flood control, and the use of water resources for irrigation and hydro-power has ignored climate adaptation, basin approaches, and the incorporation of multiple stakeholders (Dixit, 2009) In fact, the building of dams and embankments, apart from creating displacement of thousands of people, has led to greater loss of lives and property, particularly due to infrastructural accidents like embankment breaches, which led to the Koshi floods in 2008.

Transboundary river governance has ramifications beyond water resources, affecting the political economy and diplomatic relationships between states (Gyawali & Dixit, 1999, p. 553). Its governance is becoming more complex and complicated, and scholars have highlighted key challenges in the governance of transboundary river management, including competing national interests (Grey & Sadoff, 2003), power asymmetries between riparian countries (Daoudy, 2009; Zeitoun & Allan, 2008), and climate change impacts (Earle et al., 2015). There is literature suggesting that effective and innovative transboundary water governance often depends on international cooperation and effective implementation of international legal norms and frameworks like the UN Watercourses Conventions (Giordano & Wolf, 2003), integrated management approaches (Giordano & Wolf, 2003), and recognising a greater role of the epistemic community (Kibaroglu, 2008). Some of the case studies, such as the Nile Basin, the Mekong River, and the Colorado River, reiterated the importance of flexible and inclusive institutions and benefit-sharing mechanisms to advance cooperation (Deribe et al., 2024; Zeitoun & Mirumachi, 2008). Similarly, a new and emerging body of literature is urging a paradigm shift from state-centric to more collaborative approaches and building hybrid institutions in the river basin (Gerlak et al. 2021; Karkkainen, 2005, p. 73). The flexible, multi-level institutional framework which enables transboundary communities to navigate complex socio-ecological systems is particularly important in the given context of the climate crisis in transboundary river governance (Akamani & Wilson, 2011).

Scholars have attempted to explore diverse perspectives on the governance of transboundary rivers in the Indo-Nepal borderlands. Thematically, scholars have highlighted limitations of state-centric, technocratic approaches (Vasani, 2023), the hydro-hegemonic behaviour of India (Cascão & Zeitoun, 2010; Bagale, 2020; Vij et al., 2020), the legal institutional framework (Uprety & Salman, 2011), and even fragmented community-based approaches (Maharjan, 2023). They have argued for the need for more integrated, holistic, and design institutions to incorporate multiple stakeholders in transboundary river basins. In this regard, Thomas (2017) has attempted to reconceptualise the Indo-Nepal border and river in the concept of 'river-border complex', showcasing the transboundary river as synergistic, co-constitutive, and interdependent, and urges foregrounding it in sociopolitical and economic contingencies that shape their creation and continued existence. Bisht (2019) critiques the limitation of the macro-narrative of water diplomacy in Indo-Nepal by proposing the concept of 'water communities' for a scalar understanding of the river basin and highlights the vital role of communities in water governance. Drawing on the case of Narayani-Gandak and Mahakali basins, Pandey et al. (2023) reveal limited state-to-state and community-to-community cooperation on flood management between India and Nepal. In a similar line, Upadhyaya (2020) advocates for pragmatic approaches to Indo-Nepal transboundary river governance involving private actors, civil society, and participation in the decision-making process. This research urges a paradigm shift from existing transboundary river governance to more nuanced, holistic, and integrated approaches to ensure transformational impact on borderlands. The integration of diverse actors, forces, and perspectives, though competing, is indeed important for transboundary river governance in Indo-Nepal borderlands.

Thus, this study aims to address a critical gap in governing transboundary rivers by going beyond state-centric and technocratic approaches to incorporate plural rationalities—hierarchy, individualism, egalitarianism, and fatalism into the management of Indo-Nepal borderlands. While existing research has mainly centred on bilateral treaties or framed community efforts, drawing influence from works of Marco Verweij (see Verweij, 2000), this research innovatively attempts to apply cultural theory in Indo-Nepal transboundary water governance. It argues for 'clusmy solutions' that dynamically integrate competing solidarities, such as state agencies, markets, local communities, and civil societies. Using qualitative case studies of Koshi, Pandai, and Tinau, this research uncovers the limitations of the rigid hierarchical approaches and how polycentric governance can be better utilised to tackle hydrological extremes. This research not only critiques conventional water governance paradigms but also offers various empirical insights from cases for governing over hundreds of smaller and ignored rivers and rivulets in the Indo-Nepal borderlands.

STUDY AREA

The study area spans the transboundary region of Nepal-India borderlands encompassing Nepal's terai districts and their corresponding counterparts in India. This region forms a part of the Indo-Gangetic plains and falls within the larger Ganga basin, with hundreds of rivers of different sizes originating in Nepal. The region is known for its diverse physiographic zones from the Chure (Shiwalik) mountains to the fertile plains, which are also a hotspot of different hydrological disasters like floods and droughts (Adhikari & Ojha, 2021). Thus, communities in these borderlands are not only linked through socio-economic and political ties but also through the shared vulnerability of hydrological disasters.

The study incorporates all three different categories based on their hydrological scales, governance attention, and disaster impact to make the study representative of the borderlands. Major rivers in Nepal (Category I), like Koshi, Gandaki, Karnali, Mahakali, and their tributaries, originate in the Tibetan plateau or the High Himalayas of Nepal and debouch into Nepal's terai before flowing into India to meet the Ganga. These Himalayan rivers have sustained flow throughout the year, owing to the snow and glacial melting, unlike other rivers. Non-snow-fed rivers originating from the Mahabharat ranges are medium-sized rivers (Category II), which include rivers like Bagmati, Tinau, Kamala, Rapti, Babai Lakhandehi, etc. These perennial rivers are also fed by springs and groundwater and have much lower dry season flow. Finally, there are small and marginalised rivers (Category III) which originate in the Shiwalik (Chure) mountains of Nepal. Many of these rivers are seasonal in nature and can dry up during the low-flow season while being prone to flash floods during the monsoon. These marginalised rivers, such as Pandai, Hardinath, Ratu, Khando, etc., are generally ignored in hydrological disaster management and water governance and seldom form part of research and studies. Despite their marginalisation, they frequently affect the communities living across the borderlands (Dixit et al., 2023). This study of transboundary water management in the Indo-Nepal borderlands focuses mainly on three river basins: Koshi (a Himalayan River), Tinau (a medium-sized river), and Pandai (a small and marginalised river), each of which is described briefly.

Koshi River

Koshi is one of the largest river basins in Nepal, with a catchment area of around 74,030 km^2, of which 43% falls in China, 15% falls in India, and the remaining 42% falls in Nepal (Sharma, 1997). The river is one of the principal contributors to the River Ganga. The Sapta Koshi River has seven main tributaries—Sun Koshi, Dudh Koshi, Indrawati, Likhu, Tamakoshi, Tamor, and Arun. The three largest tributaries—Arun, Tamor, and Sunkoshi—meet at a place called Triveni, from where the river flows South through the Barahakshetra gorge before reaching Chatara in the Terai plains.

Tinau River

Tinau River originates on the southern slope of the Mahabharat range at Palpa district and is also fed by several small streams like Sisne, Jhumsa, Kusum, and Dobhan Khola in Western Palpa. Then the river flows through the Chure range before entering the Rupandehi district in Nepal's terai. The river is called Dano in the Southern part of Rupandehi district and is named Kunda River after it reaches Uttar Pradesh, India, where it joins the Rapti River. Tinau River has a total catchment of 3200 km^2, out of which 850 km^2 lies in India, while the rest lies in Nepal (Dhungana et al., 2016).

Pandai River

Pandai River is a small transboundary river originating in the Shiwalik (Someshwar) hills in Nepal. From Thori Rural Municipality of Parsa district of Nepal, the river flows into India to meet the Budhi Gandak River, a tributary of the Ganga (Prasad, 2020). *Thute Khola*, one of Pandai's transboundary tributary streams and the main site of water sharing between Nepal and India, has repeatedly become the centre of transboundary water conflict. Pandai is prone to flash floods during monsoons, while water volume significantly reduces during the dry season.

DATABASE AND METHODOLOGY

The study relies on a qualitative database of multiple sources of evidence about transboundary governance in the Nepal-India borderlands. Both primary and secondary data were collected.

1. ***Secondary data:*** The study mainly relies on secondary data. The secondary data collection involved a review of grey literature, newspaper reports, government reports, strategies, articles, journals, and books relevant to the topic of the study.

2. ***Primary data:*** These were collected through a field visit and observation of three different river basins in the Nepal-India borderlands: Koshi, Tinau, and Pandai. The fieldwork was conducted in September 2024. Key informant interviews (KIIs) with water experts, residents, NGO workers, and local officials were conducted, as listed in Table 12.1 and shown in Figure 12.1. Purposive and snowball sampling were used to select respondents for the interview. The use of sampling techniques, a short time frame, and limited immersion in the community constrained the depth and comprehensiveness of the data, which may also limit the generalizability of the findings.

TABLE 12.1
Key Actors in the Study

Type of Actors	Number of KIIs	Basin/Place
Local residents/farmers	30	Koshi, Tinau, Pandai
Local government officials	9	Koshi, Tinau, Pandai
Water/irrigation groups	8	Koshi, Tinau, Pandai
Business people	10	Koshi, Tinau, Pandai
Civil society/activists	9	All three basins and Kathmandu
Water experts	3	Kathmandu

FIGURE 12.1 After taking response with head of Thori Rural Municipality, Lal Bahadur Shrestha, and his staffs. (*Source:* Photo taken by authors.)

This study utilised Cultural Theory as a conceptual framework. Cultural Theory, also known as the Theory of Plural Rationality, is based on the groundbreaking work of anthropologist Dame Mary Douglas (Verweij, 2017). Cultural Theory is an interpretive approach to studying society. It posits that there are four ways of perceiving and organising social relations (rationalities): hierarchy, egalitarianism, individualism, and fatalism, which dynamically constitute almost every domain of social life from the operation of a local school to the functioning of an international regime (Thompson, 2008; Verweij, 2017). While social scientists are familiar with the solidarities of hierarchy and individualism, Cultural Theory's contribution lies in introducing two other solidarities of egalitarianism and fatalism. These four distinct cultures can be captured based on two discrete criteria: Group and Grid. The criteria of group rate the perceived importance of the group in a culture, while some cultures are built around the group, like a football team, and some tend to have greater individual orientation, like a gathering of entrepreneurs (Thompson, 2008, p. 32). Similarly, the criteria of Grid are associated with the importance of rules and roles in a culture; government bureaucracy can be highly rigid in contrast to informal clubs, which tend to be liberal (Thompson, 2008). In the High Group and High Grid context, rigid hierarchies like that of an army are seen. However, if the culture lacks concern for roles and rules (i.e. low Grid) despite having a high group orientation, we tend to see egalitarianism (i.e. the sacredness of group meetings and search for consensus). Similarly, individualist culture tends to have a low Grid and low Group orientation, often manifested in entrepreneurs, rational choice, and pro-market politicians. Finally, isolated individuals who perceive themselves as undermined by roles and rules, i.e. low group orientation and high grid rules and roles, constitute Fatalists (Thompson, 2008, p. 33).

Table 12.2 shows four styles of organisations that are briefly discussed below (Beck et al., 2011; Gyawali, 2001; Thompson et al., 2016):

1. ***Hierarchy:*** Control is the most dominant concern in the hierarchy, with unequal roles for unequal group members. Solutions are sought, conflicts are resolved, and control is exercised through upgrading, co-opting, re-segregating, and transferring. Some examples of the hierarchy are the army, the internal structure of large corporations, and bureaucracy.
2. ***Egalitarian:*** While they lack internal differentiation of roles, they are bound together by loyalty and commitment to the ethical cause. This makes conflict resolution within groups hard, as compromises can be frowned upon, thus creating a high possibility of schism. Outside threats and alarmist causes can only keep the group together in the long run. Some examples are environmental, religious, and social campaigns.
3. ***Individualism:*** Humans are seen as self-interested and atomistic, and nature as resilient. Thus, all boundaries are provisional in individualist social contexts and are negotiated freely. The risks and opportunities are viewed as open to exploitation by the enterprising ones, making it the style of markets and business people.
4. ***Fatalism:*** Based on the idea of the untrustworthiness of humans and the capriciousness of nature, this way of organising means coping with everyday life as fate dictates. Peasants and some squatters in river basins are examples.

TABLE 12.2
Four Rationalities in the Governance Landscape

Rationality	Belief System	Preferred Governance Style	Typical Actors	Key Characteristics
Hierarchical	Order, control, and expertise create stability	Centralised, top-down, rule-based	Government agencies, bureaucracies, technocrats, and military	Emphasise regulation, control, planning, authority, and social conformity
Individualism	Competition, freedom, and innovation lead to success	Market-based, incentive-driven	Private sector, entrepreneurs	Prefer risk-taking, efficiency, freedom, and deregulation
Egalitarianism	Equity, Justice, and Participation build resilience	Participatory, decentralised, consensus	Civil Society, NGOs, local communities, social movements, activists	Value environmental justice, fairness, and inclusiveness
Fatalism	Nature and life are unpredictable	Passive, resigned, and reactive	Poor and marginalised groups and households, squatters	Distrust institutions, lack agency, and have defensive adaptation

Source: Prepared by authors.

The methodology followed a case study approach grounded in the analytical framework of Cultural Theory. Three river basins (Koshi, Tinau, and Pandai) selected represent three different categories of rivers in the Nepal-India borderlands and have different levels of formal governance, sociopolitical visibility, and infrastructure presence. Cultural Theory distinguishes societal responses to environmental governance problems into four rationalities: hierarchical, egalitarian, individualist, and fatalist. This categorisation of rationalities provided the heuristics to interpret the preferences and behaviour of actors in the basin. Thematically coded data from the field observation and interviews were mapped against these four rationalities identified by the framework.

RESULT

Both states have been keen on controlling and exploiting the rivers for hydroelectricity, irrigation, and riverbed materials. This study looks into all three categories of transboundary rivers in the Nepal-India borderlands.

KOSHI

Koshi is a category-I transboundary river governed through the Koshi Agreement between Nepal and India, which led to the construction of a barrage over a kilometer long. There is a long history of attempts to tame the Koshi River (see Mishra, 1997). The agreement that led to the creation of the barrage is just one element of the hierarchical-technological paradigm that is dominant in borderland water management. The embankments in the river were completed in 1959, and the barrage in 1964, accelerating sediment deposition, significantly distorting the river morphology. Since then, eight major breaches have occurred, including the 2008 breach, which led to a devastating flood in Nepal and Bihar (Dixit, 2009). The politicians and hydrocrats in India promoted structural solutions like embankments and dams as a panacea against floods and inundation, giving people a false sense of security (Dixit, 2009). Floods continue to cause massive damage around the Koshi River in Nepal and India. The river with its high level of sedimentation—100 to 120 million cubic metres of silt annually—has elevated the riverbed or belly higher than the levees in many places of eastern embankments in the borderlands (Dixit et al., 2023). Despite increasing embankment length from 160 km in 1954 to 3465 km in 2004, flood-prone areas expanded from 2.5 to 6.89 million hectares (Pandey & Patil, 2015). Vast tracts of fertile floodplains remain waterlogged year-round, making them uncultivable.

Similarly, institutional mismanagement has also exacerbated problems in the Koshi basin. The treaty bestows Indian hydrocracy full control over the project's design, construction, operation, and maintenance, while upstream Nepal holds no rights and responsibilities. Communication between Nepali authorities and the Koshi Project under Bihar's Water Resource Department (WRD) is poor, reflecting the neglect of embankment maintenance (Shrestha et al., 2010). The project's communication mechanism is 'complex, indirect and lengthy' and lacks continuity (Shrestha et al., 2010). Nepali district officers have the authority to communicate with WRD, and even central authorities in Nepal must communicate through the Koshi local liaison office (Shrestha et al., 2010). The treaty's vagueness on maintenance and lack of clear procedures and guidelines also contribute to infrastructure failure and embankment breaches (Shrestha et al., 2010). Under the 'neo-colonial' Koshi agreement, Nepal cannot open barrage gates during floods, causing inundation of Nepali territory, while Nepali farmers face acute irrigation shortage during the dry season.

Life around Koshi is precarious. Many have lost their agricultural land to embankments and waterlogging, forcing them into poverty. While able-bodied men migrate for employment opportunities, those left behind often fall prey to diseases and destitution (Gyawali, 2001). During field visits, residents complained that during floods and inundation, people in the affected areas are deprived of clean drinking water and other basic amenities. As the Koshi agreement and its infrastructures fail to address flood issues, India has renewed its push for the Koshi high dam. Initially proposed in the 1950s at Barahakshetra in Nepal, it was dropped due to high costs and replaced by a barrage. Post-2008, it has gained new momentum. The 269 m proposed high dam would generate 3300 MW of hydroelectricity and irrigate lands in both countries (Dixit, 2020; Oza, 2014). However, it would submerge 196 sq. km of land and displace 75000 people in 80 villages in Nepal (Dixit, 2020).

Civil society, in many instances, has challenged the structural paradigm. In Bihar, *Barh Mukti Andolan* leads the opposition to the embankments and the high dam from both social justice and scientific perspectives (Dixit, 2020). Similarly, activists and groups such as the South Asia Network on Dams, Rivers, and People (SANDRP) are also challenging the large dam structures (Dixit, 2020). In Nepal, several local villages and the people's rights forum have continuously protested the Koshi high dam (see Adhikari, 2015). They advocate for alternatives such as a series of multiple small dams along the river system, preserving local communities, culture, and ecology (Oza, 2014, p. 35). Organisations like Megh Payeen Abhiyan promote sustainable water management and safe water access in the Ganga basin (Dixit et al., 2023). Such efforts have helped build resilience and adaptability among local communities in the Koshi basin, prone to hydrological stress, such as floods and droughts (Dixit et al., 2023). Civil society groups on both sides regularly engage in cross-border dialogue, underlining the need for climate-sensitive and community-oriented water governance. However, several civil society actors and water experts interviewed in the Koshi basin and Kathmandu lamented the declining environmental activism in recent years due to youth apathy.

In recent years, the early warning systems (EWS) operated by the government with the assistance of NGOs have reduced flood-related fatalities and property loss. Nepal's Department of Hydrology and Meteorology has developed a disaster management system, in which upstream community volunteers alert downstream residents based on river-monitoring data (Davison, 2022). Although both Nepal and India operate their own EWS, there is no transboundary EWS in the Koshi River basin. Existing systems are centrally managed, top-down, and lack broad participation, limiting contextual understanding (Dawadi, 2023). Geopolitical tensions, such as river data securitisation and flood origin myths, further complicate collaboration (Dawadi, 2023). During interviews, many people expressed that the information obtained from the EWS was often not timely to enable early escape and move to safety. However, field observations also showed that strong cross-border community ties enable informal flood warnings via mobile phones and digital media. FM radio from Nepal, audible across borders, also has the potential to be an effective channel. Modern digital tools can enhance the effectiveness of EWS, including cross-border alerts, ensuring critical information reaches at-risk communities.

Beyond state and civil society, contractors with transboundary and multinational ties play an important role in the region's water resources governance. While Indian contractors have been jacketing the river in Nepal and India, Nepali contractors are increasingly involved in cross-border projects. They exert influence through their network and bargaining power. Scholars have critiqued the contractor-politician-engineer nexus shaping river control efforts in the Koshi floodplain (see Dixit, 2009, 2020). The business community in the region is not just bound to the state hierarchy; they are seen actively competing against unreliable state-managed irrigation and electricity services, supplying alternative technology to farmers like portable diesel pumps and solar pumps. This dynamic private sector often steps in during floods to deliver essential goods as government aid falls short.

Tinau

Tinau is a category-II river that crosses Nepal to merge with the Rapti River in the Indian state of Uttar Pradesh. Known as Kuda in India, the river demands cross-border cooperation and communication for effective hydrological disaster response. However, the Tinau River receives scant attention in joint technical meetings between the Ministry of Water Resources of both neighbouring countries and other discourses, except during flood events and inundation (Dhungana et al., 2016). Tinau flows through the Rupandehi, one of Nepal's most populous districts, home to a major urban agglomeration. There is a high demand for Tinau's water for household, agricultural, and industrial use on both sides of the border. On the other hand, several floods and inundations have occurred at various places in the district near the Indo-Nepal border. The river has been extensively jacketed, most recently as part of the People's Embankment programme, but such structural remedies have failed to discipline the river as expected.

Riverbed mining is widespread in the Tinau, which has been an important revenue source for the local government. A strong nexus between local contractors and politicians drives riverbed extraction, amid high demand for such materials in Nepal and India. While local government officials interviewed justified the revenue source, it has caused problems like riverbed lowering and scouring of hydraulic structures and bridges. Similarly, it has made irrigation difficult, and it is hard to divert water from the deep riverbed using traditional brushwood dams. There are also reports of lowering the groundwater table due to extraction activity in Tinau's vicinity, leading to the failure of shallow tube wells to yield water. While benefiting business people and industrialists, these practices externalise costs to the general population.

The Nepal government has tried to harness the Tinau river for irrigation, hydroelectricity, and drinking water through different state-owned institutions. The basin has a rich history of farmer-managed irrigation systems (FMIS), even in state-promoted settlement areas. Many traditional FMIS have been taken over, expanded, and modernised by the state and development agencies. The ongoing decline in the traditional FMIS in Tianu is partly due to the scarcity of logs and other traditional materials required to make temporary barriers and the difficulty in mobilising voluntary labour to manage the irrigation system amidst social and economic change. In the Marchawar region along the Nepal-India border, state-led modern irrigation schemes have increasingly displaced traditional FMIS, but have failed to supply adequate water to farmers, especially tailenders. This has led to the social and economic decline of the community. Moreover, farmers find it difficult to maintain the irrigation machinery and must rely on outside technicians. These failures have pushed farmers towards individualist solutions like shallow tubewells, now even subsidised by the government.

Across the border in India, the government's embankment at Kuda for flood control turns the land between embankments into a watercourse during the monsoon. These embankments, which have been ineffective in controlling floods, do not allow water to recede, leading to waterlogging on the land, making it difficult to grow even a single crop in a year. In this hinterland of Uttar Pradesh, there is drinking water and irrigation water scarcity during the dry season, while flood ravages the area in the Monsoon.

Pandai

This river flows through the Thori and Bhikhana Thori region in the Nepal-India borderland, near Chitwan National Park, Parsa National Park in Nepal, and Valmiki

Tiger Reserve in India. Rivers like Pandai are often marginalised in state narratives and rarely in bilateral talks. In the summer, when water runs low in the *Thute Khola* stream, a tributary of Pandai, it creates conflict between the upstream and downstream communities, both critically dependent on the water of Pandai for their needs (Siddiqui et al., 2020). The last time a conflict like this emerged was in 2020 and was settled by sharing two canals of water to India while diverting one but larger canal to Nepal, as shown in Figure 12.2. Even in the upstream Nepal, which enjoys better water security, the state-led irrigation system has failed to provide water to all farmers, leaving them high and dry. Nepali farmers and local leaders during a field visit blamed India's claim and expropriation of the majority of water in the Pandai for making farming untenable due to a lack of irrigation water even in the monsoon, forcing local Nepali youths to migrate for employment opportunities.

During the last few decades, the Chure region from which rivers like Pandai originate has seen rampant deforestation, unchecked sand and gravel extraction, often abetted by local authorities. This has heightened monsoon flood risks in the monsoon while causing drying of streams and groundwater depletion in Nepal-India borderlands. While the river is not under any treaty like hundreds of other transboundary rivers and streams in the Indo-Nepal borderlands, the solution for the drought and water scarcity in the Pandai sub-basin must be sought, going beyond the state-led hierarchical means. The techno-managerial solutions, like the supply of water from far away through Nal Jal Yojna, have yet to bear any fruit and dependence on Nepali drinking water sources, particularly Rai Dhara, persists. The authorities

FIGURE 12.2 Photo of the Thute Khola water-sharing site, where the far-right canal diverts flow to Nepal, while the middle canal and remaining water continue into India. Water sharing becomes critical during the drought. (*Source:* Photo taken by authors.)

in both countries have sought to look at drought and water scarcity and the conflict it has engendered as a law-and-order problem instead of addressing the underlying causes.

Historically, people across the border who depended on the river for their lives came together to solve the water problem through an access mechanism, and the river was previously divided into different channels to supply water to dependent borderland communities. It is a sign that before the states and belongingness to a particular state became much stronger among the people, an egalitarian outlook that recognises the river as a common heritage vital to the survival of mankind across the border was much more prominent. So, when disputes occurred, village elders and informal village-level groups helped de-escalate the situation even during droughts and scarcity (Prasad, 2020). During the field visit, people generally expressed that federalism, which has empowered local government in resource governance, has played a constructive role in solving the problems, even though local politics can also possibly exacerbate the conflict. Nepali Armed Police Force (APF), stationed near water sharing sites of *Thute Khola*, plays a vital role in monitoring water sharing sites to ensure stability, and maintaining law and order if contention arises.

Beyond dry season droughts, Pandai has a high sediment load, which causes flash floods in the monsoon. Local governments in both countries have sought to control the river through competitive embankment building. Nepali residents interviewed claimed that embankments on the Indian side created riverbank erosion on the Nepali side, affecting settlements, the bus park, and local markets (Figure 12.3).

FIGURE 12.3 Erosion of the riverbank and damage to the Postal Highway on the Nepali side occurred as a result of flood-related water flow. (*Source:* Photo taken by authors.)

FIGURE 12.4 (a) and (b) Illustrate embankments constructed on the Indian side by Indian authorities. (*Source:* Photo taken by authors.)

Many locals in Nepal interviewed hoped that embankments being constructed would protect them from the floods (Figure 12.4).

DISCUSSION

Plural Rationalities and River Governance

The cases of interaction of different solidarities in the dynamics of three different transboundary rivers can provide us with different lessons for water management in the India-Nepal borderlands. All four rationalities—hierarchical, egalitarian, individualist, and fatalist—are prevalent in all three transboundary river basins, and their dynamic interaction shapes transboundary water management along the Koshi, Tinau, and Pandai rivers. The story of three rivers and the interaction between the rationalities prevalent there can be analysed using the framework provided by Cultural Theory. This study found that water governance in all three borderland river basins reflects the coexistence of all four rationalities; however, hierarchism is the most dominant one. It was found that active solidarities interact, cooperate, and form alliances across borders. Among the three rivers, the hierarchical approach was most dominant in the Koshi, followed by Tianu and Pandai. Tinau has a relatively strong individualist-egalitarian mix with a strong sense of fatalism prevalent mostly among squatters and peasants. In the Pandai basin, where the river flows through the hinterlands of Bihar and Madhesh Pradesh, near national parks and wildlife reserves, people exhibit fatalist and egalitarian tendencies. This shows that a marginalised river basin has better opportunities to bring together all solidarities for decentralised water governance compared to large river basins, which attract overwhelming state attention.

Nation States and the Hierarchical Fix: Technocratic Solutions and Their Discontents

Two nation states and their sub-national governments have their concerns regarding the management of the water in the transboundary rivers. While being deeply hierarchical and technological, their perspectives show how government machinery is organised and how the issue of water management and governance is securitised while viewed from a state-centric lens. This hard path of water management, where massive structures are seen as a wizardly demonstration of technological prowess and a means of realisation of the cornucopian dream, has buyers among the hydrocrats, business communities, and the public. This hierarchical path is strongly opposed by the grassroots social and environmental activists who share a sense of brotherhood across borders and have their behaviour dictated by this egalitarian solidarity. Egalitarians have fundamentally different values on issues of technological choices, governance, and resource use. They feel that the sense of control and command over rivers and nature is illusory and believe that cooperation, adaptation, and adjustment with nature are an appropriate way to survive.

Observation at all three river basins showed that nation states and their hydrocrats and politicians have a misunderstanding of the natural and social aspects of water and rivers. Political Science and its associated disciplines, like International Relations and International Law, demonstrate a myopia rendering the knowledge they produce seriously inadequate in a holistic understanding of the river system, particularly in the borderlands (Gyawali, 2020). They have a tendency to see the nation state as the only medium through which all social engagement happens, and the state mechanisms can take care of everything that happens. Even in large rivers of Nepal like the Koshi (discussed above), the Gandaki, and the Mahakali, bilateral treaties and mega structures created for their governance have failed to satisfy the needs and aspirations of the borderland communities. Such agreements and projects have arguably impoverished the borderland riverine communities not only in Nepal, which have been imposed with neo-colonial water treaties, but also in India. However, these patent failures have not dissuaded nation states from engaging in the folly of state centrism and high technology optimism, and generally, the tyranny of hierarchism continues unabated.

Once a society enters a resource governance pathway that is high technology, capital-intensive, requires high expertise, and has a long gestation period, such as building barrages, long canals, high dams, and embankments, it is difficult to extricate itself from it due to a high degree of technical and social inflexibility (Gyawali, 2001). In both of these countries, the failure of each high technological infrastructure has led to proposal of much larger infrastructural solution either in the form of the Koshi high dam to alleviate Bihar's water woes in the Koshi basin or the Kaligandaki-Tinau diversion project to supply water for the people in the Tinau basin. Hierarchs concerned about national or regional water rights and 'development' are indifferent to the suffering of the people at the local levels, as these projects displace thousands of people, impoverish them, and defile the sacred landscapes. This mirrors McCully (2001) and Scudder (2019), who have discussed how large hydraulic structures and projects generate social, economic, and ecological risks and Nixon's work on development and slow violence. Nixon (2010) showed that states use both direct violence and symbolic violence through appeal for sacrifice for the greater national good, hiding the human costs; they also seek to invisibilise those affected by using acronyms and euphemisms such as PAPs (Project Affected People), sanitising, and depersonalising the experience of displacement. They are also being increasingly treated as anachronisms in a rapidly modernising world. Thus, while mega dams and other megastructures serve as captivating monuments of development, they hide what Nixon calls the underlying 'violent underdevelopment'. It is actually this indifference, along with the tyranny of hierarchs, the warped political economy of the hard technological choices, and the resulting human sufferings that fuel the egalitarian solidarity, much to the chagrin of the hierarchy.

Marginality and the Fatalist Experience

In the Pandai basin, where a small, unregulated transboundary stream flows through the hinterlands of Bihar and Madhesh Pradesh, near national parks and wildlife reserves, governance neglect has generated a fatalist landscape. In the absence of a formal treaty or bilateral agreement, the river basin and riverine communities have been rendered mostly invisible in state planning and discourses. Local people in downstream India experience recurring droughts and flash floods with little expectation of state support. This has created a sense of abandonment and helplessness. This marginalisation mirrors Ahmad (2003), who argue that despite the significance of small rivers, they remain unaccounted in flood management and modelling in the Ganga-Brahmaputra-Meghna (GBM) region. However, evidence of egalitarian rationality beneath the fatalist tendencies can be found in riverine communities. Informal transboundary cooperation, historical channel-sharing agreements, and village-level conflict resolution show embedded strategies for managing shared water resources at the local level. However, the decline of these practices in recent years points to the consequence of scaling governance away from people's hydro-social relationships. Thus, in the absence of inclusive frameworks, these marginalised riverine communities risk further exclusion by the hierarchic-technocratic paradigm.

Nepal finds itself in a marginalised position in bilateral relations with India since it is at the receiving end of the technological imposition by India, a regional hydro-hegemon (see Cascão & Zeitoun, 2020; Tandan, 2021). Nepal has been forced to accept Indian infrastructure like barrages, embankments, etc. in the borderlands as a fait accompli either through formal agreement or high-handedness. However, such marginalisation and fatalism have the potential to generate conflict once the nation or the community in the borderland realises their loss. Such conflict has been simmering in Nepal-India borderlands for years now, which can explode if Indian unilateralism continues further and exacerbates the plight of the Nepali borderland communities.

Informality, Technology, and the Rise of Individualist Logics

There has been a growth of informal and market-driven technological solutions reflecting the influence of individualist rationalities in water management in

Nepal-India borderlands. As a formal state-led irrigation and flood control system has failed to deliver the promises because of inefficiency, locals have turned to private alternatives such as shallow tube wells, diesel pumps, and informal contractors for water access and emergency response. These reflect adaptive, pragmatic strategies grounded in the logic of autonomy and innovation. However, individualist rationality also manifests itself in extractive market practices, such as riverbed mining fuelled by the nexus of local contractors, politicians, and bureaucrats in different river basins. Market demand for sand and gravel in rapidly urbanising regions has turned the river into yet another extractive frontier, seriously affecting ecological stability and worsening problems such as riverbed lowering, scouring of infrastructure, and groundwater depletion. In places in the global south where formal governance is weak, we see similar patterns of technological workaround, informal extraction, commodification, and privatisation of water (see Bakker, 2007; Johnson et al., 2016). Similarly, we can also see the support of individualist contractors and business people crucial in supporting the hierarchy's hard technological path, executing those construction projects.

Civil Society, NGOisation, and the Challenge of Egalitarian Governance

Though hierarchical and individualist solidarities continue to dominate transboundary water governance in the Nepal-India borderlands, egalitarian rationalities, particularly through civil society activism and grassroots resistance, have articulated an egalitarian vision rooted in participation, socio-economic justice, and ecological integrity. Egalitarians attempt to check on the ravages of hierarchical despotism and the rapacity of unchecked market individualism. Across three river basins, we can find civil society movements actively resisting the hierarchical tyranny, focusing on the failure of its technological choices in delivering on promises of development. Transnational social movements have rapidly expanded since the 1980s and 1990s, and many Indian civil society organisations resisting hierarchical despotism and championing an egalitarian vision for water governance actively engage with their counterparts in Nepal.

People across the river basins at borderlands for years have naively believed that embankments and other hydraulic infrastructures will free them from floods and droughts and bring an economic windfall. With the failure of hierarchism and the influence of transnational civil activists, environmental activists, and religious groups, they are growing increasingly disillusioned and disenchanted with the state and its machineries. With strong social and environmental movements, the fatalism of the masses can be transformed into activism which can challenge the hierarchical despotism. One case of strong resistance against the state's hard technological choice is evident in the case of Tinau, where environmental, social, and religious activists are protesting the diversion of the holy Kaligandaki River to Tinau. Not only the civil society movements and community resist those policies which they feel are unfavourable, but there is also evidence from all the river basins of their playing a constructive role in resolving disputes, particularly if they involve communities in borderlands. In the Tinau basin, the design of embankments in Siddharthanagar, India, was changed so that it would reduce inundation in the Marchawar region of

FIGURE 12.5 Peoples from the Indian side collect water from the Raidhara for drinking purposes, which is located within Nepali territory. (*Source:* Photo taken by authors.)

Nepal, partly due to media coverage and the role of India-Nepal Friendship Society, formed by border town residents of both countries and other community initiatives (Dhungana et al., 2016). In Nepal-India borderlands, which have deep social, economic, and cultural linkages, inter-community relations can help reduce misunderstanding and reach optimal solutions (Figure 12.5).

However, it was also observed that in many instances local community mobilisations, which emerged as spontaneous, egalitarian responses to hydrological disasters and ill governance of water resources, have now been superseded by the umbrella of NGOs, which often align their goals with state-approved developmental and water governance goals and market logics for funding and visibility shifting away from structural changes. This aligns with the literature on NGOisation of civil society and social movements, criticising the donor-driven agendas and professionalisation, which often depoliticises community action, shifting focus from structural changes (Choudry & Kapoor, 2013; Kaldor, 2003). Their action often reproduces or reinforces the same top-down approach, marginalising a large section of the population, as seen in their role in championing a flawed EWS in flood-affected areas. While they were found to be working in places like Koshi, which garners more media attention, they were visibly absent in marginalised spaces like Pandai basin, where the community struggles with both flood and drought.

CONCLUSION AND RECOMMENDATION

This chapter studied the interaction of four different rationalities (hierarchical, individualist, egalitarian, and fatalist) shaping transboundary water governance in Nepal-India borderlands, focusing on three rivers: Koshi (a major river), Tinau (a medium-sized river), and Pandai (a small, marginalised river). Drawing on the Cultural Theory (Theory of Plural Rationalities), the study reveals that water governance in the borderland is marked by the existence and contestation of four rationalities. At the same time, hierarchical solidarity remains dominant and privileged. The findings showed how hierarchical state-led approaches like embankments, barrages, and dams have failed to reduce disaster vulnerability and secure a cornucopian future for the riverine community as promised. The intervention of market individualism and egalitarian civil societies and social movements has given only mixed results. However, they are trying to resist exclusion and assert their alternative ways of living.

This study addressed the research question by showing that the governance of transboundary rivers in Nepal-India borderlands, despite being shaped by institutional complexity, is still greatly technocratic and hierarchical. However, the attempts to deal with 'wicked problems' of water management through top-down infrastructural and technocratic intervention, which is a hallmark of hierarchism, are unlikely to succeed and may even intensify the local vulnerabilities if they proceed without local participation, particularly with an increasing uncertainty due to climate change. There is a need to move away from equilibrium thinking based on homogeneity, predictability, linearity, and simplification and look for more bottom-up ways of dealing with problems, being cognisant of local conditions of the borderlands. Informal mechanisms such as community communication across borders, EWS, and decentralised adaptation strategies were considered essential but under-recognised modes of building resilience.

The success of water governance in Nepal-India borderland will depend not upon elegant solutions devised by the hierarchists but on the plural solutions created through deliberative interaction and interplay of multiple world views regarding the perception of problems, risk, and ways of organising, thus creating a new, creative, and alternative response. Therefore, there is a need for statesmanship that can bring together and balance all three institutional solidarities (bureaucracy, market, and activism), helping create 'clumsy solutions' for 'wicked problems' (see Grint, 2008). While the overweening dominance of hierarchy is gradually giving way to the alliance between hierarchy and market individualism, there is an urgent need to make greater room for the egalitarian civil society motivated by things other than control and profit and can check their tendencies to some extent. The findings suggest that 'clumsy governance' that draws from all rationalities in contextually appropriate ways may be suitable for dealing with dynamic risks, conflict, and tensions of water governance in the borderland.

There are practical implications for the local riverine communities in the borderlands. Government policies and interventions that recognise ecological knowledge, adaptive practices, and the communities' lived experience will be deemed legitimate and become more effective. Communities must seek to enhance cross-border EWS, regulate riverbed mining and groundwater exploitation, strengthen traditional FMIS, and revive and use traditional water harvesting technologies used for centuries in the Ganga plains to bolster water security. India, with its long history of using high technology based on megastructures for water governance since colonial times (see D'Souza, 2006), has led to a technological lock-in. Since Nepal is a late adopter of high-modernist technologies for water management, the Nepali community still has the opportunity to embrace and expand those traditional systems and technologies for irrigation, watershed conservation, and water harvesting that better suit their reality. Modern multimedia, telecommunication, and other digital technologies have effectively reduced fatalities and other losses during disasters. Such technologies must be adopted for effective disaster management and water governance.

Policymakers and stakeholders in India and Nepal must internalise the lessons from minor and marginalised river basins—not just major treaty-bound rivers—as they reflect the full spectrum of governance failures and adaptive innovations. Stakeholders in these two countries require a democratic environment conducive to discursive contestation, interaction, and synthesis of different viewpoints to reach a 'clumsy solution' and overcome the yoke of hierarchism in water governance. Their ability to achieve such socio-economically and ecologically just solutions for the riverine communities in the borderlands is a litmus test of democracy in the two neighbouring countries.

It is recommended that transboundary river commissions/committees in Nepal-India borderlands and other rivers across South Asia, instead of focusing solely on state initiative (hierarchy), should make themselves representative, including civil society actors (egalitarian), private sectors (individualist), and local voices. The planning process must integrate local narratives of hydrological risks, entrepreneurial innovation, and expert forecasting to build flexible and robust plans for effective irrigation management and flood preparedness. The river sharing agreements should allow greater flexibility, particularly during hydrometeorological shocks and stresses. Pilot-level 'clumsy policy experiments' can be conducted to test plural governance in the borderland communities involving shared flood shelters and cross-border village collaboration for early warning. These pilot experiments can form the basis for greater institutional changes. These countries should no longer fatalistically blame each other for a share of their water woes, but engage in serious soul-searching about the technological choices appropriate in their territory to improve resilience and adaptability towards hydrological extremes based on socio-economic, political, and ecological reality.

LIMITATION

This research showed that using plural rationality theory (Cultural Theory) in qualitative research offers a robust framework to understand institutional diversity, contestation, and conflict in complex water governance systems. Further studies can build on this framework to study other transboundary and domestic resource governance and environmental issues in South Asia and beyond. However, this study had a few limitations. It relied on qualitative data from three rivers in the Nepal-India borderlands, but the findings might not be entirely generalisable. Moreover, due to time and resource constraints, the study lacked a high level of immersion in the study area. Additionally, due to language and cultural barriers in study areas, some biases and inaccuracies might have crept into the research despite deliberate attempts to minimise them. Furthermore, the study did not account for the geopolitical shifts in the region, nor quantify the socio-economic impacts in the borderland communities. Future research could use a mixed methods approach, increase the immersion and engagement time in the transboundary study area to improve the robustness of the study and findings, and expand the comparative research to include shared rivers in other South Asian countries.

REFERENCES

Adhikari, B. R., & Ojha, R. B. (2021). Geology and physiography. In R. B. Ojha & D. Panday (Eds.), *The Soils of Nepal* (pp. 29–39). Springer International Publishing. https://doi.org/10.1007/978-3-030-80999-7_4

Adhikari, D. (2015, June 2). Protests scupper India-Nepal dam. *The Third Pole*. Retrieved on November 20, 2024, from https://www.thethirdpole.net/en/energy/protests-scupper-india-nepal-dam/

Ahmad, Q. K. (2003) Regional cooperation in flood management in the Ganges-Brahmaputra-Meghna Region: Bangladesh perspective. *Natural Hazards*, *28*, 181–198.

Akamani, K., & Wilson, P. I. (2011). Toward the adaptive governance of transboundary water resources. *Conservation Letters*, *4*(6), 409–416. https://doi.org/10.1111/j.1755-263X.2011.00188.x

Alfthan, B., Gupta, N., Gjerdi, H. L., Schoolmeester, T., Andresen, M., Jurek, M., & Agrawal, N. K. (2018). "Outlook on climate change adaptation in the Hindu Kush Himalaya, mountain adaptation outlook series." *United Nations Environment Programme, the International Centre for Integrated Mountain Development.*

Bagale, D. R. (2020). Nepal–India water cooperation: consequences of mutuality or hegemony? *Water Policy*, *22*(6), 1098–1108. https://doi.org/10.2166/wp.2020.135

Bakker, K. (2007). The "commons" versus the "commodity": Alter-globalization, anti-privatization and the human right to water in the global south. *Antipode*, *39*(3), 430–455.

Beck, M. B., Thompson, M., Ney, S., Gyawali, D., & Jeffrey, P. (2011). On governance for re-engineering city infrastructure. *Engineering Sustainability*, 164, 129–142.

Bhattarai, A., Timilsina, S., Ayer, S., Paudel, G., & Maharjan, M. (2024). Evaluation of Drought in Gandaki River Basin of Nepal. *Journal of Drought and Climate change Research*, *2*(2), 1–26. https://doi.org/10.22077/jdcr.2023.6865.1046

Biggs, E. M., Duncan, J. M. A., Atkinson, P. M., & Dash, J. (2013). Plenty of water, not enough strategy: How inadequate accessibility, poor governance and a volatile government can tip the balance against ensuring water security: The case of Nepal. *Environmental Science & Policy*, *33*, 388–394. https://doi.org/10.1016/j.envsci.2013.07.004

Bisht, M. (2019). From the edges of borders: reflections on water diplomacy in South Asia. *Water Policy*, *21*(6), 1123–1138. https://doi.org/10.2166/wp.2019.124

Cascão, A. E., & Zeitoun, M. (2010). Power, hegemony and critical hydropolitics. In *Transboundary Water Management*. Routledge.

Choudry, A., & Kapoor, D. (2013). *NGOization: Complicity, Contradictions and Prospects*. Bloomsbury Publishing.

D'Souza, R. (2006). Water in British India: the making of a 'colonial hydrology'. *History Compass*, *4*(4), 621–628.

Daoudy, M. (2009). Asymmetric Power: Negotiating water in the Euphrates and Tigris. https://doi.org/10.1163/157180609X432860

Davison, C. (2022, August 2022). Nepal's early-warning system reduces flood fatalities. *DW*. Retrieved on June 19, 2024, from https://www.dw.com/en/nepals-early-warning-system-reduces-flood-fatalities/a-62981276

Dawadi, G. (2023). Understanding transboundary early warning systems and their cultural context in the Kosi River Basin. [Doctoral Dissertation, King's College, London]. https://kclpure.kcl.ac.uk/ws/portalfiles/portal/229478060/2023_Dawadi_Gaurab_1833727_ethesis.pdf

Deribe, M. M., Melesse, A. M., Kidanewold, B. B., Dinar, S., & Anderson, E. P. (2024). Assessing international transboundary water management practices to extract contextual lessons for the Nile River Basin. *Water, 16*(14), 14. https://doi.org/10.3390/w16141960

Dhungana, H., Pain, A., & Dhungana, S. P. (2016). Disaster risk management and meso-level institutions in Nepal: a case study of floods in Tinau River in western Terai. *Climate Change and Rural Institutions (CCRI) Research Project, CCRI Case Study, 6*. Case-Study-Report-Tinau-Flood-1.pdf

Dixit, A. (2009). Kosi embankment breach in Nepal: need for a paradigm shift in responding to floods. *Economic and Political Weekly, 44*(06). https://www.epw.in/journal/2009/06/special-articles/kosi-embankment-breach-nepal-need-paradigm-shift-responding-floods

Dixit, A., Rimal, N., Bhandari, D., & Sharma, S. (2023). Catalyzing change for resilience across boundaries: Challenges in Nepal's small transboundary rivers. ISET-Nepal. Report_24_May 2023 (3).pdf

Dixit, A., Upadhya, M., Pokhrel, A., Dixit, K., Rai, D. R., & Devkota, M. (2008). Flood disaster impacts and responses in Nepal Tarai's marginalised basins. https://www.semanticscholar.org/paper/Flood-Disaster-Impacts-and-Responses-in-Nepal-Dixit-Upadhya/1bb528abb1ca5dd0e238d55cee6304658c91781c

Dixit, K. M. (2020, October 27). The ill-conceived push for a high dam to curb Kosi flooding is a litmus test for democracy in India. *Scroll.in*. Retrieved on November 8, 2023, from https://scroll.in/article/976756/the-ill-conceived-push-for-a-high-dam-to-curb-kosi-flooding-is-a-litmus-test-for-democracy-in-india.

Earle, A., Cascao, A. E., Hansson, S., Jägerskog, A., Swain, A., & Öjendal, J. (2015). *Transboundary Water Management and the Climate Change Debate*. Routledge. https://doi.org/10.4324/9780203098929

Gaire, S., Castro Delgado, R., & Arcos González, P. (2015). Disaster risk profile and existing legal framework of Nepal: Floods and landslides. *Risk Management and Healthcare Policy, 8*, 139–149. https://doi.org/10.2147/RMHP.S90238

Gerlak, A. K., Karambelkar, S., & Ferguson, D. B. (2021). Knowledge governance and learning: examining challenges and opportunities in the Colorado River basin. *Environmental Science & Policy, 125*, 219–230.

Giordano, M. A., & Wolf, A. T. (2003). Sharing waters: Post-Rio international water management. *Natural Resources Forum, 27*(2), 163–171. https://doi.org/10.1111/1477-8947.00051

Grey, D., & Sadoff, C. (2003). Beyond the River: The benefits of cooperation on international rivers. *Water Science and Technology: A Journal of the International Association on Water Pollution Research, 47*, 91–96. https://doi.org/10.1016/S1366-7017(02)00035-1

Grint, K. (2008). Wicked Problems and Clumsy Solutions: The Role of Leadership. *BAMM Publications*. Retrieved from https://www.researchgate.net/publication/281357989_Wicked_problems_and_clumsy_solutions_The_role_of_leadership

Gupta, N., Dahal, S., Kumar, A., Kumar, C., Kumar, M., Maharjan, A., Mishra, D., Mohanty, A., Navaraj, A., Pandey, S., Prakash, A., Prasad, E., Shrestha, K., Shrestha, M. S., Subedi, R., Subedi, T., Tiwary, R., Tuladhar, R., & Unni, A. (2021). Rich water, poor people: Potential for transboundary flood management between Nepal and India. *Current Research in Environmental Sustainability, 3*, 100031. https://doi.org/10.1016/j.crsust.2021.100031

Gyawali, D. (2001). *Water in Nepal* (Vol. 280). Himal Books.

Gyawali, D. (2020, October 3). Rethinking Transboundary Waters. *Geopolitics & Ecology of Himalayan Waters*. Retrieved on June 20, 2025, from https://www.himalayanwaterproject.org/post/rethinking-transboundary-waters.

Gyawali, D., & Dixit, A. (1999). Mahakali impasse and Indo-Nepal water conflict. *Economic and Political Weekly, 34*(9), 553–564.

Johnson, H., South, N., & Walters, R. (2016). The commodification and exploitation of fresh water: Property, human rights and green criminology. *International Journal of Law, Crime and Justice*, *44*, 146–162.

Kaldor, M. (2003). *Global Civil Society: An Answer to War.* Polity Press.

Karki, S. (2025). *The Nepal-India Open Border: Centering Climate Security in the Political Space.* SAFF. https://www.southasianfuturesfellowship.org/analysis-2/the-nepal-india-open-border%3A-centering-climate-security-in-the-political-space

Karkkainen, B. C. (2005). Transboundary ecosystem governance: Beyond sovereignty? In *Public Participation in the Governance of International Freshwater Resources.* UNU Press. https://collections.unu.edu/view/UNU:2453

Kibaroglu, A. (2008). The role of epistemic communities in offering new cooperation frameworks in the Euphrates-Tigris rivers system. *Journal of International Affairs*, *61*(2), 183–198.

Maharjan, K. (2023). Transboundary river governance and climate vulnerability: Community perspectives in Nepal's Koshi river basin. *Geographical Research*, *61*(4), 512–524. https://doi.org/10.1111/1745-5871.12598

McCully, P. (2001). *Silenced Rivers: The Ecology and Politics of Large Dams.* Zed Books.

Mishra, D. (2008). The Kosi and the embankment story. *Economic and Political Weekly*, *43*(46), pp. 47–52. https://www.jstor.org/stable/40278181

Mishra, D. K. (1997). The Bihar flood story. *Economic and Political Weekly*, *32*(35), 2206–2217. http://www.jstor.org/stable/4405794

Nixon, R. (2010). Unimagined communities: Developmental refugees, megadams and monumental modernity. *New Formations*, *69*(1), 62–80.

Oza, J. (2014). Resisting for the River. *Third World Network.* Retrieved on November 8, 2024, from https://twn.my/twnf/2014/4160.htm

Pahl-Wostl, C., Gupta, J., & Petry, D. (2008). Governance and the global water system: A theoretical exploration. *Global Governance*, *14*, 419–435. https://doi.org/10.1163/19426720-01404003

Pandey, A., & Patil, S. (2015). *The Agony of Rivers: Floods in Bihar.* ICIMOD.

Pandey, C. L., Joseph, J., Deshar, R., & Niraula, P. (2023). Transboundary flood resilience: Insights from Narayani and Mahakali Basins. *International Journal of Disaster Risk Reduction*, *86*, 103535. https://doi.org/10.1016/j.ijdrr.2023.103535

Prakash, A., George, R., & Barua, A. (2025). Socio-hydrological frameworks for adaptive governance: Addressing climate uncertainty in South Asia. *Frontiers in Water*, *7*. https://doi.org/10.3389/frwa.2025.1556820

Prasad, E. (2020, July 10). *Pandai River.* SANDRP. https://sandrp.in/tag/pandai-river/

Ranjan, A. (2016). Contours of India–Nepal relationship and trans-boundary rivers water disputes. *Journal of International Affairs*, *1*(1), 1. https://doi.org/10.3126/joia.v1i1.22637

Rieu-Clarke, A., Moynihan, R., & Magsig, B.-O. (2016, February 16). *Transboundary Water Governance and Climate Change Adaptation: International Law, Policy Guidelines and Best Practice Application | PreventionWeb.* UNDRR. https://www.preventionweb.net/publication/transboundary-water-governance-and-climate-change-adaptation-international-law-policy

Scudder, T. (2019). *Large Dams: Long Term Impacts on Riverine Communities and Free Flowing Rivers.* Singapore: Springer.

Sharma, C. K. (1997). *A Treatise on Water Resources of Nepal.* MASS Printing Press.

Shrestha, A. B., & Aryal, R. (2011). Climate change in Nepal and its impact on Himalayan glaciers. *Regional Environmental Change*, *11*(1), 65–77. https://doi.org/10.1007/s10113-010-0174-9

Shrestha, R. K., Ahlers, R., Bakker, M., & Gupta, J. (2010). Institutional dysfunction and challenges in flood control: A case study of the Kosi flood 2008. *Economic and Political Weekly*, *45*(2), 45–53. http://www.jstor.org/stable/25663989

Siddiqui, S., Chohan, S., Khanduri, Khanduri, V., Kumar, S. (2020). Challenges for water cooperation on small rivers between India-Nepal: Lessons from the fringe villages along the Pandai River in the shared Gandak Basin. *India Water Partnership*. Challenges-for-Water-Cooperation-on-small-Rivers-between-India-and-Nepal.pdf

Tandan, P. (2021). Challenging hydro-hegemony of India: Resistance of Nepal in the Upper Karnali and Saptakoshi dam project. *Sustainable Water Resources Management*, *7*(6), 106.

Thomas, K. A. (2017). The river-border complex: A border-integrated approach to transboundary river governance illustrated by the Ganges River and Indo-Bangladeshi border. *Water International*, *42*(1), 34–53. https://doi.org/10.1080/02508060.2016.1247236

Thompson, M. (2008). *Organising and Disorganising*. Triarchy Press.

Thompson, M., Ellis, R., & Wildavsky, A. (1990). *Cultural Theory*. Boulder, CO and Oxford: West View.

Thompson, M., Gyawali, D., & Verweij, M. (2016). The dharma of development. In D. Gyawali, M. Thompson, & M. Verweij, *Aid, Technology and Development: The Lessons from Nepal*. Taylor & Francis.

Upadhyay, M. (2020). Pragmatic approaches to transboundary river governance: Exploring pathways for co-operation on shared water resources. *World Water Policy*, *6*(1), 101–114. https://doi.org/10.1002/wwp2.12030

Uprety, K., & Salman, S. M. A. (2011). Legal aspects of sharing and management of transboundary waters in South Asia: Preventing conflicts and promoting cooperation. *Hydrological Sciences Journal*, *56*(4), 641–661. https://doi.org/10.1080/02626667.2011.576252

Vasani, H. (2023). International water law and hydropolitics: An enquiry into the water conflict between India and Nepal. *Water International*, *48*(2), 259–281. https://doi.org/10.1080/02508060.2023.2176595

Verweij, M. (2000). *Transboundary environmental problems and cultural theory: The protection of the Rhine and the Great Lakes*. Springer.

Verweij, M. (2017). The remarkable restoration of the Rhine: plural rationalities in regional water politics. *Water International*, 42(2), 207–221.

Vij, S., Warner, J., & Barua, A. (2020). Power in water diplomacy. *Water International*, *45*(4), 249–253. https://doi.org/10.1080/02508060.2020.1778833

Zeitoun, M., & Allan, J. A. (2008). Applying hegemony and power theory to transboundary water analysis. *Water Policy*, *10*(S2), 3–12. https://doi.org/10.2166/wp.2008.203

Zeitoun, M., & Mirumachi, N. (2008). Transboundary water interaction I: Reconsidering conflict and cooperation. *International Environmental Agreements: Politics, Law and Economics*, *8*(4), 297–316.

13 Sustainable Water Resources Management in Brahmaputra River Basin

Key Challenges and Way Forward

Ahmad Shams Qadir, Sadia Afrose Esha, Sakib Mahmud, Md. Tawhidul Aziz, and Md. Ashraful Haque

INTRODUCTION

Freshwater resources, especially those that transcend national boundaries, are becoming a focal point of geopolitical tension and ecological concern across the globe. Rivers like the Brahmaputra, which flows through China (Tibet Autonomous Region), India, and Bangladesh, serve not only as vital freshwater sources but also as catalysts for cooperation or conflict among riparian states (Pradhan et al., 2021). In the context of the global water crisis, the Brahmaputra River Basin (BRB) represents a critical case where environmental sustainability, regional diplomacy, and human development intersect. The consequences of mismanaging such transboundary resources are profound, potentially resulting in escalated regional disputes, loss of biodiversity, heightened flood risks, and displacement of vulnerable populations (Yang et al., 2016). Hence, sustainable management of this river is inevitable. Water resource management in transboundary basins has emerged as a global research priority, particularly in regions where hydrological interdependence intersects with political sensitivities. The BRB covers an area of approximately 580,000 km^2 and supports the livelihoods of over 130 million people across three nations (Yang et al., 2016). Despite its substantial water flow, the basin remains underutilized and poorly coordinated in terms of integrated water resource management. According to the United Nations Educational, Scientific and Cultural Organization (UNESCO, 2021), over 60% of international river basins lack adequate cooperative management frameworks, underscoring the urgency of institutionalized transboundary governance mechanisms.

DOI: 10.1201/9781003599692-13

In South Asia, this challenge is intensified by climate variability, geopolitical tensions, and uneven development. The Himalayan region, source of the Brahmaputra, is one of the most climate-sensitive zones in the world. Glacial retreat, erratic monsoon patterns, and increasing frequency of extreme weather events have exacerbated water insecurity (ICIMOD, 2022). Simultaneously, the rise in unilateral hydroelectric projects, particularly by upstream states, threatens to disrupt the delicate hydrological and ecological balance of the region. In this context, the Brahmaputra represents both a test case and an opportunity for regional cooperation in line with international water law principles, including the Helsinki Rules and the UN Watercourses Convention (United Nations, 1997). Globally, a wide array of theoretical and empirical frameworks has been employed to study transboundary water governance. The Integrated Water Resources Management (IWRM) model, promoted by the Global Water Partnership (2000), emphasizes the coordinated development and management of water, land, and related resources. IWRM has been applied in river basins such as the Nile, Mekong, and Danube to align ecological sustainability with economic development and equitable access. However, criticisms have emerged regarding its applicability in politically complex regions where trust and institutional capacity are limited (Biswas, 2004). Another influential model is the hydro solidarity framework (Falkenmark, 2001), which argues for a basin-wide perspective that emphasizes shared responsibilities and ecological interdependence. The benefit-sharing approach has also gained traction, particularly in Africa's Nile Basin Initiative, by shifting the focus from water sharing to sharing the benefits derived from water use (Sadoff & Grey, 2002). In Asia, the Mekong River Commission is often cited as a partial success in regional cooperation, although challenges persist due to China's non-membership and asymmetrical power dynamics (Sneddon & Fox, 2006). These models underscore the importance of institutional design, power asymmetry, data-sharing, and stakeholder inclusion in managing shared waters sustainably.

In contrast to the extensive body of work on other major transboundary rivers, scholarly attention on the BRB remains relatively limited. Recently, some studies have begun to reveal the unique hydrological, geopolitical, and ecological characteristics of the basin. Hanasz (2017) notes that despite the abundance of water, the Brahmaputra is under growing stress due to increasing demographic pressure, deforestation, and infrastructure development. Singh et al. (2020) identified significant data gaps and mistrust among riparian states as major impediments to integrated basin management. Hydro-political literature also points to growing tensions between India and China over upstream dam construction, such as the Zangmu Dam on the Yarlung Tsangpo (the upper stream of the Brahmaputra). While China has engaged in limited data-sharing agreements, these are often seasonal and lack enforcement mechanisms (Tortajada, 2016). Moreover, Bangladesh as the lowest riparian faces challenges related to water quantity, sedimentation, and saline intrusion, particularly during the dry season. There is also a paucity of coordinated ecological assessments, despite the river's critical role in sustaining floodplain wetlands, fisheries, and endangered species such as the Gangetic dolphin and the greater adjutant stork (WWF, 2019).

This study seeks to generalize the Brahmaputra's water management challenges in terms that resonate with broader international concerns: the fragility of hydro-diplomacy, the environmental costs of unilateral development, and the limitations of current governance regimes in addressing interconnected climate and ecological threats. Despite the growing recognition of transboundary river basins as global public goods, the BRB continues to lack a cohesive framework for sustainable and equitable management. Most existing research focuses on either technical aspects (e.g. sediment transport, glacial melt) or political tensions in isolation. What is missing is a comprehensive, multidimensional analysis that synthesizes environmental, social, economic, and geopolitical factors while also offering actionable recommendations for policy and institutional reform. This study also seeks to fill this gap by offering an integrated assessment of the Brahmaputra's water resource challenges, combining a review of international best practices with an analysis of regional dynamics. The primary research questions guiding this study are: What are the key challenges facing sustainable water management in the BRB? How can regional cooperation and governance be strengthened to manage the basin sustainably? What role do climate change, hydroelectric power projects, and ecological degradation play in shaping the future of the basin? And finally, what institutional frameworks and strategic interventions can be implemented to ensure long-term sustainability and equity in water distribution?

Unlike previous literature that often isolates hydrology from governance or diplomacy, this research bridges those domains to suggest context-sensitive and regionally adaptable strategies. A key novelty of this study is its proposal for the establishment of a Joint River Basin Organization (JRBO) modelled on successful international precedents but tailored to South Asia's complex geopolitical landscape. The JRBO would serve not only as a platform for data sharing and conflict resolution but also as a mechanism to promote ecosystem-based management and inclusive development. The objectives of the study are thus threefold: (i) to identify and critically examine the key challenges of sustainable water management in the BRB; (ii) to assess the current state of regional cooperation and governance mechanisms; and (iii) to propose a strategic roadmap, including institutional innovations and policy reforms, for achieving sustainable and equitable water governance. By contributing new empirical insights and a policy-relevant framework, this research aims to inform both academic debates and practical decision-making in transboundary water governance.

PREVIOUS LITERATURE: WORK DONE SO FAR

Samaranayake et al. (2016) illustrated that the upper riparian countries like China and India focus more on the political aspects of the basin, whereas Bangladesh, being the lowest riparian, is primarily interested in the physical terms of the basin. At present, water cooperation in this basin is limited. Though each riparian has its own domestic considerations, positive interactions can be achieved in the Brahmaputra basin at both bilateral and multilateral levels. This paper depicted that it is absolutely crucial for China, India, and Bangladesh to work cooperatively to prevent future problems. A research paper titled "Transboundary Water Cooperation over the Brahmaputra River" published by the Hague Institute for Global Justice (2015) sheds light on eight action situations of water cooperation surrounding the Brahmaputra River including Track I, II, and III types of cooperation. It stated the key factors affecting each action situation, including basin-wide contexts, some of which are situation-specific to a particular action situation, formal and informal institutions, and actors and agencies. Biswas (2014) focused on the importance of diplomacy and regional cooperation in addressing water challenges. It offered practical tools and frameworks for negotiating transboundary water agreements, building trust, and fostering regional collaboration, particularly relevant for South Asia's complex water landscape, especially in the Brahmaputra Basin. He indicated that hydro-hegemonic mindset of the upper riparian countries and lack of cooperation and mistrust of the lower riparian countries are the primary reasons for not being able to reach a common platform for the stakeholders.

The research "Darkening Waters, Gathering Storm: Sino-Indian Water War on the Brahmaputra River" by Wu and Latham (2021) reflected that the prospect of a conflict in the Brahmaputra Basin is concerning not only because of the grave consequences for the region but also for the global economy and the balance of power in Asia. They expressed their concern by stating that the mistrust and non-cooperation can lead to unimaginable human suffering and further environmental degradation in the region. This research suggested that if such a conflict indeed comes to pass, it may represent the beginning of a new, even more devastating form of resource conflict. They also stated that it is crucial that policymakers in both New Delhi and Beijing recognize the gravity of the situation and also for the stability of the region and the future of Sino-Indian relations depend on it. Similarly, the research article titled "Comparative analysis of trans-boundary water governance of Ganges-Brahmaputra-Meghna (GBM) and La Plata River basins: Issues and concerns" by Rahman (2020) suggested that effective governance of transboundary rivers requires regional-level mechanisms that balance the needs and interests of all basin countries. For instance, the La Plata River basin's governance framework is often cited as a successful example of regional cooperation, where countries engage in pragmatic and mutually beneficial approaches to address their own needs while respecting the interests of other riparian states. This cooperative governance model highlights the importance of evolving governance structures that accommodate political, economic, and cultural considerations, as well as the shared nature of water resources. In the case of the GBM basin, the literature underlined that the future of effective water utilization hinges on collaboration among the riparian countries. Given the interdependence of the nations within the basin, the notion of unilateral control over water resources becomes unsustainable in the long term. Rather, transboundary water management in the GBM basin necessitates cooperation to ensure equitable and sustainable water sharing. This is particularly important considering the complex political, cultural, and security issues that often surround water governance in South Asia. Drawing on the La Plata experience, it is evident that a consistent and evolving governance framework can facilitate successful water sharing and development. Over the decades, the La Plata basin's governance structure has fostered political interrelationships among co-basin countries, providing a model for how such cooperation can lead to mutually beneficial outcomes. The lessons from La Plata suggest that the GBM basin must develop similar mechanisms that view water as a regional public good, which can ultimately benefit the people of South Asia by ensuring equitable and sustainable access to transboundary water resources.

Another study, "Transboundary Water Cooperation over the Brahmaputra River Legal Political Economy Analysis of Current and Future Potential Cooperation" by The Hague Institute of Global Justice (2016), analysed transboundary water cooperation in the BRB, focusing on sociopolitical, economic, and institutional factors influencing cooperation among China, India, Bhutan, and Bangladesh. It examined bilateral agreements and Track II/III initiatives, including the development of hydropower and cross-border trade. Using the Multi-Track Water Diplomacy Framework, the research identified the "Zone of Possible Effective Cooperation"[1] (ZOPEC), which envisions basin-wide cooperation through economic integration and sustainable resource management, factoring in environmental and geopolitical shifts. The study highlighted the potential for a collaborative future amidst complex regional dynamics. Consequently, Rahaman and Varis (2009) emphasized the need for IWRM in the Brahmaputra basin[2] to address development challenges and mitigate the risk of future conflicts. The absence of coordination among riparian countries hampers sustainable development, despite the river's significant hydropower potential. The paper explored the current state of water resources, evaluated the perspectives of riparian nations, and highlighted opportunities for regional cooperation. It also advocated for collaborative approaches to harness the basin's resources for energy security and regional growth.

STUDY AREA

The BRB is one of the most significant transboundary river systems in South Asia due to its ecological, strategic, and geopolitical aspects. Originating in the Angsi Glacier of the Tibet Autonomous Region in China, the river known as the Yarlung Tsangpo flows for over 2900 km through China, India, and Bangladesh before merging with the Ganges and eventually falling into the Bay of Bengal. The basin encompasses an estimated drainage area of around 580,000 km^2 (Figure 13.1), stretching across diverse topographies and climatic zones, ranging from the high-altitude Tibetan Plateau to the low-lying floodplains of Assam and Bangladesh.

The BRB supports a population of more than 130 million people and is a lifeline for communities that depend on it for agriculture, fisheries, transportation, and domestic water use. The river is characterized by an exceptionally high sediment load and one of the highest discharge rates in the world, with an average annual flow of about 612 billion cubic metres (FAO, 2011). Seasonal monsoon patterns contribute to frequent and severe flooding, particularly in the Indian state of Assam and parts of Bangladesh. The basin is also home to unique biodiversity, including several endangered species, and contains important wetlands, floodplain forests, and migratory bird habitats. Despite its ecological and economic importance, the Brahmaputra Basin faces mounting challenges. These include the impacts of climate change, unsustainable infrastructure development, and geopolitical tensions among riparian states. Glacial retreat and erratic rainfall patterns have led to increased flood risks and altered flow regimes. Moreover, large-scale hydropower projects, especially upstream in China and India, are raising concerns over water availability, ecological disruption, and the rights of downstream communities. The absence of a basin-wide governance mechanism further exacerbates the challenges of integrated and cooperative water resource management (Ray et al., 2015).

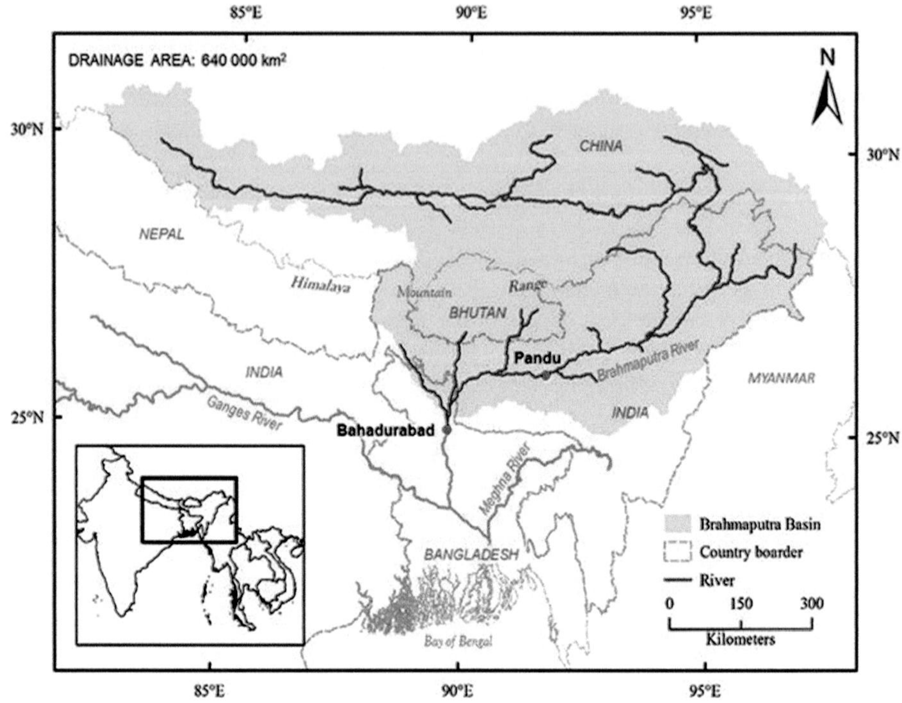

FIGURE 13.1 Map of the Brahmaputra River basin. The focus reach is located between the Pandu discharge station in India and the Bahadurabad discharge station in Bangladesh. (*Source:* Prepared by author based on Fischer et al., 2017.)

OBJECTIVES AND RESEARCH QUESTION

The overarching goal of this study is to explore the key challenges and opportunities for sustainable water resources management in the BRB, with a particular focus on transboundary cooperation, ecological conservation, and institutional governance. To achieve this, the research is guided by several interrelated objectives:

1. To identify the major water management challenges within the BRB, including environmental degradation, hydro-political tensions, and infrastructural pressures.
2. To evaluate the current state of water governance across the basin, examining both national policies and transboundary mechanisms (or the lack thereof).
3. To analyse the implications of climate change for hydrological patterns, flood risks, and long-term water security in the region.
4. To propose feasible strategies including institutional frameworks such as a JRBO for promoting cooperation, resilience, and sustainability.

These objectives lead to the key research questions as discussed in the introduction section and these questions are grounded in the unique characteristics and data emerging from the study area. For instance, Assam experiences annual floods that affect nearly 4.5 million people on average, damaging infrastructure, agriculture, and livelihoods (Government of Assam, 2023). In Bangladesh, seasonal water shortages and salinity intrusion threaten food security and public health in riverine communities (BADC, 2022). Meanwhile, China has constructed several hydropower dams on the Yarlung Tsangpo, the cumulative impacts of which remain poorly understood due to limited data sharing (Tortajada, 2016).

The absence of a formal river basin institution means that much of the basin's management occurs in silos, often driven by national priorities rather than shared regional objectives. The disparities in institutional capacity, political will, and hydrological data among the three countries further complicate efforts towards integrated basin-wide planning. Therefore, understanding the complex dynamics of the Brahmaputra Basin is essential not only for regional sustainability but also for contributing to broader international debates on transboundary water governance, climate resilience, and sustainable development.

METHODOLOGY

This research employs a qualitative approach grounded in a systematic review and synthesis of secondary data sources to explore and analyse the multifaceted dimensions of sustainable water resources management in the BRB. The methodological framework is designed to ensure a rigorous, interdisciplinary, and integrative understanding of the ecological, political, economic, and institutional challenges that affect water governance in this complex transboundary river system.

Research Design

The study is structured as a qualitative, desk-based analytical investigation, focusing on secondary data collection and interpretative synthesis. This approach is particularly appropriate given the study's emphasis on policy analysis, transboundary hydro-politics, and the evaluation of institutional frameworks – domains in which contextual interpretation and critical discourse are paramount (Creswell, 2013). The decision to adopt a non-empirical research design also reflects the accessibility and richness of existing literature and policy materials concerning the BRB.

Data Collection Strategy

The core of the research methodology is a comprehensive literature review and data synthesis, drawing upon a wide array of secondary sources. These include peer-reviewed academic literature from journals such as Water International, Journal of Hydrology, International Journal of Water Resources Development, and Environmental Politics. Books and edited volumes that focus on IWRM, hydro-diplomacy, climate change impacts, and South Asian geopolitics are also considered. In addition, theses and dissertations from universities with strong programme in water resources, international relations, and environmental policy provide valuable insights. Policy papers and reports from global organizations such as the World Bank, United Nations (UN), Global Water Partnership (GWP), International Water Management Institute (IWMI), and the International Centre for Integrated

Mountain Development (ICIMOD) are extensively consulted, alongside government publications and official reports from Bangladesh, India, and China, particularly from the ministries of water resources, foreign affairs, and environment. The study also incorporates NGO and civil society publications, including reports from WWF, IUCN, and local advocacy groups active in the Brahmaputra region. Online academic repositories such as JSTOR, Scopus, ScienceDirect, ResearchGate, and Google Scholar further enrich the data pool. Particular emphasis is placed on collecting materials published in the last two decades (2000–2024) to ensure the information remains current, while landmark theoretical works and international water law documents from earlier periods are included for conceptual grounding.

Search and Selection Criteria

To maintain objectivity and relevance, a systematic search strategy was employed, using carefully selected key search terms such as "Brahmaputra River Basin," "transboundary water governance," "hydro diplomacy in South Asia," "climate change and Himalayan rivers," "hydropower and water politics," "IWRM and institutional reform," "joint river basin organizations," and "China-India-Bangladesh water conflict." These keywords were applied across multiple databases and search engines, with the use of Boolean operators (AND, OR, NOT) to refine and optimize results. The initial search yielded over 400 documents, which were then filtered according to specific inclusion criteria. These criteria included direct relevance to the Brahmaputra Basin or to comparable international cases such as the Mekong and Nile, publication in peer-reviewed journals or by internationally recognized institutions, availability of full-text access, and publication in English or in official South Asian regional languages (translated as necessary).

Thematic Coding and Analytical Framework

The extracted data from the selected sources were categorized and coded into major thematic areas aligned with the research objectives, which included transboundary dynamics and hydro-politics, impacts of climate change and environmental degradation, hydropower development and ecological trade-offs, institutional frameworks and governance gaps, and best practices in international water cooperation. The data coding process was carried out using manual matrix tables, organized by author, year, geographic focus, theoretical orientation, empirical findings, and relevance to the BRB issues. This method enabled cross-comparison of findings across different thematic domains and geographical contexts. Furthermore, triangulation of multiple source types, including academic, governmental, and non-governmental materials, was employed to validate interpretations and ensure robustness in the synthesized outcomes (Denzin, 1978).

Analytical Models and Frameworks Employed

Several internationally recognized analytical models and theoretical frameworks were employed to interpret the data. IWRM was applied to evaluate institutional and policy coherence across sectors and scales (GWP, 2000). Hydro-solidarity and basin-wide cooperation models (Falkenmark, 2001) were used to highlight the importance of collective approaches, while the benefit-sharing model (Sadoff & Grey, 2002) was applied to examine alternatives to volumetric water-sharing by promoting cooperative economic outcomes. The hydro-hegemony framework (Zeitoun & Warner, 2006) was employed to understand power asymmetries among the riparian states, and adaptive governance theory (Pahl-Wostl, 2009) was used to assess resilience and flexibility in current water governance mechanisms. Collectively, these models provided a comparative lens for analysing the BRB in relation to other global transboundary river systems, such as the Mekong, Nile, Jordan, and Danube.

Ethical Considerations

As this study does not involve the collection of primary data from human subjects, the ethical concerns are minimal. However, due diligence has been taken to ensure proper attribution of intellectual property, and all sources have been cited in accordance with the APA (American Psychological Association) 6th edition referencing standards. Further, government reports and data have been verified for authenticity through official digital platforms or institutional archives.

LIMITATIONS OF THE METHODOLOGY

Despite its breadth, the methodology is subject to certain limitations. A language bias is evident, as the research relies primarily on English-language materials and translated documents, which may result in the omission of critical perspectives contained in vernacular publications. There are also access constraints, since not all government data, particularly from upstream regions such as Tibet, were publicly available. Furthermore, the dynamic nature of water politics poses a challenge, as the geopolitical context of the BRB is fluid, and recent developments, such as new dam projects or treaties, may not yet be reflected in the available literature. Another limitation lies in the lack of empirical verification; given the secondary nature of the data, field-level ground-truthing or stakeholder interviews were not feasible, which may restrict the contextual depth of certain policy recommendations. Nonetheless, the methodology is designed to deliver a robust, triangulated, and policy-relevant synthesis of the complex issues affecting the BRB.

RESULT AND DISCUSSION

Analysing the Context of Brahmaputra River Basin from Theory of International Relations

In the context of water scarcity and transboundary river disputes in South Asia, this perspective offers insights into the dynamics shaping the crisis. Within the BRB, shared by China, India, and Bangladesh, the competition for water resources has the potential to escalate tensions and even trigger conflict. Close monitoring of the situation reveals several defining characteristics. The first is the competition for resources.

In the Brahmaputra Basin, the scramble for limited water supplies becomes a central issue, raising the likelihood of disputes. China's construction of dams in the upper reaches, which alter downstream water flows, is often perceived as a strategic move to bolster its own capabilities while disregarding the concerns of downstream states.

Another defining feature is the focus on strategic calculations rather than cooperative approaches. States in the basin frequently engage in manoeuvres designed to maximize their own benefits and reduce vulnerabilities, often through diplomatic posturing, alliances, or covert measures. By virtue of its upstream position, China holds considerable influence over downstream access to water, thereby affecting agriculture, energy production, and broader economic development in India and Bangladesh (Adhikari, 2014). This interplay feeds into a constant security dilemma. China's dam-building activities are frequently interpreted by India and Bangladesh as direct threats to their water security. Such perceptions fuel spirals of mistrust, arms build-ups, and heightened political tensions, reinforcing the entrenched security dilemma in the region. A further complication arises from the overriding concern with national interest and sovereignty. States in the basin tend to prioritize their economic development and security needs, often treating cooperation as a zero-sum game. This rigid focus on sovereignty hinders collaborative responses to shared environmental challenges and heightens the risk of what some have called a "hydro-war 2.0." Taken together, these characteristics suggest that the complex dynamics of water resource management in the Brahmaputra Basin are best understood through a Realist lens, highlighting the role of power politics, strategic positioning, and national interest in shaping inter-state behaviour.

The BRB, spanning China (Tibet), India, and Bangladesh, is of immense ecological, geopolitical, and socio-economic importance. Yet progress towards sustainable water resources management is obstructed by a series of interrelated political, institutional, environmental, and developmental challenges. These barriers collectively undermine regional cooperation, data sharing, and integrated planning, creating a multidimensional set of obstacles that continue to hinder effective collaboration among the riparian countries.

Transboundary Political Tensions and Trust Deficit

One of the most fundamental obstacles to cooperation in the BRB is the absence of trust among the riparian countries, particularly between China and India, and between India and Bangladesh. Despite being hydrologically interconnected, these nations lack a basin-wide legal or institutional mechanism for joint water governance. China and India have long-standing border disputes, especially in Arunachal Pradesh, which China claims as part of South Tibet. This geopolitical tension spills over into water diplomacy, with China often refusing to share comprehensive hydrological data or disclose upstream development plans (Hanasz, 2017). Meanwhile, India and Bangladesh, though signatories to some bilateral water-sharing agreements such as the Ganges Water Treaty of 1996, have not succeeded in establishing a legally binding Brahmaputra agreement. Political sensitivities and persistent fears of upstream control continue to hinder meaningful negotiations. As a result, the lack of trust and institutionalized cooperation fosters unilateral water development strategies and strategic opacity, fuelling suspicion and obstructing integrated basin planning.

Absence of a Joint River Basin Organization

Unlike the Mekong or Nile River Basins, the Brahmaputra lacks a formal joint basin institution that includes all three riparian countries. Although India and Bangladesh operate through the Joint Rivers Commission (JRC), and China shares seasonal flood data, there remains no multilateral platform for coordinated planning, conflict resolution, or benefit-sharing. This institutional vacuum makes it difficult to address the cumulative impacts of upstream activities and undermines long-term sustainability goals. In the absence of a JRBO, data asymmetry continues, particularly with China, which controls the river's source in Tibet and has constructed several upstream dams. Without a neutral, multilateral body, efforts towards cooperative water allocation, early warning systems, joint environmental assessments, and inclusive stakeholder dialogue remain severely constrained.

Unilateral Hydropower Development and Infrastructure Race

The drive for energy security, particularly through hydropower development, has emerged as a significant point of tension in the BRB. China has constructed dams such as the Zangmu Dam and proposed a massive "super dam" near the Great Bend, raising concerns in downstream India about potential water diversion or reduced flows (Tortajada, 2016). In response, India has accelerated its own hydropower initiatives in Arunachal Pradesh, aiming to assert control and reinforce its "prior use" rights under customary international law. Meanwhile, Bangladesh, as the lowest riparian country, faces heightened vulnerability to diminished dry season flows, increased sedimentation, and deteriorating water quality resulting from upstream damming. This escalating race for hydro-sovereignty fosters fragmented development, deepens ecological imbalances, and marginalizes opportunities for cooperative planning or cumulative impact assessments.

Climate Change and Glacial Retreat

The Brahmaputra originates in the Himalayan cryosphere, one of the most climate-sensitive regions on Earth. Global warming is accelerating glacial retreat, altering precipitation patterns, and increasing the frequency of extreme events such as glacial lake outburst floods (GLOFs), droughts, and high-magnitude floods. These climatic shifts intensify challenges related to water availability, sediment transport, and agricultural productivity in downstream regions. Yet the variability induced by climate change remains poorly modelled and inadequately monitored, particularly in upper Tibet, where data is both scarce and politically sensitive. This deepening climate uncertainty introduces a volatile dimension to water planning and underscores the urgent need for adaptive, ecosystem-based governance. Unfortunately, such approaches remain absent due to persistent political fragmentation across the basin.

Institutional Fragmentation and Policy Incoherence

Within countries, water governance is often fragmented across multiple ministries and agencies such as those responsible for water resources, environment, power, and agriculture, resulting in a lack of cohesive oversight. This fragmentation is particularly pronounced in federal systems like India, where poor inter-agency coordination hampers effective decision-making. Conflicting objectives between development priorities, such as energy generation, and conservation goals, like biodiversity protection, further complicate governance. Additionally, national water policies frequently fail to align with regional or transboundary objectives, creating policy discontinuities that undermine integrated approaches. As a result, institutional fragmentation fosters policy myopia, leads to duplication of efforts, and squanders opportunities for cross-sectoral synergies, including integrated flood management and nature-based solutions.

Legal and Normative Gaps in International Water Law Application

Although international frameworks such as the UN Watercourses Convention (1997) and the Helsinki Rules (1966) provide legal guidance on equitable use and the principle of no significant harm, the Brahmaputra riparian countries have yet to fully commit to implementing these norms in a multilateral context. China has not ratified the UN Watercourses Convention, while India has demonstrated only selective engagement with international legal instruments. Moreover, unlike the Indus (India–Pakistan) or the Ganges (India–Bangladesh), the Brahmaputra lacks a basin-specific treaty to govern shared water resources. In the absence of binding legal mechanisms, customary norms remain largely unenforced, leaving cooperative frameworks without the legal authority needed to ensure accountability or foster sustained collaboration.

Inequitable Power Asymmetry

The geopolitical structure of the BRB is defined by asymmetrical power relations, with China occupying the position of upstream hegemon, India functioning as a middle power, and Bangladesh situated as the lowest riparian. China's strategic opacity and its reluctance to engage in multilateral negotiations underscore its dominant upstream status. India, meanwhile, straddles a dual role, downstream of China and upstream of Bangladesh, which creates conflicting interests and often results in inconsistent negotiation behaviour. Bangladesh, at the receiving end of the basin, faces acute downstream vulnerability, disproportionately affected by upstream activities while possessing minimal control or leverage. This hydro-hegemonic configuration suppresses the possibility of equitable negotiations and undermines the principle of "reasonable and equitable use" as advocated by international water law, including the UN Watercourses Convention of 1997.

Possibility of Hydro-war in the Brahmaputra River Basin

The BRB, shared by China, India, Bangladesh, and Nepal, is increasingly emerging as a hotspot for geopolitical tension. Competing water demands, unilateral dam construction, and the absence of binding treaties have heightened the risk of hydro-political conflict, particularly under the compounded pressures of climate change and regional mistrust (Alamghir & Alam, 2024). China, as the uppermost riparian and hydro-hegemon, holds an asymmetric advantage by controlling the Yarlung Tsangpo, the upper course of the Brahmaputra, and developing major hydropower projects in Tibet, including the Zangmu Dam. Of particular concern is the proposed "Super Dam" near the Great Bend, a 60-gigawatt project that could significantly alter downstream water flows. This unilateral development, coupled with China's strategic opacity, has bred deep distrust in India and Bangladesh. Given that water is both a life-sustaining resource and a pillar of national security and economic sovereignty, any perceived manipulation of river flows, such as diversion or sudden release, could provoke diplomatic breakdowns, retaliatory actions, or indirect confrontations. In a scenario of heightened tensions, China's potential use of water as a strategic tool by withholding or releasing flows during monsoon seasons might even be interpreted as a form of hybrid warfare.

India, positioned midstream, finds itself squeezed between China's upstream dominance and Bangladesh's downstream vulnerability. In response to China's dam-building, India has accelerated its own hydropower projects in Arunachal Pradesh, a region disputed by China. This infrastructure race is increasingly driven by zero-sum logic rather than cooperative planning. India has invoked the principle of "prior appropriation rights" and fast-tracked developments like the Upper Siang and Dibang Multipurpose Projects, framing them as strategic counterweights to Chinese initiatives. Without a coordinated basin-wide framework, the proliferation of unilateral dams on both sides risks unpredictable hydrological consequences. In times of crisis, manipulated water flows could damage downstream agriculture or flood vulnerable areas, escalating tensions in a tit-for-tat dynamic.

Water in the Brahmaputra Basin is also becoming securitized in political rhetoric and public discourse. Each country increasingly treats water control as an extension of national sovereignty rather than a shared regional resource. Chinese state media has emphasized "domestic hydropower sovereignty," while Indian strategic thinkers advocate for "harnessing Himalayan waters" to safeguard national interests. This rise in hydro-nationalism fosters an environment where aggressive posturing over river development is normalized. During election cycles or military stand-offs, such nationalistic narratives may push governments towards brinkmanship rather than diplomacy.

Meanwhile, the ecological and human security fallout from climate change and hydrological extremes such as GLOFs, sediment surges, and flash floods is intensifying the vulnerability of millions across the basin. If upstream dams exacerbate or fail during such events, the resulting disasters could be perceived as hostile acts, even if unintended. Historical incidents, like the 2020 glacial flood in Uttarakhand, demonstrate that perception often outweighs intent in fragile geopolitical contexts. In such scenarios, blame games or retaliatory measures may follow. A catastrophic dam breach or accidental flooding linked to upstream development could even trigger military or cyber responses under the doctrine of "anticipatory self-defense," further destabilizing an already volatile region.

The Way Forward

A JRBO is absolutely essential for establishing a unified platform that brings together both upper and lower riparian countries of the BRB (Figure 13.2). Such an institution would enable open dialogue among all stakeholders, fostering trust and improving communication on shared water resource challenges. Beyond

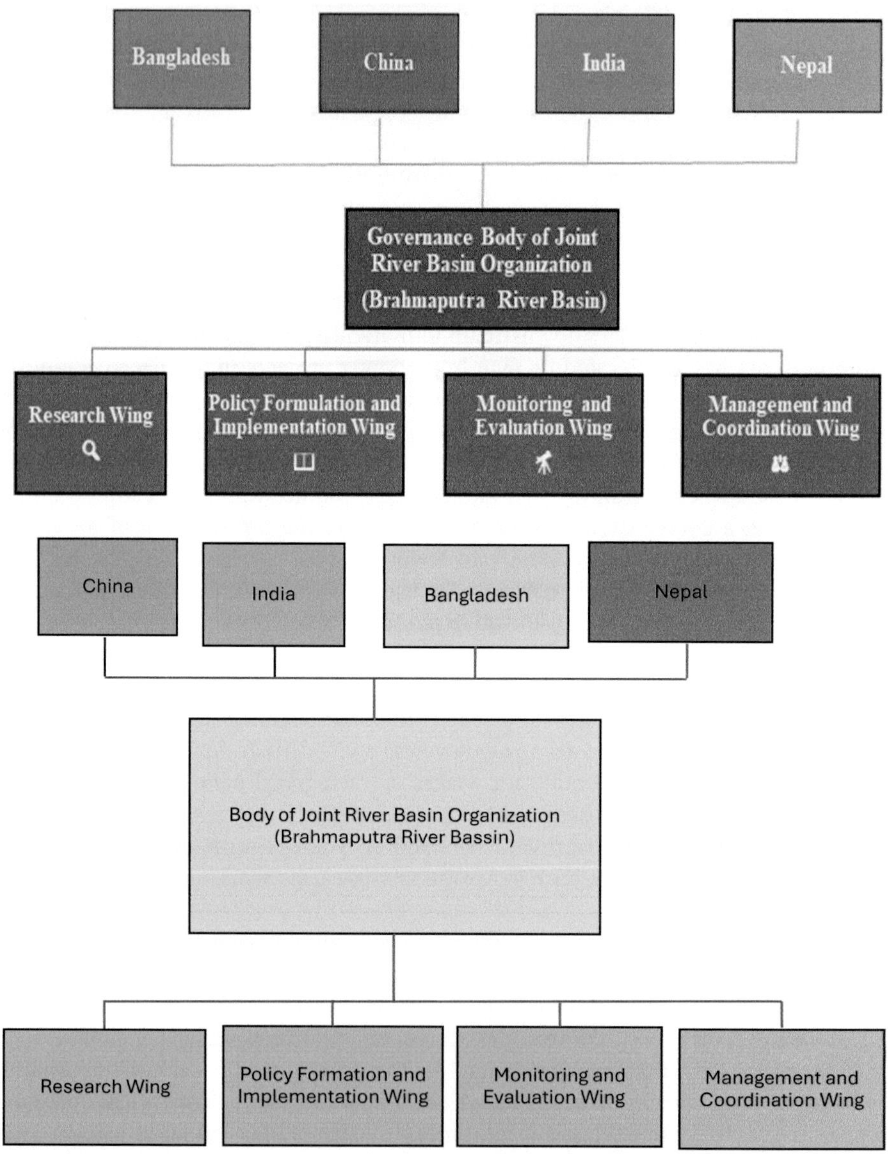

FIGURE 13.2 Conceptual framework of the proposed JRBO for the BRB. (*Source:* Prepared by authors.)

facilitating conversation, an RBO could spearhead cooperative water management by designing and implementing basin-wide strategies for flood control, drought mitigation, and equitable water allocation. It would also play a critical role in standardizing data collection and promoting transparent data sharing, thereby enhancing collective understanding of the river's dynamics. Importantly, an RBO could serve as a neutral body for conflict resolution, offering structured mechanisms for mediation and preventing water-related disputes from escalating into broader geopolitical tensions.

Governance Body of the Joint River Basin Organization for the Brahmaputra River Basin

The proposed Governance Body of the JRBO is envisioned as an inclusive, cooperative, and multi-tiered institutional mechanism representing Bangladesh, India, China, and Nepal. Its formation will be grounded in principles of consensus and sovereign equality, ensuring that each country's interests are equitably represented while prioritizing transboundary water sustainability and regional resilience. The Governance Body will be established through a multilateral treaty endorsed by all four riparian nations, anchored in international water law frameworks such as the Helsinki Rules and the UN Watercourses Convention. Each country will ratify the agreement through its respective legislative process, followed by the appointment of national representatives via an inter-ministerial panel composed of experts in water resources, environmental policy, diplomacy, and regional cooperation. To ensure balanced influence and shared ownership, the body will operate under a rotating presidency with biennial leadership transitions.

The institutional composition of the Governance Body will include several key components. At its apex will be the Council of Ministers (CoM), comprising the Water, Foreign Affairs, and Environment Ministers from each member country, serving as the principal policymaking forum. Supporting this will be the Technical Advisory Panel (TAP), made up of multidisciplinary experts in hydrology, climate science, disaster risk management, and socio-economics, nominated by the member states. A Stakeholder Forum (SF) will ensure inclusivity by engaging civil society representatives, indigenous communities, academic institutions, and private sector observers. The Secretariat, a permanent and neutral administrative unit, will be based in a jointly selected riparian city and will facilitate day-to-day operations.

The scope of work for the Governance Body will be comprehensive and action-oriented. It will develop and oversee basin-wide integrated water governance policies, facilitate data sharing, joint research initiatives, and early warning systems, and mediate disputes to ensure equitable water allocation. Additionally, it will promote climate-resilient infrastructure planning and ecological conservation, coordinate funding mechanisms including climate finance and development aid, and engage with global institutions for technical and financial support. This governance architecture embodies participatory decision-making, technical excellence, and cooperative sovereignty laying the foundation for a peaceful, resilient, and future-proof BRB.

Research Wing of the Joint River Basin Organization – Structure and Function

The Research Wing of the JRBO is envisioned as the scientific backbone of transboundary water governance in the BRB. Its core mission is to generate evidence-based insights that inform policy formulation, technical planning, and sustainable development across the four riparian nations, i.e. Bangladesh, India, China, and Nepal. Structured as a multidisciplinary, semi-autonomous body under the JRBO Secretariat, the Research Wing will consist of five thematic divisions: Hydrology & Water Quality, Climate Change & Disaster Risk, Ecology & Biodiversity, Socio-economic & Livelihood Studies, and Governance & Policy. Each division will be led by a researcher from one of the member countries, with leadership rotating to ensure equitable representation and shared ownership.

The Research Wing will engage in a wide range of scientific and policy-relevant activities, including integrated hydrological modelling and forecasting, sediment transport analysis, and studies on river morphology. It will assess climate change impacts on river flow and glacial melt and develop strategies for transboundary disaster risk management, particularly in the context of floods and landslides. Additional focus areas include ecosystem services and aquatic biodiversity, the "water–energy–food" nexus, basin-wide livelihood assessments, and the integration of social equity, indigenous rights, and gender considerations into water governance. The wing will also contribute to institutional harmonization and policy innovation across the basin.

To fulfil its mandate, the Research Wing will require a diverse team of experts, including hydrologists, glaciologists, remote sensing analysts, environmental economists, sociologists, ecologists, GIS specialists, and policy researchers. Each thematic division will be staffed by a balanced mix of senior scientists, postdoctoral researchers, and technical associates working collaboratively across disciplines and borders. Recruitment will be conducted through a merit-based, transparent, and country-inclusive process. Open calls for applications will be disseminated via academic platforms, national research councils, and the JRBO website. Selection committees will include independent international experts and national nominees to ensure impartiality. Gender balance, regional diversity, and interdisciplinary expertise will be mandatory criteria in final appointments. Contractual terms will accommodate both short-term consultancy roles and long-term research fellowships, fostering a dynamic and inclusive research environment.

Policy Formulation and Implementation Wing of the Joint River Basin Organization

The Policy Formulation and Implementation Wing of the JRBO will serve as the strategic nucleus of the institution, tasked with translating scientific research, diplomatic agreements, and stakeholder contributions into actionable, basin-wide policies. Its central role is to ensure alignment between regional priorities and national water strategies across Bangladesh, India, China, and Nepal. Composed of senior policy analysts, legal advisors, transboundary water governance specialists, institutional development experts, and liaison officers from each member country, the wing will operate under the leadership of a Director of Policy. Supporting this leadership will be four thematic sub-units: Water Allocation and Regulation, Legal and Institutional Harmonization, Stakeholder Engagement and Equity, and Monitoring & Evaluation (M&E), each with designated focal points from all member states to guarantee inclusivity and shared ownership.

The scope of work for this wing is comprehensive and integrative. It will develop basin-level protocols, guidelines, and policy frameworks addressing water sharing, flood management, sediment control, and ecological protection, all in accordance with international conventions and regional needs. Legal harmonization will be a key function, with the wing recommending reforms to align national laws with basin-level agreements to ensure enforceability and consistency across jurisdictions. Strategic integration with the Research Wing will ensure that all policies are grounded in robust scientific evidence and data, while also embedding climate resilience, gender sensitivity, and indigenous rights into every policy domain. The wing will actively engage stakeholders, i.e. local communities, civil society, and the private sector, through structured dialogues that incorporate bottom-up perspectives into decision-making. It will also establish a basin-wide performance framework to monitor policy implementation, track compliance, and recommend corrective actions. To remain responsive to evolving conditions, the wing will implement adaptive policy mechanisms, including periodic reviews to revise policies based on new data, shifting hydrological patterns, or emerging regional challenges. Ultimately, this wing will act as the vital bridge between knowledge, diplomacy, and action ensuring that policy is not only thoughtfully crafted but also effectively implemented, continuously monitored, and dynamically adapted to meet the complex and changing realities of the BRB.

Management and Coordination Wing of the Joint River Basin Organization

The Management and Coordination Wing of the JRBO will serve as the operational nerve centre, responsible for the seamless execution of the organization's vision, programmes, and institutional mandates. It will provide essential administrative, logistical, and intergovernmental coordination support across all JRBO units and member countries, Bangladesh, India, China, and Nepal. Led by a Chief Operations Coordinator, this wing will encompass units dedicated to Administrative Management, Programme Coordination, Finance and Budgeting, Information and Communications, and International Cooperation. To promote transparency, inclusivity, and shared ownership, representatives from each member state will be embedded in key functional roles. Recognizing the linguistic diversity of the basin, the wing will maintain a bilingual or multilingual secretariat staff to ensure effective communication.

The scope of work for this wing is both comprehensive and integrative. It will oversee operational planning and monitoring of annual work plans and budgets, ensuring alignment with strategic priorities set by the CoM. Programme coordination will involve synchronizing activities across the Research Wing, Policy Wing, and stakeholder forums, facilitating joint project implementation, data exchange protocols, and coordinated field operations in transboundary areas. As the primary institutional liaison, the wing will manage communication between JRBO and national agencies, resolving inter-institutional issues in a timely manner. It will also be responsible for financial and administrative management, including procurement, human resources, and reporting, all conducted in accordance with jointly agreed standards and donor requirements. Information management will include maintaining a centralized basin knowledge portal, coordinating public communications, and supporting real-time data dissemination for early warning systems and project tracking. In times of crisis, such as floods, landslides, or infrastructural failures, the wing will facilitate rapid inter-country coordination to ensure an effective response. Through structured, impartial, and efficient coordination, the Management and Coordination Wing will act as the backbone of the JRBO, harmonizing planning, execution, and diplomacy in real time to advance sustainable and resilient basin management.

Basin-wide Holistic Approach

The basin-wide holistic approach has been proved largely successful and effective in various water-stressed regions of the globe. The successful implication of the Mekong River Agreement, Cooperative Management Programs of the Senegal River and the Niger River Basin Water Charter, and SADC Revised Protocol on Shared Watercourses can be a guiding force for the basin-wide approach for the BRB (Table 13.1). These approaches, as well as the procedural standards developed in line with the codified laws of international watercourses, should be strongly followed for a fair share for the upper and lower riparian countries of the BRB. A fundamental component of effective basin management is the formulation and execution of a comprehensive basin action plan. This process involves establishing systematic procedures to explore various water resource management options, selecting the most appropriate among them, and ensuring their implementation. A crucial aspect of this approach is adaptive learning where insights gained during implementation are continuously integrated back into the planning cycle to refine strategies and improve outcomes. The basin action plan functions as a strategic guide for the sustainable management of water resources within a river basin, typically covering a planning horizon of three to six years. Its development relies on inclusive dialogue and consensus-building among a wide range of stakeholders, including government agencies, local authorities, the private sector, farmers, civil society organizations, and individual water users. Once the plan has been deliberated and agreed upon, it is officially sanctioned by the basin organization, which assumes a coordinating role.

This action-oriented blueprint delineates the specific responsibilities of various actors, establishes frameworks for cost-sharing and financial contributions, sets lines of accountability, and defines channels for effective communication and information exchange. Rather than serving as a static list of desired projects, the basin action plan reflects a dynamic, coordinated, and multi-sectoral effort aimed at achieving long-term sustainability and resilience in water governance. Rather than being a mere compilation of potential projects, the action plan integrates a balanced combination of infrastructure initiatives, maintenance activities, and non-structural interventions. These may include legislative reforms, regulatory updates, pricing strategies, institutional capacity-building, training programmes, and other supportive measures aimed at fostering effective water governance.

TABLE 13.1
Proposed Tools for Integrated Basin Action Plan for Brahmaputra River Basin

Tools for Basin-Wide Planning & Water Governance	Tools for Conflict Mitigation & Resolution	Tools for Cross-Border Communication & Engagement
Basin-scale hydro forecasting and scenario modelling	Joint fact-finding missions and hydrological audits	Real-time data sharing via cloud-based platforms
Integrated Decision Support Systems (DSS) using GIS & Remote Sensing	Third-party facilitation for dispute resolution	JRBO-wide internal communication system (intranet)
Joint river gauging and climate monitoring stations	Rapid response committees for transboundary emergencies	Regular video conferencing and technical webinars
Harmonized Environmental Impact Assessment (EIA) and Strategic Impact Assessment (SIA) reviews	Dispute resolution panel with rotating national chairs	Monthly bulletin with basin updates and technical briefs
Transnational clearance mechanisms from the Joint River Basin Organization for major infrastructure	Multi-stakeholder conflict mediation involving basin communities	Public-facing web portal with multilingual support
Joint funding mechanisms and cost-sharing protocols	Sub-basin peace forums and cross-border dialogues	Community radio and mobile alerts in flood zones
Intergovernmental basin planning boards	Cultural exchange programmes to foster mutual trust	Social media engagement and knowledge-sharing campaigns
Synchronized flood release protocols and dam safety agreements	Local-level grievance redress and traditional dispute practices	Virtual knowledge repository and open-access data hub
Joint training for water engineers, planners, and technicians	Legal harmonization workshops on water laws	Stakeholder mapping and participatory outreach tools
Shared project planning units embedded in national water ministries	Civil society and NGO coordination mechanisms	Participatory app for citizen science and feedback

Source: Prepared by authors.

Functioning as the central framework for water management within the basin, the action plan provides a structured guide for coordinated efforts. While the plan assigns clear responsibilities for implementation, it is essential to recognize that the basin organization does not directly execute all components. Instead, its principal role lies in facilitating collaboration and coordination among various actors. For example, in the context of a pollution reduction strategy, responsibilities might be distributed among municipal councils, industrial waste managers, agricultural stakeholders, environmental planners, regulatory agencies, conservation bodies, and recreational authorities. The basin organization, in this scenario, would act as the orchestrator, ensuring that each entity knows its role, the timing and location of its actions, and the shared objectives of the initiative. The development of a basin management plan must be approached with the intent of ensuring its active implementation, rather than allowing it to become a neglected or symbolic document. To prevent this, it is essential to integrate a robust system of reporting and accountability within the plan itself. For instance, both national and transboundary basin management frameworks should clearly define the nature of the information to be reported to national authorities, along with the frequency and format of these reporting cycles. Furthermore, to uphold transparency and credibility, the basin organization should be subject to periodic audits conducted by an independent body. These audits should critically evaluate the outcomes and effectiveness of the management plan and, where necessary, propose adjustments or reforms to enhance performance and ensure that the objectives of sustainable water governance are being effectively pursued.

Effective implementation of a basin management plan requires the basin organization to coordinate the diverse actions of multiple stakeholders, each with distinct roles and responsibilities. Consider the case of irrigation: ensuring that water reaches the farmer involves a chain of coordinated efforts between the reservoir operator, the water distribution authority, the farmers, and potentially an environmental regulator. However, these functions are often distributed among various government departments or, particularly in developing countries, may even fall outside formal government oversight. As a result, communication among these entities is frequently limited or entirely absent. In expansive river basins, certain actors involved in the process may be unaware of others' roles or existence, leading to a fragmented approach to water management. This lack of integration often results in isolated, localized solutions that fail to consider broader basin-wide impacts or synergies. To mitigate such fragmentation, it is essential for basin managers to embed coordination mechanisms within the overall basin plan. Additionally, individuals and institutions responsible for implementing these tasks must possess the necessary skills to work collaboratively across sectors and disciplines. Strengthening institutional and human capacity through training, knowledge sharing, and cross-sectoral planning is therefore critical to fostering effective, integrated basin management.

Diplomatic Initiatives on Various Levels

Establishing a platform for ongoing diplomatic dialogues and collaborations among the basin countries to foster understanding, trust, and cooperation, Track II and III diplomacy offer alternative approaches to address transboundary water challenges, complementing traditional government-to-government (Track I) diplomacy. These non-state actors and informal processes can play a crucial role in fostering dialogue, building trust, and developing cooperative solutions. Building trust and fostering dialogue among the researchers, academicians, and gradually among the government officials and politicians can be an efficient step for equitable water resources management. As a turbulent river, the Brahmaputra provides a potential platform for generating electricity from hydropower. The riparian countries have the right to use water potential to the fullest. But it is to be noted that the dams built on the rivers should be built through joint collaboration and ensuring that these dams do not pose harm to the lower riparian countries. If the countries can reach a suitable location for the dams where building a dam would be useful for all and harmful to none, and the electricity can later be shared among all the riparian countries, it would be a win-win situation for all. Joint ventures in hydropower can provide clean energy for all countries involved, promoting economic development while minimizing negative impacts on downstream regions.

Flood is a common hazard for the riparian countries of the Brahmaputra. But investing in advanced flood management systems and early warning mechanisms by these riparian countries to mitigate the impact of floods in downstream areas is a key to sustainable and cooperative riverine flood management. This will eventually increase joint venture initiatives, reduce mistrust, and promote cooperation among the riparian countries. Promoting the construction of environmentally friendly infrastructure projects that consider the ecological health of the river basin is essential as sustainable infrastructure development ensures that economic growth does not come at the expense of environmental degradation, benefiting all countries in the long run. The borders can only partition the geographical boundary, not the ecosystem. All the riparian countries should realize that ecosystem of the Brahmaputra will only be preserved when a development plan is initiated integrating all these countries. Involving local communities in decision-making processes and ensuring that the benefits of development projects are equitably distributed. Engaging local communities fosters a sense of ownership and ensures that development projects are aligned with the needs and aspirations of the people directly affected. Realizing the needs of the people living beside and depending on the Brahmaputra River should get the highest priority while initiating any development projects on the river.

Scientific Research and Data Sharing

Encouraging collaborative scientific research on the Brahmaputra's hydrology, ecology, and climate patterns and facilitating data sharing among the basin countries should be promoted as informed decision-making relies on accurate data. Shared scientific knowledge can lead to better policies and management practices. First, year-round information sharing. The exchange of hydrological data round the year will help in improving the quality of flood and drought forecasting for all four riparian states. Second, develop flood forecasting and early warning systems through the exchange of real-time data on river flow from upstream areas of the basins. The benefits of data-sharing among the riparian countries have been demonstrated and well observed in the Danube basin (Rasul, 2015). The Comprehensive Data Sharing Framework visualized in the concept map (Figure 13.3) is designed to facilitate seamless, secure, and inclusive data exchange among stakeholders in a transboundary river basin context, such as the Brahmaputra. It is structured around three core pillars: Common Website, Data Governance, and Stakeholder Engagement.

The Common Website functions as a centralized digital platform, providing real-time data access, user-specific controls, interactive data visualization tools, and a

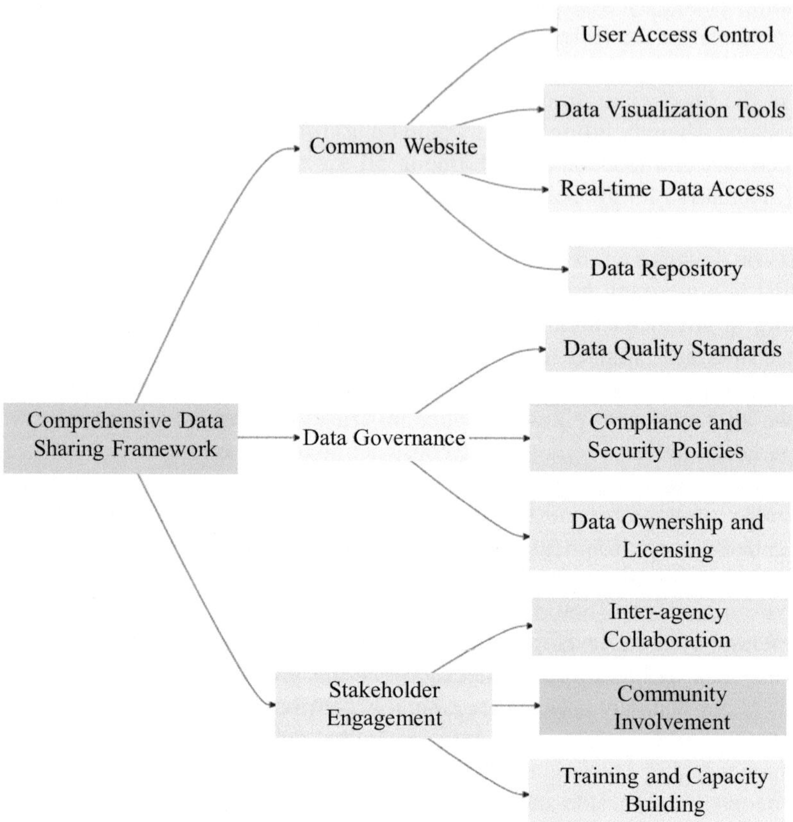

FIGURE 13.3 Efficient data sharing framework for Brahmaputra River Basin. (*Source:* Prepared by authors.)

unified repository for hydrological and environmental data. This ensures transparency and accessibility for both technical and policy users. Data Governance forms the institutional backbone, ensuring that shared data meets uniform quality standards, complies with security and confidentiality policies, and respects ownership and licensing rights of contributing countries or agencies. This ensures trust, accountability, and legal clarity in data use. Stakeholder Engagement focuses on participatory inclusion through inter-agency collaboration, community involvement, and sustained capacity building. This promotes ownership at all levels, encourages local data contributions, and ensures long-term sustainability of the system.

Overall, this framework fosters multi-level integration, reduces information asymmetry, and enhances regional cooperation, making it essential for data-driven, conflict-sensitive water governance in complex river systems. Addressing the challenges of the BRB requires a multifaceted approach that prioritizes collaboration, sustainability, and inclusivity. By implementing these creative and effective solutions, the countries sharing the basin can work together towards a fair and equitable distribution of resources.

CONCLUSION

The BRB stands at the crossroads of opportunity and peril rich in water resources and ecological diversity, yet constrained by entrenched geopolitical tensions, fragmented governance structures, and accelerating climate vulnerabilities. As demonstrated throughout this study, achieving sustainable water resources management in such a transboundary setting demands a deliberate and strategic pivot away from unilateralism and short-term development priorities towards an integrative, adaptive, and cooperative paradigm. This research has shown that the challenges of managing the Brahmaputra's waters cannot be isolated within disciplinary or national silos. Rather, they are inherently transboundary, multi-sectoral, and deeply political. The absence of a multilateral institutional framework, such as a JRBO, perpetuates a vacuum in data transparency, joint planning, and equitable benefit-sharing, thereby reinforcing power asymmetries and downstream vulnerabilities. Similarly, the proliferation of unilateral hydro-infrastructure projects and the inconsistent application of international water law norms only serve to deepen mistrust and magnify ecological risks.

Yet the path forward is not without precedent or promise. International experiences from the Mekong, Nile, and Danube basins reveal that even in regions marked by asymmetry and contention, cooperative water governance can be built incrementally through inclusive dialogue, shared data regimes, and co-designed action plans. The proposed establishment of a Brahmaputra Basin Authority, underpinned by both state and non-state diplomatic tracks, offers a feasible institutional entry point to catalyze such a transformation. Furthermore, the findings of this chapter call attention to the urgency of embedding climate resilience and ecosystem-based approaches into the heart of policy formulation. The Brahmaputra's future cannot be secured by engineering solutions alone. It requires a values-based rethinking of water not as a contested commodity but as a shared ecological and cultural heritage. Strengthening local participation, especially among indigenous and marginalized communities, and aligning basin development with the Sustainable Development Goals (SDGs) are essential to legitimizing and sustaining any governance arrangement. Ultimately, the Brahmaputra is more than a hydrological system; it is a geopolitical reality, an ecological artery, and a sociocultural entity. Its management must therefore be approached not merely as a technical challenge but as a test of political imagination and regional solidarity. By acknowledging interdependence, respecting asymmetries, and institutionalizing cooperation, riparian nations can transform the Brahmaputra from a site of latent conflict into a model of transboundary sustainability and resilience. The stakes are high, but so too is the potential for shared prosperity and peace in the region.

NOTES

1. https://climate-diplomacy.org/magazine/cooperation/transboundary-water-cooperation-over-brahmaputra-river-legal-political-economy
2. https://onlinelibrary.wiley.com/doi/abs/10.1111/j.1477-8947.2009.01209.x

REFERENCES

Adhikari, K. N. (2014). Conflict and cooperation on South Asian water resources. *IPRI Journal, 14*(2), 45–62.

Alamghir, M., & Alam, K. (2024). *Hydro-politics in South Asia: Nature, current status and overcome policies.* https://doi.org/10.13140/RG.2.2.12974.01606

BADC. (2022). *Annual Report 2021–2022.* Bangladesh Agricultural Development Corporation. Ministry of Agriculture, Government of Bangladesh. https://www.badc.gov.bd/

Biswas, A. K. (2004). Integrated water resources management: A reassessment. *Water International, 29*(2), 248–256. https://doi.org/10.1080/02508060408691775

Biswas, A. K. (2014). *Water Diplomacy: A Primer for the 21st Century.* Routledge.

Creswell, J. W. (2013). *Qualitative Inquiry and Research Design: Choosing among Five Approaches* (3rd ed.). Thousand Oaks, CA: Sage.

Denzin, N. K. (1978). *The Research Act: A Theoretical Introduction to Sociological Methods* (2nd ed.). New York, NY: McGraw-Hill.

Falkenmark, M. (2001). The greatest water problem: The inability to link environmental security, water security and food security. *International Journal of Water Resources Development, 17*(4), 539–554. https://doi.org/10.1080/07900620120094073

FAO. (2011). *Aquastat: Brahmaputra River Basin.* Food and Agriculture Organization of the United Nations. http://www.fao.org/nr/water/aquastat/basins/index.stm

Fischer, S., Pietroń, J., Bring, A., Thorslund, J., & Jarsjö, J. (2017). Present to future sediment transport of the Brahmaputra River: Reducing uncertainty in predictions and management. *Regional Environmental Change, 17,* 515–526.

Global Water Partnership. (2000). *Integrated water resources management.* TAC Background Paper No. 4. https://www.gwp.org/globalassets/global/toolbox/publications/background-papers/04-integrated-water-resources-management-2000-english.pdf

Government of Assam. (2023). *Assam State Disaster Management Authority Flood Report 2023.* Assam State Disaster Management Authority (ASDMA). https://asdma.assam.gov.in/

Hague Institute for Global Justice. (2016). *Transboundary Water Cooperation over the Brahmaputra River: Legal Political Economy Analysis of Current and Future Potential Cooperation.* The Hague Institute for Global Justice. https://thehagueinstituteforglobaljustice.org

Hanasz, P. (2017). The politics of water governance in the Brahmaputra basin: An Indian perspective. *Observer Research Foundation Occasional Paper, 126.* https://www.orfonline.org/research/politics-water-governance-brahmaputra-basin-indian-perspective/

Pahl-Wostl, C. (2009). A conceptual framework for analysing adaptive capacity and multi-level learning processes in resource governance regimes. *Global Environmental Change, 19*(3), 354–365.

Pradhan, N. S., Das, P. J., Gupta, N., & Shrestha, A. B. (2021). Sustainable management options for healthy rivers in South Asia: The case of Brahmaputra. *Sustainability, 13*(3), 1087.

Rahaman, M. M., & Varis, O. (2009). Integrated water management of the Brahmaputra basin: Perspectives and hope for regional development. *Natural Resources Forum, 33*(1), 60–75. https://doi.org/10.1111/j.1477-8947.2009.01208.x

Rahman, M. A. (2020). Comparative analysis of trans-boundary water governance of Ganges-Brahmaputra-Meghna (GBM) and La Plata River basins: Issues and concerns. *Water Policy, 22*(S1), 59–74. https://www.biiss.org/storage/uploads/pdfs/6a921edfc513abd9abdd7ecbf4c367c2.pdf

Rasul, G. (2015). Water for growth and development in the Ganges, Brahmaputra, and Meghna basins: an economic perspective. *International Journal of River Basin Management, 13*(3), 387–400. https://doi.org/10.1080/15715124.2015.1012518

Ray, P. A., Yang, Y. C. E., Wi, S., Khalil, A., Chatikavanij, V., & Brown, C. (2015). Room for improvement: Hydroclimatic challenges to poverty-reducing development of the Brahmaputra River basin. *Environmental Science and Policy*, *54*, 64–80.

Roy, J. et al. (2019). Exploring Futures of the Hindu Kush Himalaya: Scenarios and Pathways. In: Wester, P., Mishra, A., Mukherji, A., & Shrestha, A. (eds.), *The Hindu Kush Himalaya Assessment*. Springer, Cham. https://doi.org/10.1007/978-3-319-92288-1_4

Sadoff, C. W., & Grey, D. (2002). Beyond the river: The benefits of cooperation on international rivers. *Water Policy*, *4*(5), 389–403.

Samaranayake, N., Limaye, S., & Wuthnow, J. (2016). *Water Resource Competition in the Brahmaputra River Basin*. Arlington, VI, USA: CNA Analysis and Solutions. https:// https://www.cna.org/reports/2016/IRM-2016-U-013097.UNSECURED.pdf

Singh, A., Thakkar, H., & Thomas, V. (2020). Transboundary water governance in the Brahmaputra basin: Towards a basin-wide approach. *Water Policy Research Working Paper Series*. South Asia Consortium for Interdisciplinary Water Studies (SaciWATERs). https://saciwaters.org

Sneddon, C., & Fox, C. (2006). Rethinking transboundary waters: A critical hydropolitics of the Mekong Basin. *Political Geography*, *25*(2), 181–202. https://doi.org/10.1016/j.polgeo.2005.11.002

The Hague Institute for Global Justice. (2015). *Transboundary Water Cooperation over the Brahmaputra River*. The Hague Institute for Global Justice. https://thehagueinstitute-forglobaljustice.org

Tortajada, C. (2014). Water infrastructure as an essential element for human development. *International Journal of Water Resources Development*, *30*(1), 8–19. https://doi.org/10.1080/07900627.2014.888636.

UNESCO. (2021). *Transboundary Waters: Sharing Benefits, Sharing Responsibilities*. United Nations Educational, Scientific and Cultural Organization. https://unesdoc.unesco.org

United Nations. (1997). *Convention on the Law of the Non-navigational Uses of International Watercourses*. Adopted by the UN General Assembly. https://legal.un.org/ilc/texts/instruments/english/conventions/8_3_1997.pdf

Wu, A., & Latham, A. (2021). Darkening waters, gathering storm: Sino-Indian water war on the Brahmaputra River. *The Diplomat*. https://thediplomat.com/2021/02/darkening-waters-gathering-storm-sino-indian-water-war-on-the-brahmaputra-river/

WWF. (2019). *Securing a Living River: A Strategy for Conservation and Sustainable Development in the Brahmaputra Basin*. World Wide Fund for Nature. https://www.wwfindia.org

Yang, Y. E., Wi, S., Ray, P. A., Brown, C. M., & Khalil, A. F. (2016). The future nexus of the Brahmaputra River Basin: Climate, water, energy and food trajectories. *Global Environmental Change*, *37*, 16–30.

Zeitoun, M., & Warner, J. (2006). Hydro-hegemony: A framework for analysis of transboundary water conflicts. *Water Policy*, *8*(5), 435–460.

14 Conflict, Cooperation, and Climate Resilience in the Governance of South Asia's Transboundary Rivers

Mehebub Sahana and Sk Ajim Ali

REFRAMING TRANSBOUNDARY WATER GOVERNANCE IN SOUTH ASIA

The Ganga–Brahmaputra–Meghna (GBM) and the Indus River systems serve as more than simply water supplies; they also help to define the identities of the communities that live along their banks, sustain economies, and influence cultures. Riverine people have depended on these waterways for generations for transportation, trade, agriculture, and fishing, incorporating their customs and religious beliefs into the rivers' organic rhythm. However, these lifelines are under severe threat from floods, droughts, water stress, landslides, and more, stretching from the Khyber Pakhtunkhwa region in Pakistan to Arunachal Pradesh in India and the Chittagong Hill Tracts in Bangladesh. Recently, multiple catastrophic glacial lake outburst flood (GLOF) events have raised alarm across the Himalayan region (Ahmed et al., 2021; Allen et al., 2016). For example, the South Lhonak GLOF in Sikkim, India (2023) led to a massive Teesta flood, washing away many bridges, towns, and the ₹16,000 crore Teesta-III dam project (Saha et al., 2025). Similarly, the Mustang and Humla GLOFs in Nepal (8 July 2025) washed away a key China-built friendship bridge and crippled hydropower projects, while the Dharali, Uttarakhand cloudburst (2025) caused further devastation. Warming temperatures are increasing the risk of similar events, including GLOFs, flash floods, and landslides in the Himalayan region, while riverine floods downstream have become a regular phenomenon during every monsoon. The delicate balance between people and rivers has been disrupted by climate change, excessive population growth, and poor water governance and cooperation among riparian countries. Uncontrolled riverbank erosion displaces thousands of communities annually (Kaiser, 2023), while irregular monsoon patterns, protracted droughts, and more intense flash floods have made water supplies uncertain. Although they have historically been crucial in maintaining and safeguarding these water systems, women, indigenous peoples, and small-scale fishers are frequently the most at risk, and their voices are rarely heard in decision-making processes. Local community leadership, participatory governance, and regional collaboration

 DOI: 10.1201/9781003599692-14

are essential to conserve these rivers as the effects of climate change worsen. By enhancing cooperation and information exchange, the GBM and Indus rivers can continue to sustain life for future generations.

In this regard, the edited book *Rivers Beyond Borders: Environmental Sustainability in South Asia's Transboundary Waters* brings together an array of viewpoints on the complex relationship between politics, the environment, and water governance throughout South Asia's shared river basins. The Indus, GBM, Koshi, and Mahakali and Teesta, and their tributaries are lifelines that transcend political boundaries, cultural landscapes, and natural zones. However, as a result of climate change and extreme weather events these shared waterways are increasingly the subject of conflict, fragmented cooperation, and vulnerability. This edited collection brings together several perspectives to analyse the complex relationships among hydro-politics, water diplomacy, legal frameworks, stakeholder involvement, and environmental sustainability across the region. Together, the chapters highlight how important the region's rivers are to millions of people, while also being intricately linked to ecological, sociopolitical, and historical issues. Asymmetries of power, knowledge, and access influence the region's water politics, from colonial legacies rooted in hydro-social changes in the Bengal delta between India and Bangladesh to geopolitical tensions between China and India over the Brahmaputra River and between India and Pakistan over the Indus River. One recurring theme in all the contributions is that conflicts, unilateralism, and diplomatic impasses continue to shape transboundary water governance in South Asia more so than cooperative basin-wide approaches.

In order to situate the challenges within a framework of disputed sovereignty, unequal power relations, and limited regional integration, the opening chapters examine transboundary disputes, cooperation, and sustainability pathways throughout South Asia. The possibilities and limitations of institutional systems for conflict resolution are demonstrated by case studies from Bangladesh and India (via the involvement of the NRCC and the Teesta discussions). Analysis of the Brahmaputra issue and the river conflicts between China and India also reveals how unequal power dynamics and geopolitical mistrust complicate prospects for hydro-diplomacy. A *longue durée* perspective is provided by the Bengal Delta case, which shows how water politics and hydro-social connections have been permanently impacted by colonial land regimes and sociopolitical shifts. Ethics and governance frameworks are the subject of other chapters. The discrepancy between official treaties and actual practices is emphasised in the critical study of hydro-diplomacy in South Asia, the appraisal of India's ethical position in managing its waterways, and the assessment of the legal frameworks between India and Bangladesh. They demonstrate that rather than being integrated, flexible, or sustainability-oriented, current agreements are frequently chaotic, issue-specific, and reactionary. In addition, the chapters on the Brahmaputra River Basin and the borders between Nepal and India highlight the many rationalities and stakeholder complexities that can make governance "clumsy" and difficult to reconcile with basin-wide collaboration.

More emphasis is placed on environmental and ecological aspects in several chapters, including the Ganga Basin's cryosphere-delta connections and the Brahmaputra's sustainable water management. These contributions emphasise how the Himalayan cryosphere serves as both a water source and a location for emerging hazards, particularly the possibility of GLOFs and glacier retreat. However, they also highlight that most transboundary frameworks fail to adequately incorporate adaptation strategies and climate risk assessments into governance mechanisms, leaving significant gaps. River regimes are undergoing rapid transformation due to climate change, placing both upstream and downstream communities at risk from glacial retreat, increasingly unpredictable flows, and catastrophic hazards such as GLOFs, floods, droughts, and water stress. Yet most transboundary frameworks still lack the necessary tools to address these cascading risks. Furthermore, formal negotiations frequently overlook local contestations, gendered experiences, and indigenous knowledge systems. When considered collectively, these contributions highlight the fundamental contradiction of South Asia's transboundary rivers: despite their inherently shared nature, which necessitates collaborative management, governance strategies continue to be fragmented, unilateral, and short-term in focus.

RIVERS BEYOND BORDERS AND FUTURE RESEARCH IN TRBs

TRB research in South Asia is highly critical and complex, owing to politics, differing interests of neighbouring countries, issues of cooperation, and the challenges of recent extreme environmental changes and disaster resilience. Despite the comprehensive assessments provided in this book, there remain notable research gaps in the study of transboundary river governance in South Asia (Figure 14.1). The impact of climate change and the ways in which ongoing disasters are linked to upstream and downstream rivers are still not sufficiently addressed by governance structures. In particular, the consequences of snowmelt variability, and GLOFs, while widely acknowledged, are rarely incorporated into joint management, knowledge share, basin-level planning, or national adaptation initiatives. Similarly, disaster risks such as floods, droughts, riverbank erosion, and extreme hydrological events are still largely treated as domestic concerns rather than as regional and transboundary issues requiring coordinated solutions. Although the importance of institutions has been recognised, limited research has been undertaken on innovative governance strategies, such as basin organisations, cooperative monitoring programmes, or benefit-sharing arrangements that extend beyond volumetric water allocation. Likewise, despite their substantial influence on governance outcomes across scales, the roles of local and subnational actors, provincial governments, community organisations, and civil society remain marginalised in formal diplomacy.

Resilience and catastrophe governance represent another critical gap. Insufficient research has been undertaken on the transboundary effects of cryospheric hazards, particularly GLOFs along the Nepal–China–India corridors, and there are notably no integrated early warning systems (EWS) or adaptation strategies in place. With little emphasis on community-led resilience initiatives or catastrophe governance frameworks, the potential of disaster diplomacy as a complement to hydro-diplomacy remains largely untapped. Furthermore, chronic asymmetries in data sharing, especially between India, Pakistan, Bangladesh and China, continue to undermine trust and regional cooperation. Collective preparedness is further restricted by the lack of open-access systems for real-time hydrological monitoring and forecasting. Furthermore, future research, policy, and governance of South Asian transboundary

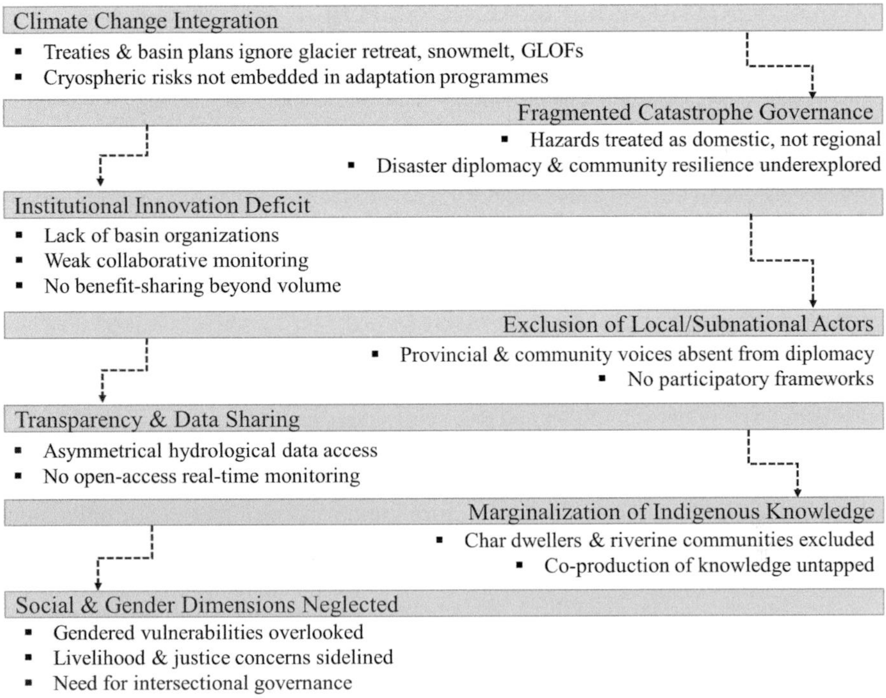

FIGURE 14.1 Research gap in transboundary water governance in South Asia. This figure is intended to help readers understand what has been looked into and what more needs to be explored in policy briefs and academic presentations.

rivers should explicitly consider small tributaries and distributaries, as well as their contributions to the larger river systems and associated ecosystems. The long-term health and resilience of the GBM and Indus basins depend on these smaller waterways, making their environmental sustainability crucial for ensuring the ecological and socio-economic resilience of the entire region.

In addition, there are no existing frameworks within transboundary river governance for supporting the resilience of local and indigenous communities who have lived with, and adapted to, riverine ecologies for thousands of years across boundaries. Institutional governance frameworks also continue to underrepresent indigenous and local knowledge systems. Opportunities for co-produced knowledge that brings together scientific expertise and local practices are generally overlooked, as the perspectives of riverine residents, char dwellers, and indigenous groups—who hold invaluable insights on resilience and adaptability—are not adequately incorporated.

Finally, the social dimensions of water governance require urgent attention. Policy and practice rarely account for the gendered aspects of climate vulnerability, water access, and decision-making, as well as wider issues of livelihood, culture and social justice. Future research must adopt an intersectional approach that meaningfully integrates climate risks, disaster diplomacy, local knowledge, data transparency, and gender equity into collaborative frameworks in order to achieve inclusive and sustainable transboundary water governance in South Asia.

TOWARDS SUSTAINABLE SOLUTIONS FOR TRANSBOUNDARY RIVER CONFLICTS IN SOUTH ASIA

South Asian transboundary river disputes have their roots in the region's unequal hydrological access, ecological interdependence, and geopolitical complexity. More than 60% of the freshwater on Earth is found in transboundary rivers, and 153 nations share at least one of the 286 transboundary rivers and lakes (Sahana et al., 2024). Every nation in South Asia shares two or more main rivers and a number of tributaries with one or more of its neighbours (Bakhsh, 2022). Despite having access to only 4% of the world's freshwater resources, South Asia is home to around 24% of the world's population (Birendra et al., 2021). Stressors brought on by climate change, worsen this imbalance by escalating conflicts over water-infrastructure development and water share or distribution, excessive agricultural and industrial water uses. For instance, the Indus Water Treaty between India and Pakistan, signed in 1960, remains one of the few enduring frameworks, yet it faces strain from upstream dam projects and shifting rainfall patterns (Khatoon et al., 2025). Recent literature on hydro-political studies shows a sharp rise in unilateral water infrastructure initiatives post-2000, with India alone initiating over 60 hydropower projects in the Brahmaputra Basin, raising alarms in downstream Bangladesh (Baruah et al., 2023). Meanwhile, data asymmetry continues to undermine trust: India and China, for example, have limited real-time hydrological data exchange, despite the increasing frequency of transboundary disaster events.

The book *Rivers Beyond Borders: Environmental Sustainability in South Asia's Transboundary Waters* offers a timely lens into these dynamics, emphasising the need for cooperative basin-level governance, collaboration, and environmental sustainability (Table 14.1). It argues that without integrating environmental sustainability

TABLE 14.1
Chapter Summaries: Key Themes and Insights

Ch. No.	Chapter	Theme/Focus	Key Takeaways
1	Transboundary River Conflicts, Cooperation, and Environmental Sustainability	Regional overview	Examines transboundary issues in South Asian countries by identifying the primary sources of disputes and their impacts, while exploring how cooperative management can promote pathways toward environmental sustainability.
2	Transboundary Water Dispute between Bangladesh and India	Institutional role	Explores the potential of the Bangladesh National River Conservation Commission (NRCC) to shape equitable water governance and looks at the governance opportunities and challenges inherent in the current institutional arrangements, paying special attention to the vertical aspects of water disputes between Bangladesh and India
3	India–China Water Disputes	Geopolitical tensions	Uses international legal frameworks and current literature to analyse the legal, environmental, and geopolitical aspects of India–China transboundary water relations in order to map strategic interests and diplomatic bottlenecks

(Continued)

TABLE 14.1 (*Continued*)
Chapter Summaries: Key Themes and Insights

Ch. No.	Chapter	Theme/Focus	Key Takeaways
4	Hydro-Social Transformations in the Bengal Delta	Historical lens	Examines how littoral Bengal was imagined as a land-based, revenue-producing frontier, where paddy cultivation displaced a more fluid hydro-social landscape
5	Challenges of Integrated River Basin Management	Stakeholder engagement	Examine how different geographical, socio-economic, and political situations affect stakeholders' opinions about river management
6	Hydro-Diplomacy in South Asia	Regional cooperation	Examines how competition for water resources affects regional politics in order to assess the function of hydro-diplomacy in South Asia
7	Interplay between Dispute and Diplomatic Deadlock	Climate lens	Explores the factors contributing to the lack of progress in river resource diplomacy among the four principal co-riparian states of South Asia (India–Pakistan to the west; India–Bangladesh–Nepal to the east)
8	Ethical Considerations in India's Water Management	Normative analysis	Analyses ethical considerations with both upstream and downstream states, mainly Bangladesh, Bhutan, China, Nepal, and Pakistan
9	Legal Frameworks for Sustainable Transboundary Water Governance in India–Bangladesh	Law and policy	Critically analyses issues pertaining to water, interacting with national and international legal systems and laws
10	Bhutan's Water Governance	Case study	Offers a model of sustainability rooted in national ethos
11	Navigating Transboundary Dynamics of the Ganga River Basin	Glacial-deltaic linkages	Connects upstream cryosphere dynamics with downstream vulnerabilities
12	Water Governance in Nepal–India Borderlands	Plural rationalities	Examines how four distinct rationalities (i.e. hierarchical, individualist, egalitarian, and fatalist) interact to shape transboundary water administration in the border between India and Nepal
13	Water Resources Management in the Brahmaputra River Basin	Strategic basin planning	Identifies the key challenges of the basin and proposes strategies for a sustainable future
14	India–China Brahmaputra Conflict	Strategic complexity	Examines bilateral engagements, water resource management practices, hydroelectric projects, ecological impacts, and geopolitical tensions between India and China

and social equity into hydro-diplomatic efforts, the region risks deepening ecological fragility and political discord. South Asia must move from fragmented water politics to collaborative stewardship if its rivers are to remain sources of life rather than conflict. This book summarises case studies from South Asian countries, particularly India, Bangladesh, Nepal, and Bhutan, highlighting historical legacies and colonial continuities in shaping hydro-politics.

HYDRO-DIPLOMACY AND COOPERATIVE MECHANISMS: BETWEEN PROMISE AND PARALYSIS

Hydro-diplomacy in South Asia stands at a critical crossroads, caught between the promise of cooperative water governance and the paralysis of entrenched geopolitical rivalries. The region continues to face challenges due to fragmented institutional frameworks, asymmetrical data access, and unilateral infrastructure development, despite decades of negotiations and treaty-making, such as the Teesta and Kushiyara Water Sharing Agreements, the Ganges Water Treaty, and the Indus Water Treaty. Although there have been some positive outcomes from shared water management, the Geneva Water Hub's roundtable on hydro-diplomacy highlighted that new models incorporating financial incentives, intersectoral coordination, and inclusive governance are still needed to fully utilise water as a tool for consolidating peace (Gharios and Farajalla, 2022). Similarly, the Stimson Centre's international hydro-diplomacy initiative emphasises how the securitisation of water resources and lack of financing for joint river management obstruct meaningful cooperation (Aamer, 2023). In 2025, Bangladesh became the first South Asian country to accede to the UN Water Convention, advancing the protection and sustainable management of transboundary rivers and lakes. This development could further complicate India–Bangladesh water management strategies, bilateral relations, and existing frameworks such as the Joint River Commission (JRC), as well as pending agreements like the Ganga Water Treaty (2026) and the Teesta Water Agreement.

The contributing authors of this edited book argue that collaborative procedures need to advance beyond volumetric allocations to include benefit-sharing, resilience-building, and participatory monitoring. However, data trends indicate a continued increase in hydropower projects and unilateral dam development without open discussion between upstream and downstream nations. This exacerbates social and ecological vulnerabilities and undermines trust. Furthermore, disaster preparedness and adaptive governance are hindered by the absence of real-time data-sharing mechanisms across riparian governments, including China, Nepal, Bhutan, Pakistan, Bangladesh, and India. Although regional forums and multi-stakeholder discussions have increased, their effectiveness is limited by the lack of legal enforcement and inclusive participation, particularly for subnational actors and indigenous communities. Thus, hydro-diplomacy in South Asia remains suspended between aspirational frameworks and operational inertia, a space where the rhetoric of cooperation must urgently translate into resilient, equitable, and climate-conscious action.

INTEGRATED TRANSBOUNDARY RIVER BASIN MANAGEMENT: STAKEHOLDER ENGAGEMENT AND GOVERNANCE CHALLENGES

A comprehensive framework known as Integrated Transboundary River Basin Management (ITRBM) aims to strike a balance between socio-economic development, ecological sustainability, and geopolitical realities across shared river basins. The implementation of ITRBM is particularly complex in South Asia, where significant small and large rivers within the GBM and Indus River systems traverse several national and subnational borders. Although basin-wide planning, equitable sharing, and ecological health are key components of IWRM, fragmented governance and conflicting sovereign interests frequently limit transboundary implementation. National governments typically dominate negotiations, leaving provincial authorities, local communities, and indigenous groups marginally involved despite their direct dependence on riverine ecosystems. In addition to undermining inclusive decision-making, this exclusion neglects important indigenous and local knowledge systems that could enhance resilience to hazards caused by climate change, drought, and flooding. Governance challenges are further compounded by asymmetries in data sharing, institutional fragmentation, and the absence of regional mechanisms for collaborative monitoring or benefit-sharing beyond water allocation.

This book emphasises the urgency of moving beyond state-centric diplomacy to embrace inclusive, multi-level governance. It identifies research gaps in understanding how riverine and agricultural communities, fisherfolk, and women—often the most vulnerable to hydrological shocks—are rarely consulted in formal basin planning, despite their deep ecological knowledge and lived experience. Data asymmetries, jurisdictional overlaps, and the absence of basin-wide organisations with enforcement capabilities all exacerbate governance issues. Furthermore, the potential for cooperative monitoring and benefit-sharing is undermined by trust deficits between riparian governments, which are exacerbated by a lack of data sharing and unilateral infrastructure projects. ITRBM must evolve into a participatory framework that acknowledges the legitimacy of multiple stakeholders and encourages knowledge co-production to address these issues. This entails establishing basin organisations represented by local governments, civil society, and the scientific community; developing open-access hydrological data platforms; and integrating equity and gender considerations into water governance. Without such changes, the potential of integrated management risks being undermined by reactive planning and exclusionary politics.

This book demonstrates that the future of South Asia's rivers depends on rethinking them as areas for collaboration, sustainability, and collective resilience rather than solely for conflict resolution and need-based diplomacy. The region can begin transforming its rivers from sources of conflict into platforms for ecological and regional security by recognising research gaps and adopting innovative approaches. The synthesis of this volume highlights the urgent need for South Asia to move beyond narrow, state-centric, and bilateral approaches towards a more holistic, integrated, and climate-resilient vision of transboundary water governance. This transition necessitates the integration of ecological sustainability, disaster risk reduction, and climate change adaptation into diplomatic processes.

ETHICS, EQUITY, AND ENVIRONMENTAL JUSTICE IN TRANSBOUNDARY WATER GOVERNANCE

Environmental justice, equality, good governance and ethical responsibilities are fundamental but frequently disregarded aspects of transboundary water governance in South Asia. The river systems in the region are vital lifelines that support a variety of ecosystems, cultures, religious faiths, and communities, in addition to being hydrological features. The ethical requirements of justice, inclusivity, and historical accountability are often neglected in favour of state-centric interests, technical efficiency, and volumetric allocations in governance frameworks. For example, it is common for upstream dam construction and diversion projects to proceed without sufficient downstream community involvement, which contravenes principles of distributive justice and prior informed consent. In South Asia, fewer than 20% of transboundary water agreements address the rights of vulnerable groups, including women, indigenous peoples, and riverine and char communities, and even fewer incorporate measures for equality or environmental protection (Jolly et al., 2023). In South Asian water diplomacy, most agreements, treaties, and related disputes or discussions do not fully adhere to international principles of good governance, including 'equitable and reasonable utilisation' and the prevention of 'significant harm' to downstream communities.

The book raises these issues and argues that social justice and ethical accountability are the foundations of sustainable water governance. It challenges the prevailing hydro-diplomacy paradigm, which views rivers as strategic resources rather than as commons, and advocates a shift to frameworks that consider ecological integrity, intergenerational equity, and the lived realities of marginalised communities. Disparities in water availability across rivers reveal glaring inequalities: per capita water availability in certain transboundary districts of India, Pakistan, Bangladesh, Bhutan, and Nepal is less than half that in upstream metropolitan areas. Women in flood-prone communities frequently suffer most from water shortages, displacement, and exclusion from decision-making, exacerbating these discrepancies.

This book collectively recommends several measures to promote more sustainable pathways for water diplomacy in South Asia. These include enhancing multilevel discussions and knowledge sharing, establishing dialogue platforms from the central ministry level to local government, and fostering people-to-people connections. Engagement should also involve NGOs, think tanks, and local authorities, particularly in border areas, to address transboundary water challenges affecting smaller tributaries and rivers. Bureaucratic, administrative, academic, and non-academic engagement—through debates, conferences, workshops, and other events—should be strengthened. Additionally, efforts should promote constructive and positive activism to address water-related environmental issues and support effective, equitable water governance.

In this context, environmental justice calls for more than technical solutions; it requires a rethinking of governance that prioritises ecological stewardship, human rights, and cultural values. This includes incorporating moral principles into treaty development, guaranteeing open and inclusive procedures, and recognising the validity of indigenous and local knowledge systems. Without such a shift, transboundary water governance risks perpetuating structural inequalities and ecological harm, rather than fostering resilience and cooperation.

CONCLUSION AND PATHWAYS FOR SHARED FUTURES

This book highlights that climate change is not solely an environmental issue but also a profound geopolitical stressor for TRB governance in South Asia. Post-colonial transboundary river governance has shifted from a focus on hydropower generation, irrigation, and resource utilisation towards addressing disaster risk, multi-hazard events, and loss and damage, which exacerbate existing challenges. A new era of water diplomacy has emerged, shaped by climate change and the growing emphasis on disaster resilience. Consequently, there is a pressing need for data-driven research, spatial and empirical analysis, and the adoption of advanced technologies, including satellite remote sensing and AI-based modelling, to support TRB research and develop sustainable governance pathways. The nexus of climate, conflict, and cooperation reveals a complex interplay of ecological vulnerability, geopolitical tension, and diplomatic opportunity. Climate change acts as a threat multiplier, accelerating glacial retreat, increasing monsoonal variability, and raising the frequency and intensity of extreme hydrological events, including floods and droughts. These phenomena disproportionately affect downstream communities, exacerbating inequalities in water access and often triggering inter-state and intra-state disputes. For instance, satellite observations indicate a 15% increase in flood intensity across the Brahmaputra basin over the last two decades, while Himalayan glacial melt has accelerated by nearly 30% since the 1990s (Devrani et al., 2022; Nie et al., 2021). Such trends heighten anxieties over water security and sovereignty, particularly in sensitive corridors such as Nepal–China–India and India–Bangladesh.

Yet these climate-induced stresses also present opportunities for collaboration. Shared vulnerabilities can catalyse joint EWS, basin-wide adaptation planning, and disaster diplomacy. This book demonstrates that climate threats, when approached constructively, can foster dialogue, trust-building, and cooperative governance. Tools such as collaborative monitoring stations, data-sharing platforms, and inclusive basin organisations can transform reactive, conflict-prone politics into proactive resilience strategies. However, entrenched power asymmetries, limited transparency, and the marginalisation of local actors often impede this potential. Ironically, regions most exposed to climate shocks tend to have the weakest institutional safeguards and least inclusive governance mechanisms. Therefore, climate diplomacy must be fully integrated into hydro-diplomatic frameworks that prioritise sustainability, equity, and co-benefits. This includes the meaningful involvement of civil society, indigenous communities, and subnational actors, whose expertise is essential for adaptive governance. Ensuring that South Asia's rivers remain sources of life rather than instruments of division is both an ethical and strategic imperative.

Transboundary river governance in South Asia embodies a fundamental paradox: rivers sustain populations across national boundaries yet remain sites of tension, mistrust, and fragmented governance. Climate change exacerbates these challenges through cryospheric risks, hydrological variability, and extreme events, exposing the limitations of existing treaties and institutional frameworks. Nevertheless, these challenges offer opportunities to reimagine cooperative futures. An integrated approach is required—one that treats rivers as shared natural systems rather than fixed territorial assets. Building trust and resilience among riparian states can be advanced through basin-wide organisations, cooperative monitoring, and open-access data platforms. Equally important is the inclusion of subnational actors, women, and indigenous communities, whose lived experiences and traditional knowledge are vital for equitable and adaptive governance.

To move beyond fragmented diplomacy and ecological precarity, this volume advocates:

- **Regional Climate-Water Compacts:** Legally binding agreements combining disaster risk reduction, climate adaptation, and fair water distribution.
- **Transboundary Knowledge Platforms:** Jointly developed spaces for community engagement, scenario planning, and shared data.
- **Justice-Centred Governance:** Institutionalising ethical principles, equality, and indigenous rights within legal frameworks.
- **Participatory Hydro-Diplomacy:** Extending diplomacy beyond states to include local communities, civil society, and academic stakeholders.

By implementing these pathways, South Asia can transform its transboundary rivers from sources of tension into platforms for ecological resilience, social equity, and sustainable regional cooperation.

REFERENCES

Aamer, F. (2023). Water security: Diplomacy, global cooperation, and effective management of shared rivers. Konrad-Adenauer Foundation & Stimson Center. Accessed from https://www.stimson.org/2023/water-security-diplomacy-global-cooperation-effective-management-of-shared-rivers/

Ahmed, R., Wani, G. F., Ahmad, S. T., Sahana, M., Singh, H., & Ahmed, P. (2021). A review of glacial lake expansion and associated glacial lake outburst floods in the Himalayan region. Earth Systems and Environment, 5(3), 695–708.

Allen, S. K., Linsbauer, A., Randhawa, S. S., Huggel, C., Rana, P., & Kumari, A. (2016). Glacial lake outburst flood risk in Himachal Pradesh, India: An integrative and anticipatory approach considering current and future threats. Natural Hazards, 84(3), 1741–1763.

Bakhsh, M. M. (2022). Analysis of water conflicts: A case of South Asian region. Pakistan Journal of Applied Social Sciences, 13(1), 79–94.

Baruah, T., Barua, A., & Vij, S. (2023). Hydropolitics intertwined with geopolitics in the Brahmaputra River Basin. Wiley Interdisciplinary Reviews: Water, 10(2), e1626.

Birendra, K. C., McIndoe, I., Schultz, B., Prasad, K., Bright, J., Dark, A. et al (2021). Integrated water resource management to address the growing demand for food and water in South Asia. Irrigation and Drainage, 70(4), 924–935.

Devrani, R., Srivastava, P., Kumar, R., & Kasana, P. (2022). Characterization and assessment of flood inundated areas of lower Brahmaputra River Basin using multitemporal synthetic aperture radar data: A case study from NE India. Geological Journal, 57(2), 622–646.

Gharios, G., & Farajalla, N. (2022). Dams, Floods and Cuts: Diverging Waters Between Governments and Civil Societies – Examples from The Levant. In First Annual Report of the Global Observatory for Water and Peace (p. 68). Geneva Water Hub. Accessed from: https://discovery.dundee.ac.uk/en/publications/dams-floods-and-cuts-diverging-waters-between-governments-and-civ/.

Jolly, S., Singh, S., & Saraswat, A. (2023). Mainstreaming gender in transboundary water governance: A South Asian perspective. Asia Pacific Journal of Environmental Law, 26(2), 221–243.

Kaiser, Z. A. (2023). Analysis of the livelihood and health of internally displaced persons due to riverbank erosion in Bangladesh. Journal of Migration and Health, 7, 100157.

Khatoon, S., Taj, S., Fayaz, H., Naseer, M. Z., & Tanveer, M. R. (2025). The hydro-politics of Indus River Basin: The role of water in Pak-India relations. Journal of Social Signs Review, 3(06), 200–218.

Nie, Y., Pritchard, H. D., Liu, Q., Hennig, T., Wang, W., Wang, X. et al (2021). Glacial change and hydrological implications in the Himalaya and Karakoram. Nature Reviews Earth & Environment, 2(2), 91–106.

Saha, S., Bera, B., Sengupta, D., Mukhopadhyay, U., Ghosh, D., Tamang, L. et al (2025). Multiple drivers of the recent South Lhonak glacial lake outburst flood in Sikkim Himalaya and its aftermath on Teesta River Valley. Geosystems and Geoenvironment, 4(2), 100375.

Sahana, M., Dhali, M. K., & Lindley, S. (2024). Global disparities in transboundary river research have implications for sustainable management. Communications Earth & Environment, 5(1), 786.

Index

For Product Safety Concerns and Information please contact our
EU representative GPSR@taylorandfrancis.com Taylor & Francis
Verlag GmbH, Kaufingerstraße 24, 80331 München, Germany